T0202377

What Makes Time Special?

As we navigate through life we instinctively model time as having a flowing present that divides a fixed past from open future. This model develops in childhood and is deeply saturated within our language, thought and behavior, affecting our conceptions of the universe, freedom and the self. Yet as central as it is to our lives, physics seems to have no room for this flowing present. *What Makes Time Special?* demonstrates this claim in detail and then turns to two novel positive tasks. First, by looking at the world "sideways"—in the spatial directions—it shows that physics is not "spatializing time" as is commonly alleged. Even relativity theory makes significant distinctions between the spacelike and timelike directions, often with surprising consequences. Second, if the flowing present is an illusion, it is a deep one worthy of explanation. The author develops a picture whereby the temporal flow arises as an interaction effect between an observer and the physics of the world.

Using insights from philosophy, cognitive science, biology, psychology and physics, the theory claims that the flowing present model of time is the natural reaction to the perceptual and evolutionary challenges thrown at us. Modeling time as flowing makes sense even if it misrepresents it.

Craig Callender earned his PhD with research on the direction of time at Rutgers University. He then worked at the London School of Economics before moving to the University of California, San Diego. He has interests in time and physics, the interpretation of quantum mechanics, quantum gravity, philosophy of science, and environmental ethics. He is editor of *the Oxford Handbook of Philosophy of Time* (2011).

Praise for *What Makes Time Special?*

"Time is a big invisible thing that will kill you' (p. 1). I cannot think of a more striking opening sentence to a work of philosophy in recent times—or at any time, for that matter. What follows is a comprehensive tour of philosophy of time from Callender's perspective, written with great insight, as well as wit and flair. . . . Callender has written a survey of issues in philosophy of time from a broadly naturalistic perspective. It is rich in detail and argument. . . . Anyone interested in understanding time will be rewarded by further digging."

Steven F. Savitt, *British Journal for the Philosophy of Science*

"Craig Callender's *What Makes Time Special?* is a special book in at least four dimensions. It is the most original work on the philosophy of time for many years. It has a good claim to be the best recent book on the physics of time, too, if we read the phrase to stress Callender's title question: What is it in physics that distinguishes time from space? It is wonderfully clear. And—most remarkable of all, given the subject matter-it is often fun."

Huw Price, *Philosophical Review*

"[A]mbitious and highly original contribution to the philosophy of time . . . displays nothing short of profound insight into the way physics informs old debates about time . . . densely-argued, fascinating treatment of the problem of time, that breaks new ground . . . will be compulsory reading for anyone interested in the topic [of time], not just philosophers of physics"

Comments from selectors of the 2018 Lakatos Award

"This is a golden age for the philosophy of time. . . . Callender's book is a novel and engaging contribution to this positive development, driven by a desire to understand the emergence of "manifest time" from a physical world initially hostile to it, with the help of disciplines as different as hardcore theoretical physics and experimental psychology (and much in-between). . . . I wholeheartedly recommend his new book to everyone interested in time and its puzzles."

Yuri Balashov, *Notre Dame Philosophical Reviews*

What Makes Time Special?

Craig Callender

OXFORD
UNIVERSITY PRESS

Great Clarendon Street, Oxford, OX2 6DP,
United Kingdom

Oxford University Press is a department of the University of Oxford.
It furthers the University's objective of excellence in research, scholarship,
and education by publishing worldwide. Oxford is a registered trade mark of
Oxford University Press in the UK and in certain other countries

The moral rights of the author have been asserted

First published 2017
First published in paperback 2023

Published in the United States of America by Oxford University Press
198 Madison Avenue, New York, NY 10016, United States of America

British Library Cataloguing in Publication Data
Data available

Library of Congress Cataloging in Publication Data
Data available

ISBN 978-0-19-879730-2 (Hbk.)
ISBN 978-0-19-288746-7 (Pbk.)

To the ones who make life special, Isla, Lily, Ewan, and Lisa

Contents

Acknowledgments   xi
List of Figures   xv
List of Boxes   xix

1. The Problem of Time   1
   1.1 Manifest Time   4
        1.1.1 Now   7
        1.1.2 Flow   10
        1.1.3 Past/Future asymmetry   12
        1.1.4 Is manifest time universal?   14
   1.2 Physical Time   19
        1.2.1 No manifest time   20
        1.2.2 What makes time special?   22
   1.3 The "Two Times" Problem   23
   1.4 From There to Here   26

2. Lost Time: Relativity Theory   31
   2.1 Classical Physics   31
        2.1.1 Recovering space and time   34
        2.1.2 Classical ideal clocks and manifest time   37
        2.1.3 Trautman–Cartan theory   40
   2.2 Relativity   42
   2.3 Where's Time?   49
   2.4 Minkowski Spacetime   52
   2.5 Lorentzian Time   57
   2.6 Outside Minkowski: Domes, Donuts, and Diamonds   59
   2.7 Conclusion   66

3. Tearing Spacetime Asunder   67
   3.1 Cauchy Time   67
   3.2 "Unique" Time Functions   72
   3.3 Time, Stuff, and Laws   76
   3.4 Conclusion   80

4. Quantum Becoming?   81
   4.1 Quantum Mechanics   82
   4.2 Popper's *Experimentis Crucis*   84
        4.2.1 Quantum preferred frames and time   89
        4.2.2 Caveats and alternatives   89
        4.2.3 The coordination problem   90
   4.3 Quantum Becoming via Collapses?   94
   4.4 Conclusion   96

5. Intimations of Quantum Gravitational Time 98
  5.1 The Best of Times: "Asynchronous Becoming" in Causal Sets 99
    5.1.1 The basic kinematics of CST 100
    5.1.2 Taking growth seriously 103
  5.2 The Worst of Times: Disappearing Time in Canonical Quantum Gravity 110
    5.2.1 Semiclassical time 112
    5.2.2 Justifying the approximations 116
  5.3 Conclusion 118

6. The Differences Between Time and Space 119
  6.1 The Project Reconceived 120
  6.2 Time in Physics 122
    6.2.1 The metric 122
    6.2.2 Dimensionality 126
    6.2.3 Mobility asymmetry 130
    6.2.4 Direction of time 132
    6.2.5 Natural kind asymmetry 133
  6.3 The Fragmentation of Time 135
  6.4 Conclusion 137

7. Laws, Systems, and Time 138
  7.1 System Laws and Time 140
  7.2 Time is the Great Informer 142
  7.3 Binding Time 144
    7.3.1 One-dimensionality 144
    7.3.2 Closed timelike curves 145
    7.3.3 The direction of time 148
    7.3.4 Natural kind asymmetry 148
  7.4 Metaphysical Variations 149
  7.5 Questions and Connections 153
  7.6 Conclusion 156

8. Looking at the World Sideways 157
  8.1 Strength and Well-posed Cauchy Problems 158
  8.2 The Worlds 164
  8.3 Proposal 166
  8.4 The Argument 167
  8.5 Illustration 170
  8.6 Is It Time? 171
  8.7 Turning Pages in Non-temporal Directions 173
    8.7.1 Pages of light 174
    8.7.2 Pages of time 177
  8.8 Conclusion 179

9. Do We Experience the Present? 180
  9.1 Metaphysics of Time 181
  9.2 The Problem of the Presence of Experience 182

9.3	The Temporal Knowledge Argument	185
9.4	From Metaphysics to Psychology: Perceived Synchrony	189
	9.4.1 Temporal ventriloquism	193
	9.4.2 Temporal recalibration	194
	9.4.3 Comments	196
9.5	Interlude: Measuring Subjective Simultaneity	197
9.6	Exploding the Now	201
9.7	Does Synchrony Pop Out?	203
9.8	Conclusion	205

10.	Stuck in the Common Now	206
10.1	Disagreement and the Case of PH	210
10.2	Manufacturing the Now: Signals, Speed, and Stamps	213
	10.2.1 Time stamps not needed	213
	10.2.2 The common now	217
10.3	Wiggling in Time vs Wiggling in Space	221
10.4	Conclusion	223

11.	The Flow of Time: Stitching the World Together	226
11.1	Sharpening Focus	228
11.2	Meet IGUS	231
11.3	Getting IGUS Stuck in Time	234
11.4	Outfitting IGUS	237
	11.4.1 Sensing motion and change	238
	11.4.2 Specious present	240
	11.4.3 Felt duration	242
11.5	Memories and Flow	243
11.6	The Enduring Witness	247
11.7	From Flowing Selves to Animated Time	252
11.8	Temporal Decentering and the Self	255
11.9	The Acting Self	259
11.10	The Explanation of Passage	261
11.11	Conclusion	262

12.	Explaining the Temporal Value Asymmetry	264
12.1	"Thank Goodness That's Over"	266
12.2	The Proximal/Distant Asymmetry	270
12.3	The Humean Solution	272
12.4	The Knowledge Asymmetry	274
12.5	The Affect Asymmetry	275
12.6	Explaining the PF Asymmetry	279
12.7	Other Temporal Biases	282
12.8	Explaining Other Time Biases	285
12.9	Conclusion	288

13.	Moving Past the ABCs of Time	290
13.1	Analytic Philosophy of Time: A Potted and Biased History	290

13.2	The Explanatory Challenge	294
13.3	The ABCs of Physics	300
13.4	Eliminating Tense?	302
13.5	Conclusion	303
14.	Putting It All Together	304
14.1	Common Structure	305
14.2	A Unified Flowing Now	306
14.3	Animals	309
14.4	An Illusion?	310
14.5	Conclusion	311
Bibliography		313
Index		337

Acknowledgments

I was bitten by the topic of time during my second year of graduate school. I didn't know it then, but the resulting infection would prove long-lasting. I succumbed because the topic of time has it all. Time is deep and mysterious, but it also matters to daily life—all those time clichés are on to something. Philosophy runs the danger of getting trapped in verbal disputes, but some questions about time are definitely genuine ones. The topic is not worked to death, as plenty of promising and untapped projects exist. Its study is burgeoning in the sciences, from physics to neuroscience. And best of all for someone of my temperament, one can attack time with a variety of intellectual weapons. Logic, metaphysics, epistemology, physics, cognitive science, developmental psychology, evolution, and more can and should be used.

My fascination couldn't have landed in a more fertile environment. I went to Rutgers initially to study Spinoza and maybe metaphysics, but I benefited from the most amazing stroke of luck. Unlikely as it may seem, New Jersey turned out to be the ideal location for studying time, philosophy, and physics. Bob Weingard, my supervisor, was an expert in the field. He generously taught me physics lessons most afternoons for years, exposing me to ideas that I never would have learned through coursework. Bob's knowledge and kindness was incredible, and his premature death is still painful. Now, sadly, a lot of the physics I learned from Bob has faded away, but I hope two other things he taught me rubbed off, namely, his nuts-and-bolts approach to philosophy— philosophical posturing and unnecessary complexity was alien to Bob—and his sense of awe at how amazing the physical world is. New Jersey also offered two of the top philosophers of the science in the world, Barry Loewer and Tim Maudlin. Ying to the other's yang philosophically, they nonetheless shared a vision of how philosophy and science should work together, each weaker without the other. As if all that were not enough, as external advisors I had David Albert, Huw Price, and Steve Savitt. All three gave me early versions of manuscripts on time that were crucial to my thesis. Huw in particular has never stepped out of his advisor role, although now the advice is as much about coffee as philosophy. Throw in the other outstanding faculty at Rutgers and dozens of exceptional graduate students and friends, and it's hard to imagine getting luckier.

Fast forward to 2005. Perfectly content in my career to go from problem to problem, article to article, I never dreamed of writing a long manuscript. Yet when I moved to UC San Diego in 2001 I began to learn something about time and the science of the mind. No doubt my subconscious struggled to reconcile the time of physics with the time of experience for years. At some point in 2005 the tension percolated up to the surface in an epiphany: I was a self-loathing philosopher of time! While I learnt a great deal from philosophy of time, I felt that it had stolen the best problem—what I here call "the" problem of time, namely how scientific and manifest temporal images relate—and turned it into a language and logic-soaked enterprise disconnected from the rest of science. No disrespect to language or logic—I try to use both often—but these tools didn't seem capable of *answering* the problem of time, as opposed to sharpening it and outlining possible solutions. An answer requires explanations based upon our best sciences, yet they weren't brought to the table in philosophy. Outside philosophy, the question wasn't being asked. Hence I began this book.

A short decade later, what seemed a stray research topic back in 1992 has become a good portion of my career. Looking back I see innumerable influences to whom I owe gratitude. The biggest are probably the institutions I've worked at and the students I've taught. I benefited from fantastic environments at both the London School of Economics and UC San Diego, two schools that pride themselves on fostering substantive interdisciplinary work. My students have taught me an enormous amount, and I thank in particular those classes at UC San Diego and the University of Sydney who studied early drafts of some of this material. And while teaching is a great privilege, a break isn't bad; hence I'm tremendously grateful to the Foundational Questions Institute and UC San Diego's Center for Humanities, each of which gave me grants releasing me from a course. I also thank Kristie Miller, Sam Baron, and Dean Rickles for proposing me for an International Collaborator Award that supported my fun two month stay at the University of Sydney.

Interactions with colleagues were also crucial. Many on this list may not have a clue why they are on it. But for each I recall a conversation, question, or other form of communication—perhaps even just a raised eyebrow at the right moment—that improved this book in a specific way: David Albert, Mark Barber, Jeffrey Barrett, Thomas Barrett, Gordon Belot, Dean Buonomano, Lisa Callender, Nancy Cartwright, Patricia Churchland, Paul Churchland, Mauro Dorato, Heather Dyke, John Earman, Nina Emery, Matt Farr, Mathias Frisch, Bob Geroch, Carl Hoefer, Alex Holcombe, Nick Huggett, Nat Jacobs, Barry Loewer, Christian Loew, David Malament, Tim Maudlin, Tarun Menon, David Meyer, John Norton, Naomi Oreskes, Laurie Paul, Oliver Pooley, Huw Price, Agustin Rayo, David Rideout, Steve Savitt, Michael Stoeltzner, Chris Suhler, Paul van der Wagt, Daniel Vitkus, Jim Weatherall, Steve Weinstein, Robert Westman, Alastair Wilson, Jeff Wincour, Eric Winsberg, and Dean Zimmerman. I apologize to those I have forgotten. With ten years gestation, it's inevitable. A few people deserve special mention. Jenann Ismael, whose fantastic "side on" look at time coincides with mine, gave me the backbone to pursue the project the way I have. Chris Wüthrich, a glutton for punishment, attended my seminar on time and also provided written feedback. Jonathan Cohen, my philosophical sounding board, helped smooth the way for me when it came time to learn about philosophy and cognitive science. My recent students Casey McCoy and John Dougherty also have been invaluable. Motivated by the shame that would unjustifiably be cast upon them if their supervisor wrote a horrible book, they each helped materially and intellectually.

Although almost everything has been seriously reworked, portions of this book have appeared in print or are at least based on what is in print elsewhere. I need to thank the publishers and coauthors (Chris Suhler and Chris Wüthrich) for permission to use that material. Section 5.1 appears as part of "What Becomes of a Causal Set" with Chris Wüthrich in the *British Journal for the Philosophy of Science*; re-worked parts of Chapters 8 and 9 began life in "The Common Now" in *Philosophical Issues*; Chapter 11 is a modified and updated version of "Thank Goodness That Argument is Over: Explaining the Temporal Value Asymmetry" with Chris Suhler in *Philosophical Imprints*; and parts of the Epilogue overlap with "Time's Ontic Voltage" in Adrian Bardon's *The Future of the Philosophy of Time*.

For ten years I've joked to the family that whereas most authors thank their loved ones for helping them complete their book, I would do the opposite. I now fulfill that promise: Lisa, Ewan, Lily, and Isla, *despite your best efforts, the book is now done*. In fact, although neither can read, let me cast the net a bit wider and include in the mix of book-delayers the dog and tortoise (who, speaking of time, will outlive us all). What distractions you all were! Dinners, camping, films, barking (okay, mostly the dog), jokes, sports, choir, art shows, holidays, trips to the beach, games, walks, and vacations. There is no question that the book is finished later than it otherwise would have been. That said, looking over the list just now, I grant that most of those distractions weren't so bad. So what if the book is delayed? In exchange I have a life. That is the best deal anyone could ask for. Thanks for being such wonderful nuisances. For that I owe you everything.

List of Figures

1.1	Clichéd T-shirt	8
1.2	Dividing the world in two	9
1.3	An updating now	11
1.4	The past/future asymmetry	13
1.5	Some possibilities for time	20
1.6	Eddington's two tables	24
1.7	Iggy thrown into spacetime	26
1.8	Iggy models time as manifest time	27
2.1	Newton–Cartan spacetime	34
2.2	Foliation of M into time and space	36
2.3	Time as a "base space"	38
2.4	Path-independence of classical duration	39
2.5	Lightcone structure	45
2.6	Path-dependence of duration	48
2.7	Relativity of simultaneity	53
2.8	Stein's *definite for* relation	56
2.9	Lightcone presents	61
2.10	Puny presents	62
2.11	Donut presents	62
2.12	Private presents	63
2.13	Diamond presents	64
3.1	Domain of dependence	68
3.2	Tearing spacetime into space and time	69
3.3	Friedman–Lemaître–Robertson–Walker time	74
4.1	EPR–Bell experiment	88
4.2	Bohmian configuration space	91
4.3	Coordination problem	93
5.1	Two different maximal antichains	102
5.2	Alice and Bob's birthdays come into being	107
5.3	Post growth	109

5.4	Semiclassical time	115
6.1	Two times	129
6.2	Mobility asymmetry	131
6.3	One or two balls?	134
7.1	Wormhole: no self-interaction	146
7.2	Wormhole: with self-interaction	146
7.3	Conservative metaphysical variation	149
7.4	Radical metaphysical variation	150
8.1	Transcendental determinism	160
8.2	Cauchy problem	162
8.3	Sideways Cauchy problem	164
8.4	Which direction?	166
8.5	Defining spacelike in a toy example	171
8.6	Varieties of initial null surfaces and boundaries	175
8.7	Telling the world's story sideways	177
9.1	Temporal integration window	193
9.2	Simultaneity judgments	198
9.3	Temporal order judgments	199
9.4	Stream–Bounce illusion	200
9.5	Pop out	204
10.1	Odie	214
10.2	No time stamp needed	215
10.3	Our common now	216
10.4	Payoff matrix	218
11.1	IGUS	232
11.2	IGUS and IGUS*	235
11.3	IGUS near IGUS*	236
11.4	Reichardt-like motion detector	239
11.5	IGUS's memories	245
11.6	The self in time	253
11.7	Ego-moving perspective	254
11.8	Time-moving perspective	254
11.9	Undifferentiated past and future	255
11.10	Timeline	256

11.11	Causal order	256
11.12	Temporal decentering	257
12.1	Care versus time	266
12.2	PF asymmetry	280
12.3	The Gertrude case	282
13.1	Tensed theories	294
13.2	Growing block	298
13.3	Shrinking block	299
14.1	Explanatory schema	306
14.2	Intentional binding	308

List of Boxes

1   Einstein–Poincaré "Radar" Synchronization   51
2   Spacetime Behavior Chart   72
3   A Note on Mental Mechanisms and Evolution   278

"I wasted time, and now doth time waste me."

William Shakespeare
Richard II

Time is a great teacher, but unfortunately it kills all its pupils.

Louis Hector Berlioz
1856 Letter

Some people kill time; but this time, time is going to kill you.

The Clock King, *Batman*
"The Clock King's Crazy Crimes"

1

The Problem of Time

Time is a big invisible thing that will kill you. For that reason alone, one might be curious about what it is. Yet it is also something that matters to us throughout our daily lives. We represent time from a very early age. These representations arrive and evolve in fits and starts with the development of new abilities and distinctions. While young children we stumble with serial order recall—say, remembering the order of a bunch of flash cards. Our concept *yesterday* tends to include all past events, not merely the day before today. We quickly improve. By age four or so, we may know that one's birthday is close and we might embed the start of school in a temporal sequence. Gradually we abstract away from particular events—birthday parties, holidays, a classmate eating a bug—and employ temporal frameworks such as an ordered timeline into which we locate events. Days, weeks, and years frame events. Soon we distinguish the history of the world from the history of the self. We decenter from our own temporal perspective and adopt others. Lily can now see that when Ewan lost his first tooth, Isla's birth was in his future. Sooner or later, our temporal representations develop to the point where we can detach from particular perspectives and form a somewhat unified model of time itself.

Etched into this proto-theory of time are features central to what makes a human life recognizably human. Unlike with spatial representation, we treat as objective our own current temporal perspective on the world. We possess an almost irresistible temptation to believe that the present moment—right now, as you read this—is the moment that is *really* happening. Conscious decisions and actions happen in the present, we think. The present is not simply one perspective among many, like up-side-down and right-side-up are two spatial perspectives on one world, but we regard it as something objective and common amongst us. This amazing slice of time divides the world in two. We invest the two sides of this division with importantly distinct properties. Past events are fixed whereas future events are ripe with possibility. These differences are crucial guides in how you live your life. Unlike future headaches, past headaches mean little to you now, for they are "over and done." As the Epicureans noticed, our pre-natal non-existence doesn't concern us in the least, but our future deaths are the source of great angst. Naturally, this tripartite classification of the world into past, present, and future is not itself fixed. Present moments are continually replaced with new ones, updating the tripartite division accordingly. Time seems to flow.

These features—that the present is special, that time flows, that the past is fundamentally different from the future—are core components of the model of time we
develop. Dub this somewhat rough-around-the-edges common conception *manifest
time*. Manifest time is not a reflection of time as we directly experience it. It is informed
by experience, to be sure, but it is instead time as it arises from a kind of regimented
common sense picture of the world, a theory that psychology suggests we come to
in late childhood. This conception is no less important for emerging from untutored
theory. Quite the opposite: as the philosopher D. H. Mellor (2001) writes, it is the *time
of our lives*.

Despite its importance, our best science of time suggests that manifest time is
more or less rubbish. We don't know the true nature of time, but we do have strong
hints from the physical sciences that manifest time may portray a deeply mistaken
picture.

We arrive at manifest time no differently than we do many of our other models of
the world, through a mix of innate endowment, experience, and cognitive inference.
Science as an enterprise is a kind of refinement of these inferential techniques, improving on common sense by increasing the data—via experiments, data collection,
etc.—and the rigor of the inference—via advanced statistics, randomized control trials,
meta-analyses, etc. Something funny happens to our understanding of time as these
methods are ratcheted up: time gets deflated to the point where it hardly seems to count
as time anymore. Science, and fundamental physics in particular, develops its own
image of time, *physical time*. Whereas manifest time works in ordinary life, physical
time is what works in doing physics. If we assume that physical time is our best
theory of time, then it turns out that manifest time paints a very misleading image
of time.

By itself this is hardly surprising. Science is constantly informing us that our pre-
scientific representations of the world are wrong. Regimented common sense tells us
that space is Euclidean when it is really (as an aspect of spacetime) non-Euclidean.
Folk physics tells us that objects dropped when we're in motion fall straight down, that
the Sun rotates around the Earth, and so on. That a conception of the world succeeds
in some aspects of ordinary life doesn't guarantee that it will succeed when exported
elsewhere. We ought not expect time to be different.

The special problem in the temporal case is that what we're apparently wrong about
really matters to us, and to add insult to injury, we have no good story about why we're
wrong. If folk physics is wrong about the geometry of space and the trajectories of
falling objects, well, we can live with that. We weren't so invested in those beliefs. By
contrast, whether something is past, present, or future affects our language, thought,
behavior, and even social policy in countless ways. Worse, we lack an understanding
of why manifest time arises in us. It's easy to explain why we treat space as Euclidean,
falling objects as falling straight, and the Sun as rotating around the Earth. No mystery
there: spacetime curvature's deviation from flatness in our local region of the universe
is small and therefore unlikely to be noticed by creatures like us. A similar story can
be told in the case of objects falling, and the Earth's motion explains why we thought

the Sun orbits the Earth. In the case of time, we have no such story. We're wrong and we don't know why.

The philosopher Rudolf Carnap reports that this same problem disturbed no less than Albert Einstein:

> Once Einstein said that the problem of the Now worried him seriously. He explained that the experience of the Now means something special for man, something essentially different from the past and the future, but that this important difference does not and cannot occur within physics . . . That this experience cannot be grasped by science seems to him a matter of painful but inevitable resignation. (1963, pp. 37–8)

The worry continues to this day, both in academic journals and the popular press:

> The [physical] equations make sense, but they don't satisfy those who ask why we perceive a "now" or why time seems to flow the way it does. *Science*, July 2005 (Anonymous, 2005)

> [T]his coldblooded mathematical formulation doesn't do justice to the experience we all have of being in time. *New York Times*, July 2013 (Overbye, 2013)

The choice before us appears bleak. Either physics is right about time or it isn't. If correct, then it seems we must throw away all that is near and dear to us in our temporal representation, chalking up the now and flow and everything that hangs upon them to an elaborate and unexplained illusion. If it's incorrect about time, then it's incumbent upon us to show where it errs and to devise a more successful theory. Having discovered the lion's share of contemporary physics, Einstein no doubt recognized the enormity of the second task and reconciled himself to the first outcome.

Unlike Einstein, Carnap didn't see resignation as inevitable:

> I remarked that all that occurs objectively can be described in science: on the one hand the temporal sequence of events is described in physics; and, on the other hand, the peculiarities of man's experiences with respect to time, including his different attitude toward past, present and future, can be described and (in principle) explained in psychology. But Einstein thought that scientific descriptions cannot possibly satisfy our human needs; that there is something essential about the Now which is just outside of the realm of science.
> (1963, pp. 37–8)

Unfortunately Carnap didn't pursue this project of explaining "the peculiarities" of our experience of time, yet I think it's important to do so.

Central parts of the time we live by, the structured time of life, have no counterpart in the fundamental science of our world. We seem to have, to echo the consciousness debate, an "explanatory gap" between time as we use it in life and as we find it in science. Resignation in the face of this gap is intolerable. Maybe the now and flow are illusions in some sense. But if so, they are remarkably deep and universal ones that demand explanation. Time is connected to far too much—causation, freedom, identity—to leave mysterious.[1] We shouldn't abandon hope of explaining consciousness or free

---

[1] "I soon discovered that the 'problem of time' is rivaled only by the 'mind-body problem' in the extent to which it inexorably brings into play all the major concerns of philosophy" (Sellars (1962b), p. 527).

will, nor should we manifest time. Rather than live with Einstein's resignation, I want to explain the now and associated temporal features. Reconciling these two images of time is for me the principal goal of philosophy of time. It's also the goal of this book.

1.1 Manifest Time

Watching a gull fly over the local lagoon, I see a continuous shift in position, two wings moving in harmony, one flap after another. As it approaches, it cries out, perhaps as a warning. My impression is of a smooth process. No gaps, no dislocations of order, no flicker. I experience the sound from the cry as happening when the gull's mouth opens. These impressions and others are bound together as simultaneous, part of my subjective present, each set of simultaneous impressions sewn together with the "next" in what appears to be a continuous temporal sequence. My experience overall roughly matches the external reality.

This peaceful scene belies the commotion lurking underneath and the work needed to overcome it. The world is teeming with activity. Objects are blooming and buzzing by, regularly altering their locations and properties while bombarding us with a discordant cacophony of signals. The light reflecting off the gull arrives at my eyes much faster than its cry does to my ears. Tonight is humid, so sound is especially slow. Some of the gull is in shadow, other parts illuminated. While admiring the gull, I also see a child running, hear the dog barking, feel the sand's texture . . . Despite the assault from all sides by a chaotic mix of information, we are able to form mental representations of the temporal sequence of events that—especially for those events that matter to us—more or less matches the objective temporal sequence of events. We hear the gull's cry when we see its mouth open, notwithstanding the signal discrepancies. As we'll see in Section 1.1.1, we get it "wrong" to get it "right": our brain and perceptual system "fix" the information so that properties of the same event are experienced together.

Getting the timing approximately right is only the beginning of our relationship with time. We also experience duration—how long events last. Here too we can distinguish between the objective temporal properties and our experiences of them. The film *Pan's Labyrinth* may have seemed to you an eternity, but for me it flew by. Perhaps you're not a fan of dark fantasies, or maybe you had a hard day and weren't very alert. Science tells us that the subjective sense of duration depends on an indefinitely long suite of variables, including wakefulness, interest, mood, caffeine intake, and much more. We therefore disagree about how long the film seemed. But *seemed* is the key word here, for when presented with the evidence we do not disagree that objectively the film took 112 minutes, that in that interval the Earth rotated approximately 7.8% of its daily rotation, and so on. Similarly, time may have "frozen" for me during a car accident, but I do not expect it to have done so for those safely at home.

We distinguish between subjective and objective duration. The overall pattern of covariations make plausible the idea that some aspects of our experience are due to us and some due to the outside world. Though we remark on the discrepancies when watching a film, what's again interesting is that the experience of duration more or less matches the objective duration for most macroscopic (observable) events. If I ask you to match tones that are fairly different in duration, odds are that you won't make many mistakes. For longer durations, some studies suggest that minus temporal cues each objective hour subjectively is judged on average to be 1.12 hours (Campbell, 1990). We employ many biological clocks, and some of them are quite good at estimating durations of salient macroscopic events. We make mistakes judging durations at the level of milliseconds, but even here we tend to be within 10% of the objective duration (Mauk and Buonomano, 2004).

Temporal experience, in the above respects, more or less gets time right. Or slightly more accurately, often it gets the duration and objective temporal sequence of events approximately right when those events are macroscopic and salient to us. These experiences, when coupled with some of our other abilities—crucially, memory—allow us to develop a kind of rough-and-ready map or timeline of events in time. This map provides information about the temporal sequence of events—what events come before other ones—and also the duration of events—how long they last.

Manifest time contains much more than a map of temporal order and duration, and not all of it coincides with what is directly experienced. That's the reason why I echo the phrase "manifest image" from the philosopher Wilfrid Sellars (1962a), for the manifest image is a kind of regimented common sense picture of the world—not what is directly experienced. Separating our actual experience from what we cognitively infer is tricky. For now, I want to put experience to the side and instead get our theory or proto-theory of manifest time out in the open. Some of these features are not based on "experience" in any normal sense of the word. Later, when we dig deeper into manifest time we'll try to determine which of its aspects, if any, are supported by conscious phenomenology—as, say, seeing red supports the existence of redness. Conflict with common sense is forgivable (and often what makes science exciting), but conflict with experience is unacceptable.

The picture of time we inherit from childhood is a vast, complicated tapestry, one strewn with confusing, vague, and perhaps even inconsistent threads. The resulting model of time has *dozens* of aspects to it. It is, I claim, ordered, metrical (events have durations), flowing, global, one-dimensional, shared, continuous, absolute, productive, and much more. Tackling all of these properties would make our investigation unwieldy. As a result, although I will stray slightly here and there, I want to narrow our focus to the temporal counterparts of what psychologists of spatial representation call *egocentric* or *perspectival* or *internal* representations. In the case of space, there are spatial frameworks in which one locates objects by reference to one's own spatial location or *here*. Examples are claims such as that the fork is on the right and that the car is in the front. *Nonperspectival* or *allocentric* spatial frameworks, by contrast, locate entities independently of any particular perspective, e.g., Kansas City is between San Diego and Providence. The difference is easy to see: spin around and the fork is on your left and the car behind you, but Kansas City is still between those same two cities.

Psychologists, linguists, and philosophers also make the temporal version of this distinction. Egocentric temporal frameworks are those that locate events by reference to one's current temporal location, e.g., Socrates' death is past, and the first mile-high building is future. McTaggart (1908) famously points out that we regard events as arranged in a so-called *A-series*, where events move from the past to the present and to the future. Crucially, this is done with respect to the distinguished now mentioned above, for the now is what defines past, present, and future. In linguistics and cognitive science the A-series is referred to as *deictic time*. This is an attribution to time based on a temporal reference point; so long as that point coincides with McTaggart's now—typically the time of utterance—then the two are the same. Philosophers and linguists will also refer to such attributions as "tensed." Allocentric temporal frameworks, by contrast, locate events without reference to one's current temporal location or a now. The relations before and after serve this function. Calendars and history accomplish this, as do clocks. McTaggart called the series of events running from earlier to later the *B-series* whereas psychologists may call it *sequence time*. These types of temporal attributions are "tenseless." One way to notice the difference between egocentric and allocentric temporal representation is to appreciate that it's "always" true that Socrates drank hemlock *before* he died, no matter when I claim this, yet that Socrates *will in the future* drink hemlock is not true if uttered now. The truth of a remark about the A-series varies with the changing now, unlike the truth of a remark about the B-series. March 1968 is later than December 1967 no matter where the now is. The later than relation obtains independently of one's current temporal location. The key to understanding the B-series is to note that it is devoid of any deictic center; therefore, minus a point anchoring the now, there is no past and future in this series, only before and after.

Developmentally, there is some reason to think that we adopt egocentric temporal representations by about the age of 4 or 5. Much is unknown and interpreting experiments is a delicate task. However, I suspect that temporal decentering is crucial to the development of the idea that all events are unified in an A-series. Taking our cue again from space, where more work has been done in developmental psychology, spatial decentering is the ability to adopt a spatial frame on an object from a location that isn't where the subject is located. Although the fork is on my left, I can take your perspective and see that it's on your left. When I can understand that these are all perspectives on the same spatial map, then I have achieved spatial decentering. Similarly, temporal decentering is the ability to adopt a temporal frame on a time that isn't the time when the subject is located. Without it, it's hard to see how we could understand the idea of the past, present, and future updating itself and flowing by.

A child may understand that breakfast happened in the past and dinner will happen in the future. But when do they learn to decenter? When do they also appreciate that today's breakfast is to the future of yesterday's dinner? When can they "lift" their now and move it along a temporal sequence? In probably the first work on temporal decentering, Cromer (1971) told children a story of a child who climbs a ladder up to a slide and then slides down, hitting the bottom a bit too hard. Illustrated in a series of pictures arranged from left to right in accord with the later than relation, children were asked to point to the card representing a child who might say, "I will go down the slide." Cromer found that by age 4 or 5 children could answer correctly. He interpreted the task as displaying mastery of decentering. To be successful at the task, after all, children had to place the now at a different time than their own.[2] Spatial decentering is prompted by the most ordinary of active perceptions—just turning one's head. But to learn to temporally decenter it seems memory of the self in past times is crucial (McCormack and Hoerl, 2008). Unless one can recall that what is now past once was present, one will have little experience supporting decentering.

The adult conception of time is one in which we have detached temporal properties from particular events and is fully decentered. We refer to moments of time themselves without reference to any particular event. And we can easily slide the tripartite division of the world into past, present, and future up and down a unified timeline. Although this division {past, present, future} is an egocentric as opposed to allocentric representation, don't let the terms confuse: manifest time treats these distinctions as *objective*, not egocentric. That is a key difference between spatial and temporal representation (see Chapter 9). As we navigate through life, we take these distinctions to be basic features of time and invest great importance in them.

Although manifest time marks many distinctions besides what I'm calling the egocentric, in what follows we'll focus on the now, the flow, and the past/future asymmetry. Dozens of additional features could be selected for study. I concentrate on these three because they play a central role in the way we live our lives, are more or less special to time, and yet seem to conflict with physical time.

1.1.1 Now

Inspirational t-shirt messages exhort us to "live in the now" (Fig. 1.1). Tattoo designs in Japanese characters commonly proclaim the same insight. Self-help guides promise ten-step guides to living in the moment. And today, we are told, is Pooh's favorite day.[3]

While one can appreciate the advice not to dwell upon events outside one's control, there is a sense in which these clichéd exhortations are annoying—not just for being clichés—because taken literally, we have no choice but to live in the present. That is when all the living—conscious actions, decisions—seems to happen. Events in our lives don't seem like they're simply strewn along a four-dimensional manifold. We think that there is a further fact of the matter about when they *happen*. When we

---

[2] With its reliance on language comprehension, the experiment has many weaknesses. Still, it fits with other data (Friedman, 2004; Weist, 2002) suggesting that the ages of 4 and 5 mark a crucial transition in the development of our temporal representations of the world.

[3] "What day is it? It's today, squeaked Piglet. My favorite day, said Pooh." A.A. Milne, *Winnie-the-Pooh*.

Figure 1.1 Clichéd T-shirt

conceptualize time and our lives, we regard one set of events as 'lit up,' as especially vivid or happening or real. In a sense you are stuck in the now and to be told to stay there seems unnecessary.

The nature of the psychological present has been controversial in psychology and philosophy for well past a century. Obviously, we process temporal information over many time scales. To localize sounds, one needs to be able to discriminate temporal intervals as short as 600 μs (the time it takes sound to travel the distance from one ear to the other). Longer periods in the tens to hundreds of milliseconds range are needed for motion detection, speech recognition, motor coordination, multisensory integration, and music perception. Predicting events, such as a skateboarder darting across your path, might be in the range of a second or so. Circadian rhythm tracks information through a day. The mechanisms subserving these processes typically differ: circadian rhythm occurs due to a complicated protein translation–transcription feedback loop, sound localization occurs thanks to axonal delay lines, and yet neither is the mechanism necessary for the other processes listed. Amidst all of these processes, how do we determine what is part of the subjective present and what isn't?

Discussion of this question has been confused for a long time. People take their present perceptual experiences to include non-instantaneous temporal properties such as succession and duration. Watching the second hand on a clock just now, it seems to be in motion—yet motion takes time. This nonpunctate "present" was dubbed the *specious present* by E. Robert Kelly in the late nineteenth century.[4] Research has been directed ever since at establishing the specious present's duration. Due to the

---

[4]   Kelly published under the pseudonym "E. R. Clay." For an excellent discussion of the history and philosophy, see Andersen and Grush (2009).

vehicle/content distinction sometimes being elided (i.e., mistaking how long is the experiential percept versus how long the percept represents), estimates tend to vary widely, from in the milliseconds range to many seconds.

The problem at bottom is that there is no clear experiential boundary delimiting the present. Absent that, we can define it in terms of various operations (e.g., sensory integration, attention). The psychologist Michon describes the present as "a time interval in which sensory information, internal processing, and concurrent behavior appear to be integrated within the same span of attention" (Michon, 1978). That is helpful but still somewhat vague because these operations vary depending on target and much else. The English concept *present* may not be a natural kind of psychology; that is, the science may have no use for the notion. It may be more accurate to simply specify certain time scales and say what is happening then, with some of these scales better latching onto our concept of the present than others. For convenience's sake, however, I'll still speak of the *subjective* or *psychological present*, by which I'll mean the immediate conscious experiences typical of these vague but very short time scales.

No doubt inspired by these short experiential "presents," we attribute to the world itself a distinguished present moment (Fig. 1.2). It's not simply that moment by moment information comes to us through different channels. It's that the world itself is lit up by presentness. When it's so illuminated, we think, we're conscious and we can *do* things. The past is lost to us—time cliché warning: what's done is done—but changes enacted during this special moment can ripple into the future. We have *power* in the now. The now is so central to our lives that it's almost hard to appreciate its importance.

If one looks, however, one finds its mark all over our language, thought and behavior. In English one can scarcely say anything without implicitly marking a

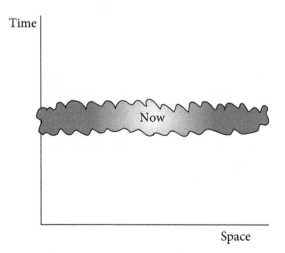

Figure 1.2 Dividing the world in two

relationship to the present. If I wish to express my connection to happiness, English forces me to decide: am I happy, was I happy, or will I be happy? The choice hangs on happiness' relationship to the present. For other languages this information gets conveyed differently, but it's always there. The anthropologist Alfred Gell reports that all known languages denote presentness in one way or other (Gell, 1992).

When we talk about things happening now, we often feel free to drop any reference to clock or calendar time. "We're going now," I tell my children, without any mention of a clock. We'll even drop the word "now." "Look out!" means look out now. Those who yell back "When?" are people who get hurt. As with other indexicals, the time being now can motivate people to action. Learning that *now* is 11:45 motivates one to rush for a noon meeting in a way that merely being told that 11:45 is 15 minutes prior to noon doesn't (one already knew that). As we'll see in Chapter 12, our preferences, attitudes, and desires are also all finely tuned to the present.

So confident are we that the present is shared by everyone, we assume that it is objective and global. Snap your fingers. That snap picks out a present event for you, but you also assume this present extends indefinitely. If there are aliens living on some extrasolar planet, then their present is the same as yours when you snapped your fingers. Events don't escape the reach of the present. We probably don't explicitly model the extent of the present. But it's being indefinitely extended is what enables the now to slice the world into unambiguous past and future regions.

Treating our present as objectively special stands in sharp contrast to our practice with the spatial counterpart of the now, the spatial *here*. We are not tempted to objectify our egocentric divisions with respect to space. Here, there, back, front, left, right . . . we know that these attributions hang on a choice of reference frame. That is why we say the fork is on *your* left, *his* right, and so on. We don't treat the spatial here as objective. Carving the world up with a distinguished present slice and not doing the same with any spatial counterpart is a critical difference between manifest space and time.

1.1.2 Flow

The present is not "stuck" on one set of events. Time is thought to flow or pass. Poets, philosophers, and musicians from ancient times to now all tell us that time is like a river. Some of this attribution we readily take as subjective, as when we say that time flies when we're having fun. We don't mean that time itself sped up because we were enjoying ourselves. But some of this attribution is directed at time itself. We commonly speak of time itself as moving, dynamic, changing, passing, and so on. All sorts of genetically and geographically distinct language communities commonly ascribe motion to time, including English, Japanese, Chinese, and the Niger-Congo language Wolof (Evans, 2003, p. 14).

There are many senses in which time is said to flow. What I'll regard as the primary sense is the idea that the tripartite structure {past, present, future} is continuously

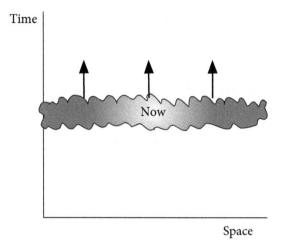

Figure 1.3 An updating now

updating itself in a temporally directed manner: future but not past events *become* present. Time flows in the sense that the tripartite structure is continually refreshed with each new present (Fig. 1.3). Given that we intuitively feel that the now is something objective, it is not surprising that we believe in a kind of flow. Northern Californians may be fixated on 1968, but only a few think that it's *still* 1968. The now updates itself. Characterized as I have, the flow of time is predicated on the existence of an objective now. I won't assume that the now is prior to the flow in any non-logical sense, however, for it's probably better to think of them as two aspects of a package. We probably wouldn't be tempted to regard the subjective now as picking out something objective if it didn't change content.

Whether the flow of time is given in phenomenology is a hotly contested topic. No doubt our experiences and their contents change. The question is whether these experiences (or others) are a reflection of time itself flowing. Since flow hangs on presentness, the answer is tied up with the similarly controversial issue of whether presentness is a phenomenological property.

The flow of time is often connected with change. Distinguish between change and *souped-up change*. Change is merely the having of different properties at different times. The moving ball occupies different locations at different times. To change one's mind one adopts new beliefs. Playing basketball changes one's life by making its practitioners happier but also orthopedically damaged. This ordinary understanding of change doesn't generate conflict with physical time. I don't wish to deny that the metaphysics of physical movement is a rich ongoing topic, nor that cutting edge physics like quantum gravity may demand some readjustment in our picture of change. My point is simply that physics has been successfully handling the alteration of physical objects for hundreds of years. We won't generate an interesting conflict with physical time by simply pointing out that there is change. Souped-up change, by contrast, is connected with time flow. The moving ball is not merely occupying different locations at different clock times, but it is *now* at a certain location. Change in this sense is the alteration of properties with respect to the present. It is sometimes thought to be an additional difference between temporal variation and spatial variation. Leaves turning color in the autumn is *real* change; the distribution of colors across my plaid shirt is *mere* spatial variation. What is the difference? If we make reference to the now, we have a difference between the two, one parasitic upon the flow as currently understood.

The updating of the now is both ignored by physics and without a spatial counterpart. True, we do update the spatial *here* as we walk around. Yet because we don't treat it as objective in the same way as the now, we do not think that space itself flows.

1.1.3 Past/Future asymmetry

> Araman said, The past to you is the dead past. If any of you have discussed the matter, it's dollars to nickels you've used that phrase. The dead past. If you knew how many times I've heard those three words, you'd choke on them, too.
>
> Isaac Asimov, "The Dead Past"

As time flows, we don't merely note a special present whooshing by, but we also invest important differences between the future in front of it and the past left behind. Imagine that you're making an important decision . . . whether to get married, move, quit your job, order the flourless chocolate cake or the peach cobbler, etc. You agonize over the selection, evaluating all the pros and cons of each possible future outcome. The future, it seems to you, is truly *open*. Anything is possible, even the flourless chocolate cake. The past, by contrast, is determinate and settled, leaving you with no hope of past events being undone. After lengthy deliberation, the moment of choice arrives and you opt for the cake. After devouring it, you reflect on the decision: no regrets.

Notice two temporal asymmetries in the above story. First, the present whooshed by in one direction and not the other, consistent with the above flow. Second, observe that at any given moment we regard the future as open and the past as fixed. Now, the dinner's fate is sealed. Openness attaches to the future, and the future is only future with respect to a changing tripartite division of reality. This second asymmetry between the past and future is what I'll dub the *past/future asymmetry* (Fig. 1.4).

The direction of openness is intimately associated with what psychologists call our *sense of agency*. As in the chocolate cake example, we have an irresistible urge to believe that future options are open to us in a way past ones are not. The future seems to have branches—possible futures—and we enjoy the power to prune these branches to the actual one. We'll discuss this sense in Section 11.2. Right now, let me point out that the past/future asymmetry shapes many aspects of our lives.

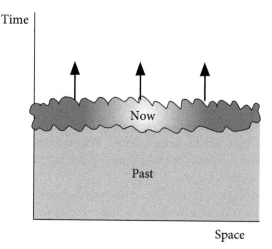

Figure 1.4 The past/future asymmetry

Psychologists have recently unearthed an unexpected range of temporally asymmetric judgments. In Chapter 12 we'll look at some of the more surprising ones, such as that whether we regard an act as intentional may hang on whether it is past or future. Focus for now on the asymmetry that caught the attention of philosophers of time, the *preference asymmetry*. One and the same event will induce in us desires and attitudes that differ systematically with our temporal perspective on it.

This can manifest itself in two ways, so there are really two preference asymmetries. In one, desires and attitudes differ depending upon whether the event is past, present, or future. We dread future pains and feel relief when they are over, but we don't dread past pains and feel relief that we're not in the future. Had you ordered the peach cobbler, you would now regret it (flourless chocolate cake is better); you wouldn't have regretted it before ordering it. In the other asymmetry, desires and attitudes differ depending upon whether the future event is close or distant. We're horrible procrastinators. We prefer distant future pain to proximal future pain. Arguably, this difference in valuation is one reason for many of our personal and social misfortunes. (Think about your credit card behavior for the first and society's response to climate change for the second.) The philosopher Baruch Spinoza held that insofar as we're rational, we ought to direct our attitudes equally toward the past and future. Perhaps Spinoza is right that we *ought* to retrain our attitudes. Even if possible, doing so would constitute a radical departure from normal human behavior.

Both of these preference asymmetries rely on the tripartite division of the world into past, present, and future. This dependence is clear in the first preference asymmetry, for the difference is triggered by whether (say) a headache is past or future. It's equally the case for the second asymmetry. Granted, we—all else being equal—prefer our pain to be in 2027 and not 2017. But when 2027 rolls around, we prefer our pain in the distant future again. We don't say, "good, I'm getting the pain just when I wanted it, 2027!" This asymmetry between proximal and distant initially strikes one as independent of a preferred present, but it isn't.

Don't confuse the past/future distinction with related asymmetries that don't depend on the tripartite division of reality. Consider, for example, the *asymmetry of causation*, the fact that causes typically precede their effects. Your decision to go for the cake, we can agree, caused the cake to be delivered. That is true from any temporal perspective, unlike the openness of the cake choice. When later asked about the decision, you won't think that the choice is still open, but you will continue to believe that the decision caused the cake to arrive. The causal asymmetry attaches to the *earlier than* relation, not the past/future asymmetry, whereas the openness asymmetry attaches to the past/future asymmetry, not the *earlier than* relation. The world contains many directed philosophical processes, e.g., entropy increase, cosmological expansion, as well as many general asymmetries such as the causal asymmetry. I believe they underlie the past/future asymmetry, but it's important not to conflate the two.

Like the now and flow, the past/future asymmetry couldn't be more significant for the way we navigate through life. Yet physics doesn't posit an open future, nor does our picture of manifest space have a counterpart. Take a three-dimensional plane intersecting your here that isn't the present. No one is tempted to think that the "east" side of this plane is fixed and the "west" side open.

1.1.4 Is manifest time universal?

I speak of "manifest time" as if it were a single representation of time, common to all people. However, time has many features and surely we have many representations of these features that differ from person to person. Influenced by relativistic physics, I personally conceive of time as a special set of directions on a spacetime endowed with a semi-Riemannian metric structure. We can be reasonably confident that Shakespeare did not share this conception. What do Shakespeare and I share when representing time? Clearly not our high-level theoretical beliefs about time. Theoretical beliefs about time may flat out contradict the tenets of manifest time. Many cultures embrace cosmological myths implying that time is circular in some sense; similarly, physicists sometimes propose models of circular time. In either case we find theoretical beliefs that contradict (for example) a strict past/future asymmetry.

Manifest time is the proto-theory of time we develop before coming to these heavily theory-laden views about time. What I suspect remains more or less invariant in us are the core aspects of manifest time that we've been discussing, including (but not limited to) the felt specialness of the present, the flow of time, and the past/future asymmetry. These concepts (past, future, etc.) are products of high-level cognition too, but they are developmentally prior to higher-level beliefs about cosmology and more grounded in bodily experience. These core aspects appear to be natural responses to a set of very basic challenges most organisms face when trying to organize temporal experience and communicate with others about it—or so I'll argue.

Support for this comes from those who study cross-cultural variation in temporal concepts. Not only have they yet to find evidence that these core features vary, but they tend to treat these aspects as "core" temporal concepts too (Nùñez and Cooperrider, 2013), as we'll see in a moment. Manifest time remains stubbornly persistent in the face of contradicting high-level beliefs. Suppose I embrace a model of cosmological inflation that posits a closed time. That still won't give me a sense of agency over past events. Shakespeare didn't regard time as a special direction on a semi-Riemannian metric space. Yet when Lady Macbeth tries to sooth her husband by saying "what is done, is done," Shakespeare is reaching across cultures and time to a shared representation of the past/future asymmetry.

I admit that I'm engaging in a bit of rational reconstruction when characterizing manifest time. The distinction between "high-level" cognition and "lower-level" cognition is not sharp. Maybe our temporal concepts never fully escape theory. That's fine; then the question is just one of degree and not kind. I've also cleaned up the model in ways nature probably didn't. If I am consistent in my attributing distinct ontological attributes to the past and future, the present had better slice the world exhaustively in two; otherwise there exist regions that are neither past nor future. My language, behavior, and thought thus commit me to the idea that the present is global. Yet few are actually asked about, or reflect upon, whether the now extends all the way to Andromeda. Consistency may demand it, but perhaps we're not consistent? That fuzziness is fine. To the extent that manifest time is committed to less than my reconstruction of it, so much the better. It's easier to recover less than more. Not knowing how messy matters here are, convenience suggests using the clean version.

Do people speaking different languages, living in different cultures, places, and times all share these core components of manifest time? I've said that there is no evidence of cultural or other variation on these aspects, but this claim is commonly denied in fields as diverse as linguistics, cognitive anthropology, and cultural anthropology. Let's briefly probe these challenges. While I'm a big fan of many of the findings in these fields and of course open to empirical falsification, I do believe that the evidence points to the core aspects of manifest time being surprisingly widespread. Too often the more radical conclusions to the contrary are not warranted by the evidence. The judgments are based on either a philosophical error, not clearly separating the many different temporal concepts we employ, or overly enthusiastic reporting.

Societies measure time in different ways, and in different cultures the pace of life varies. We invent clocks, metrology, and impose standardizations upon duration. Time gets chopped into weeks, days, hours, minutes, and so on. Often these actions have wrought huge changes upon the world, playing a role in the story of industrialization and labor. The introductions of the church bell tower and the punch-clock are prime examples in this narrative. Historians and sociologists have recently shown how the extension of time standardization also fell in step with imperialistic ambitions, as we learn via a study of telegrams, semaphore lines, phones, and railroad timetables.[5]

Likewise, interesting work in social psychology demonstrates that the "pace of life" affects us in surprising ways. How fast we walk, talk, and operate business may vary with the size of the city we live in, the average temperature of the city, the country's culture, and much more (Levine, 1998). In certain precise senses, the pace of life is slower in Sicily than in New York City. There is no doubt that what society or culture we live in affects the way we think about time and how we go about living. Yet none of these important observations threaten the widespread common components of manifest time. Industrialized or not, we have no evidence that people don't or didn't regard the present special, the past settled, and so on.

My claim flies in the face of at least one important tradition. The great sociologist Emile Durkheim argued that the philosopher Immanuel Kant was wrong, that time couldn't be a universal category of thought (Durkheim, 1912). Feeling that time must be specific, that it must come in definite sub-categories of minutes, hours, and so on, he insisted that our understanding of time must accordingly be society or culture relative. In Durkheim's wake various schools of thought in anthropology have assumed that this social relativity about time is right. Surely it is for time standardization. How we chop up durations is mediated by society, culture, and language. But that doesn't mean that all temporal concepts are so heavily mediated. The Durkheimian project seems based on a conflation between the units we use to measure and express temporal facts with the facts themselves. A milk carton may be a gallon to me and about 3.8 liters to you, but we'll agree on how many fit into a volume twice that size. Similarly, the calendars used by Mayans differ from those used by Aztecs, but still there is no evidence that they disagree on (say) there being a past.

Gell, summing up his study of time and culture, writes:

> There is no fairyland where people experience time in a way that is markedly unlike the way in which we do ourselves, where there is no past, present and future, where time stands still or chases its own tail, or swings back and forth like a pendulum.   (1992, p. 315)

I take it that Gell is saying that there is no evidence that people differ much over the core features of manifest time. Throughout history and across cultures and languages one finds the same attitudes towards these core features expressed again and again. High-level cosmological beliefs vary, as do standardizations and the pace of life, but people still share the idea of a deictic time, a now changing along a timeline.

Perhaps the most famous type of relativity associated with time is the claim that language mediates one's conception and perception of time. One can trace this idea to the famous—and now discredited—claim that members of the Native American Hopi tribe don't share our conception of time because they don't have counterparts to some of our temporal terms and other linguistic devices. This is the temporal

---

[5] See, e.g., Aveni (2002), Kern (2003), Galison (2004).

version of the Sapir–Whorf hypothesis, the claim that the structure and lexicon of one's language shapes how we conceive and even perceive the world. With regard to color, for instance, the idea is that one can't see or conceive colors for which one has no color term. Before the modern period, Japanese used one word, *Ao*, for both the colors blue and green. Contrary to Sapir–Whorf, that didn't mean that Japanese couldn't distinguish the two colors, couldn't learn their names, perform well on color tests, and so on. There were no significant experiential and conceptual differences between the Japanese and Western world as regards color.

The Sapir–Whorf hypothesis also comes in a temporal form. Because the Hopi language lacks various linguistic markers of tense, Whorf concludes that the Hopi

> [have] no general notion or intuition of time as a smooth flowing continuum in which everything in the universe proceeds at an equal rate, out of a future, through a present, into a past. (1956, p. 67)

As with color, this claim has proven suspect (Malotki, 1983). The Hopi in fact have plenty of linguistic resources to express tense, units of temporal measurements, and much more. Even if they didn't, the greater sin lay in thinking that they necessarily have a different understanding of time as a result. The Hopi language contains a term for *yesterday*. Yet if it didn't there still wouldn't be reason to believe that the Hopi didn't have the concept. A great success of cognitive science is that it has falsified many of the more extreme versions of the Sapir–Whorf hypothesis.

That statement is compatible with the truth of some weaker forms of the hypothesis. Just as it would be surprising if language exerted an *overriding* influence on our understanding and experience, it equally would be surprising if it exerted *no* influence on our understanding. A muted form of the Sapir–Whorf hypothesis, according to which there is some influence from language on our concepts, seems likely to be true in many domains. Currently, there is evidence that the language one speaks affects various subtle features associated with our thinking. In word association tests, for instance, speakers in languages where the word *bridge* is male associate more traditionally masculine adjectives with bridges than speakers of languages wherein *bridge* is female (Boroditsky et al., 2003). And in the domain of color, having a color term might allow one to discriminate it from other colors faster than if one doesn't have a color term. Perhaps Japanese people with words for blue and green would perform better or faster at discriminating the two than those with only the word *Ao*. In this case and others it's not that the experience or concepts are entirely dictated by language. No, far from dramatically Whorfian, these fascinating dependencies may result from simply associative learning or some other well-understood process. To steal a nice paper title, there is no reason to be afraid of the big bad Whorf (Casasanto, 2008).

That said, cultural and linguistic relativity as regards time may be much greater than is commonly thought—even if not about the "core" aspects. Probably the best and most exciting evidence of relativity comes in the recent work of Lera Boroditsky, Rafael

Nunez, and other cognitive scientists who focus on cross-cultural variation, much of it with an eye on how we recruit spatial metaphors and gestures when communicating about time. Consider the spatial embodiment of time. Readers of this book probably describe the future as in *front* of them and the *past* behind them. While that seems natural, it is not so to many other people. The Aymara of the Andes do precisely the opposite: the future is behind them and the past in front. Perhaps more surprising, for the Australian Pormpuraaw the past is east and the future west. That means the future is to the right if one is facing north but to the left if one is facing south. The gestures employ geocentric rather than egocentric spatial reference. Even more unexpected is the recent discovery (Núñez et al., 2012) that a remote Papua New Guinean tribe, the Yupno, point downhill to the mouth of the local river to indicate the past, no matter how they are oriented (so long as geographical cues are present). The future, by contrast, is represented as lying in the direction of the uphill source of the river. This is so, apparently, no matter their outdoor orientation. A Chinese proverb states that time flows like a river. For the Yupno there is a sense in which that is correct.

What can we infer from this diversity? An otherwise fine *New Scientist* piece reports, "Time for the Yupno flows uphill and is not even linear" (Ananthaswamy, 2012), suggesting that the Yupno understanding of time is very different from ours. Was that shown? Not exactly. It's easy for media outlets to conflate *construals of time* with temporal concepts themselves. To indicate temporal direction the Yupno apparently developed a system of gesturing in space whose frame of reference is based on a river. To get from gestures in space to concepts and beliefs about time, however, is a large gulf to cross. Your "thumbs up" gesture probably does not imply that your concept of the positive is tied to the local gravitational gradient. Similarly, the Yupno's gesturing uphill to indicate the future may not mean that their concept of time is any different than ours. Systematic gesturing conveys much information. Advocates of cognitive metaphor theory and conceptual blending in psychology are often criticized for being willing to make the inference to the concept of time too quickly (Gleitman and Papafragou, 2005). Those in these schools respond with ever more surprising evidence of dependency between construals of time and culture and language.

Fortunately, I have no stake in this controversy. The reason why is that even if we leap from spatial gestures and language to temporal concepts, the evidence doesn't suggest that the Yupno or anyone lacks the core features of manifest time. In fact, quite the opposite: for the Yupno uphill/downhill corresponds to future/past, the spatial here corresponds to the now, and so on. The Yupno and other groups may construe tensed or deictic time with different spatial gestures than we do, they may speak about that time with languages quite different than ours, but at bottom, they still hold fast to the tensed or deictic series, the core feature of manifest time. *The very evidence produced in favor of cultural variation turns out to be some of our best evidence that the core features don't vary.*

If I am right, there is a reason for the ubiquity of manifest time. Manifest time is to a large extent the natural by-product of solving certain problems common to our species. In order to successfully communicate with others, to navigate throughout the world, to organize our rich and confusing sensations, and so on, we must adopt certain strategies. Perhaps other species share aspects of manifest time too, if they confront the same challenges.[6] The explanation for the widespread adoption of manifest time hangs on deep features of our psychology, biology, and physical environment. To the extent that these conditions hold—and they hold quite widely—I expect the core features of manifest time to appear. I don't deny that language, culture, and environment may affect our conception of manifest time in various ways. Yet given our current state of knowledge, it seems that the evidence sits firmly with the claim that the core features of manifest time are widely shared amongst the individuals of our species.

1.2 Physical Time

No one conception of time emerges from a study of physics. Like any other body of knowledge, physics changes—sometimes through wholesale theoretical revolutions, other times via modest alterations or even new interpretations—and our conception of physical time changes accordingly. Time itself doesn't change, but our understanding of it does. Currently we lack a unified picture of physical time. Our best theory of the small, quantum theory, is not consistent with our best theory of the very large, relativity. Worse, we know that they clash in their treatment of time.[7] For this reason, physics doesn't provide us with a single unambiguous conception of time.

Despite all the flux and controversy in our understanding, physical time—so long as there has *been* physics—has *always* clashed in various ways with manifest time. I'll support this claim as concerns current physics in later chapters. Fortunately, it's fairly easy to characterize the clash in general terms.

At a very coarse level, physics tends to model systems by describing them with a state in an abstract *state space*. The state space describes all the physically possible states the system can be in. If we were modeling a ball on a table, the state might be a particular location on the table and a particular velocity, for example. The state space would include all the other possible locations and velocities for the ball. The actual location in this possibility space, however, won't tell us how the ball moves. For that we require a *dynamics* that dictates the transition probabilities of moving from a state in this state space at one time to another state at another time. In sum, for the ball, the theory describes what states are available to the ball, what state it's actually in, and its chances of moving to another state given its current one.

---

[6] Experiments have been done to establish how much animals discount the value of distant future preferences being satisfied. See Stephens and Anderson (2001), for instance, for experiments on blue jays (but also see Hayden (2015) for caution on lessons from some of the animal studies). Any discount rate requires indexing via a present. That suggests that some animals invoke at least some aspects of an A-series-like representation of the world. How much I dare not guess. But given common evolutionary challenges, some overlap is perhaps not too surprising.

[7] Relativity tells us that time is dynamical, i.e., that time as an aspect of spacetime has its properties altered by the presence of matter-energy. Quantum theory standardly treats time as non-dynamical.

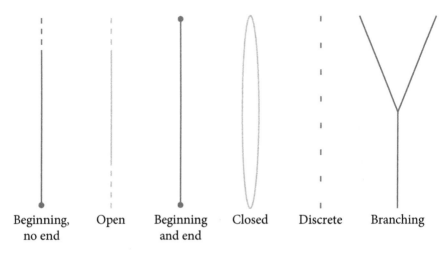

Beginning,   Open   Beginning   Closed   Discrete   Branching
no end   and end

Figure 1.5 Some possibilities for time

To devise a theory suitable for the ball, we need to do elaborate empirical invest-igation and theorizing to find the best combination of properties, state spaces, and dynamics. This demands scores of decisions about time. Are there instants? Ordered? Partially or totally? What is the topology (Fig. 1.5)? Is time continuous, dense, or discrete? Open or closed? One-dimensional? Nor is topology enough. Physics requires metrical structure. Metric structure allows us to define distances amongst events, or in the present case, the duration between instants. If I want to be able to predict how long the ball will be up in the air, we'll need a measure of duration. Less obvious but also true: velocities, accelerations, and much else are defined using the limits of various temporal intervals, which requires metric structure and also topological "neighborhoods" of instants around each instant.

The dynamics will also require choices. Is it Markovian? That is, do we need to input information only about the current state of the world, or do we also require information about, say, three hours ago? Is it time-reversal invariant, i.e., does it "care" about the direction of time? Is it time-translation invariant, i.e., does its state depend on what time it is if everything else is the same? One can see that physics, in devising good theories, makes many highly non-trivial assumptions about the nature of time.

One can then study the resulting commitments about time. In this book we'll engage in positive and negative projects, investigating what physics needs and what it doesn't need.

1.2.1 No manifest time

Physical time departs dramatically from manifest time. The important point here is not *how much* physics requires to represent the temporal aspects of the world, but *how little* it needs. No physical theory has ever required the properties characteristic of manifest time. Physics has always sufficed without positing an A-series. Aristotelian physics, Newtonian physics, Einsteinian physics, and quantum mechanics all manage without positing a special now. Lacking a now, they don't posit a temporal flow or a past/future asymmetry either, as both are dependent upon a now. This claim shouldn't be controversial. Look in any physics text and you won't find "now" in the index or anywhere else.

Where matters get more controversial is on the question of whether physics posits structure suitable for "animating" into manifest time. Relativity, our best theory that takes time as a target of study, not only doesn't distinguish any moment as present, but it also fails to provide structure that could serve as a good present if one wanted to "add" one to the theory. Suppose that you felt physical time was incomplete and were intent on adding a flowing now to it. To do so, you would have to grab a present and animate it, i.e., regard it as evolving along whatever time function the physics provides. Arguably, relativity doesn't have any good candidates for doing that. Relativity, I'll contend in Chapters 2 and 3, barely has structure worth dubbing the "present" at all. After carefully explaining time in classical and relativistic physics, I'll defend a loose dilemma, namely, that to the extent a structure corresponds to manifest time's present it is not relativistically invariant, and vice versa. We'll investigate some special "time functions" and the prospects for boosting one of these up as Time. Here we'll meet some interesting methodological choices facing relativity. Still the conclusion will be the same. Not only does relativity ignore the now, but in the language of Hollywood thrillers, it ignores the now with *extreme prejudice*.

Relativity's core insights may survive, but it is not the only putatively fundamental theory we have. Many think that quantum mechanics, although itself lacking a now, nonetheless is more friendly to manifest time than relativity. The idea is that it provides reasons for positing structure better suited for animation than relativity. Sir Karl Popper and many other distinguished philosophers and physicists argue that the quantum revolution released manifest time from the constraints of relativity. In Chapter 4 I'll examine ways in which it does and doesn't, asking whether we're being told to "bend" science in ways that don't optimize its theoretical and empirical virtues.

Not only does relativity have a competitor for current best fundamental theory, but it is also not likely to be our final theory. *Quantum gravity* is the name given to a set of research programs hoping to make relativity consistent or even unified with quantum field theory. As matters now stand, we have only a hazy view into the future because none of the programs are complete and none have been rigorously tested. Still, it's interesting to see how the fate of time hangs in the balance, as different research programs promise diametrically opposed consequences for time (Chapter 5).

As physics has progressed it has demoted time from its high throne. Ludwig Boltzmann suggested that the direction of time is not a feature of time itself but only of the local asymmetries in material process. Einstein then demoted time to an aspect of spacetime. The program known as canonical quantum gravity—plus more recent ones in the same spirit—promises to carry this dethroning to its logical end, namely, by *eliminating* claiming time altogether. By contrast, the program known as causal set theory boldly declares that it will *reinstate* manifest time to its rightful position. Advocates explicitly advertise the theory as a way of rescuing genuine time flow from the clutches of relativity. Without in any way offering the final word on the subject, I'll show that neither the restoration of manifest time nor the elimination of physical time is so easily achieved.

1.2.2 What makes time special?

Having assessed the prospects for animating structure in physics to vindicate manifest time, one can turn to the positive project of investigating time in our current physical theories. I'll largely focus on relativistic physics, as the "time" of relativity is neither as easy to eliminate or improve upon as is commonly thought; plus, its properties are spectacularly well confirmed. I'll begin by asking an old-fashioned question, namely, what is the difference between time and space? This query will take us down an interesting and largely untrodden path.

Physics is often said to "spatialize" time. That's not true, and I'll be at pains to rebut this impression. Yet as is sometimes the case with rumors, there is some residual truth lingering nearby. The truth underlying the idea that physics treats space and time alike is that physics treats the spatial counterparts of our three core manifest temporal notions the same way: it ignores them too! The counterpart of the now is the spatial *here*, and there is no *here* in physics. Without a here, there is no flow of space analogous to the flow of the now, nor do the right/left, up/down, back/front divisions mark ontologically significant distinctions in space. No one thinks that the here flows from east to west, leaving behind a fixed east. In these three crucial respects physics does treat time and space alike, not by spatializing time (as Henri Bergson suggested) or by temporalizing space (as Milič Čapek responded) but by shedding the same manifest properties. It's in this sense—and almost only this sense—in which time is spatialized by modern physics. It is only because these spatial counterparts aren't part of "manifest space" that their absence in physical space passes unnoticed. Once this is realized, one sees that there is plenty of room to treat time differently than space.

Although spacetime absorbs "space" and "time" in relativity, relativity still draws a remarkably deep and all-important distinction between the timelike and the spacelike directions on a four-dimensional manifold. I'll explore this distinction and other features that distinguish time from space such as the direction of time (as opposed to the direction of space), the natural kind asymmetry, a difference in dimensionality, and a difference in mutability. I'll then ask a question that I find natural but haven't previously encountered: why do all the logically detachable "temporal" features hang together? Why, for example, is our mobility hampered in the same directions as the set of directions distinguished by the spacetime metric?

In answering these questions, we'll pursue the intriguing idea that there is a special connection between physical modality and time, and in particular, between the laws of nature and time. Why do laws seem to govern in the timelike directions? Do they? Many have suggested an intimate link. Rather than simply assert that there is or isn't a link, I want to look under the hood, as it were, and see what connections can be found. In Chapter 6 I'll provide an informal and philosophical answer to this question inspired by the so-called "best system" theory of laws of nature. Using this answer, we'll be able to tackle the novel question of what unites all the otherwise detachable features associated with physical time. Providing an answer here will justify the novel claim of this chapter, namely, that *time is that set of directions in which our physical theory can tell the best stories.*

Then in Chapter 8 we'll tackle this question more formally. The laws of nature seem to be *productive*. One can take information across a complete three-dimensional spatial slice of time and march it forward to produce the distribution of matter and energy on the "next" slice of time. Even in relativity this is true. A good way to investigate the question of whether time is connected to the laws is to ask whether the laws can be productive in the non-timelike directions. By providing a precise sense of "productive," we'll be able to ask and answer this question, as well as rigorously illustrate the claims of the previous chapter.

"Tilt" the world over a bit. Put information on surfaces spanned by the entire lifetimes of light and march it "forward." Are the laws still productive? Now "tilt" the world on its "side." Place data on surfaces that are partly temporal and march it forward into spatial directions. Can we, compatible with the laws, narrate the story of the world from, say, "east" to "west"? I'll answer these questions and provide a surprising formal argument that illustrates how informative strength—good stories—can pick out temporal directions and bind together many of their otherwise detachable fragments. By looking at the world sideways, we'll learn something about what makes time special.

Learning about these differences between space and time is of independent interest, but it also fits in with the grand project of the book. These differences matter to creatures like us. Organisms evolving in a world with these differences will be shaped by them. They provide "hooks" onto which organisms may fix, potentially helping explain why we model the world with manifest time.

1.3 The "Two Times" Problem

Peering into physical time is illuminating, but no amount of focusing will bring manifest time into view. It's not there. How should we respond? Most physicists dismiss the flow of time and other features associated with manifest time as a grand illusion. In some loose sense this reaction may well be correct. But the effect of this dismissal is deeply unsatisfactory. Manifest time is deeply entrenched in us, and its core aspects seem almost universal. If not responding to physical time itself, still it must have an explanation. By dismissing it as illusory, one removes the project of explaining manifest time and places it on the desks of psychologists. The psychologists, however, don't know it's on their desk. The end result is that manifest time remains unexplained.

The project I recommend is analogous to the famous "two tables" problem in the beginning of Sir Arthur Eddington's *Gifford Lectures*:[8]

> we cannot touch bedrock immediately; we must scratch a bit at the surface of things first. And whenever I begin to scratch the first thing I strike is—my two tables. One of them has been familiar to me from earliest years. It is a commonplace object of that environment which I call the world. How shall I describe it? It has extension; it is comparatively permanent; it is coloured; above all it is substantial. . . . Table No. 2 is my scientific table. It is a more recent acquaintance and I do not feel so familiar with it. It does not belong to the world previously mentioned—that world which spontaneously appears around me when I open my eyes, though how much of it is objective and how much subjective I do not here consider. It is part of a world which in more devious ways has forced itself on my attention. My scientific table is mostly emptiness. Sparsely scattered in that emptiness are numerous electric charges rushing about with great speed; but their combined bulk amounts to less than a billionth of the bulk of the table itself.   (1928, pp. xi–xii)

Eddington presents us with two tables, one manifest table, the other scientific. The first is substantial, solid, colored, textured, heavy, and rigid, whereas the second lacks these features (Fig. 1.6). It is composed of mostly air, is hardly rigid, and so on. Which, we might ask, is the real table? Eddington answers, asking:

> You speak paradoxically of two worlds. Are they not really two aspects or two interpretations of one and the same world?

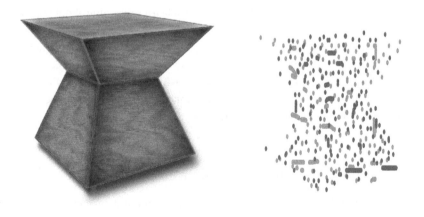

Manifest Table   Physical Table

Figure 1.6 Eddington's two tables

[8] See Savitt (2012).

To which he replies:

> Yes, no doubt they are ultimately to be identified after some fashion.

I'm sure most would agree. Perhaps the situation is best described as follows. Neither "image" is the *real* table. There is indeed one object there—it's just modeled in two different ways. For navigating through the macroscopic world of ordinary life, we represent it one way, and for doing physics, we represent it another way. The most useful model in the first context ascribes solidity to it and the best model in the second context doesn't (and so on).

That dissolves the official puzzle, yet a more difficult one lingers. Both models, despite ascribing incompatible properties to the one table, are marvelously successful, the first in our macroscopic dining pursuits, the second in making predictions about the microworld. If both are successful, then there must be something *right* about both the claim that the table is solid and the claim lacking attributions of solidity. Can we explain how it is that they are right, or nearly so? The tables need to be "identified after some fashion."

One way to do this is to begin with the scientific table and see if approximations to some aspects of the manifest table can emerge at the scales of interest to creatures like us. Assume that the physical model of the table is more or less correct. (Of course, we expect that our current physics is not the final word on what's going on, but that is beside the point.) Then we can appeal to facts about inter-particle forces, shielding at various scales, and much more in an attempt to explain why swarms of charges will resist other swarms of charges and generally behave at macroscopic scales like substantial, solid, and rigid objects. Physics will not explain all of the manifest table's properties, however. For properties like color and texture, we'll need to describe the human perceptual system and interactions between the surface and this system. For such properties, we'll need to embed a subject in the world and factor in its actions and abilities if we're to understand the manifest table.

We know—to a pretty good approximation—how to get from one of Eddington's tables to another. With that problem we are explorers who have found the signs, major trails, and even some of the transport to our destination. While acknowledging that the science is ongoing (e.g., materials science is hardly finished), armed with enough knowledge of physics and visual and tactile perception, one could tell a pretty convincing story that makes the manifest table less surprising from the perspective of science.[9] If the two tables do not yet enjoy a totally harmonious relationship, at least a kind a *détente* between the two parties exists thanks to this story.

---

[9] Let me be the first to admit that solving the "two tables" problem is *not* easy. It is enormously difficult to get even a hint of macroscopic physics from microscopic physics. Certainly there are plenty of non-trivial assumptions used in the "tight-binding" model that is used in solid state physics to explain why wavefunctions will be "repulsed" from the same atomic state—which will play a crucial role in the story of solidity. Still, I think one could at least begin a reasonable sketch of (say) how solidity emerges, fleshing out bits and pieces of this story in rigorous detail, precisely as one does in the tight-binding model. Perhaps another way to put the point: at least for solidity we have the tight-binding model, whereas we have nothing of the sort yet available for time.

No such *détente* exists between our two times. Einstein's worry is a genuine one. Beginning with the physical model of time, we do not know how to get to manifest time or anything like it. Here we are explorers deep in the mist. Perhaps the dim outlines of some landmarks are almost visible, but in general we are lost. Hence we have a "two times" problem. Unlike Eddington's problem, however, the present problem is much harder and more pressing. It is harder because physics succeeds in building explanatory and predictive theories of the world without appealing to the features of manifest time and also without providing any obvious counterparts. It is more urgent because manifest time, unlike the manifest table, occupies a deep and central place in our lives. We can't afford to dismiss either "image" of time. To understand our place in the universe, a reconciliation is badly needed.

1.4 From There to Here

The rock band R.E.M., mimicking the stereotyped answer one gets when asking for directions in rural areas, sang that one "can't get there from here." While that may be, it leaves open the reverse path, going from there to here. That is the course we'll follow. "There" in this case is physical time, the foreign conception of time we discover in physics, and "here" is the close and familiar manifest time. I agree with Einstein that time's manifest structure is deeply significant to us, but like Carnap, I do not wish to resign myself to its inexplicability. Against a background where many thinkers suggest that physics is incomplete or even inaccurate due to its omission of important temporal features, I think that it is important to give Carnap's strategy an honest try.

Consider Iggy. "Iggy" is short for an IGUS, an information gathering and utilizing system, discussed in Chapter 10. Drop Iggy into a relativistic spacetime, as in Figure 1.7.

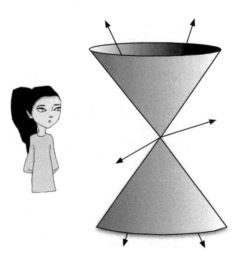

Figure 1.7 Iggy thrown into spacetime

Figure 1.8 Iggy models time as manifest time

Using all the differences between time and space in such a spacetime, plus facts about her and the interaction between her and her physical environment, can we explain why Iggy might model time as manifest time? Can we get from Fig. 1.7 to Fig. 1.8? That is the *modus operandi* of the second half of this book, and it requires some explanation and defense.

As with my treatment of Eddington's table, we'll begin with the *assumption* that the scientific image is approximately correct, i.e., we begin by dropping Iggy "there" in a world roughly as physics says it is. Why physics and not the times of other scientific theories? The answer is that physics is really the only science we have that explicitly takes *time itself* as one of its targets of study. Not only is it studied, but the physical theory of time is repeatedly subjected to rigorous and severe testing too. The test of relativistic time dilation by Hafele and Keating (1972) using synchronized cesium clocks on board commercial flights is probably the most famous, but it's worth noting that atomic clocks have been refined to where they are accurate to within 100 quadrillionths of a second. If you own a smartphone, then you are regularly "testing" relativistic time dilation because your phone is correcting for relativistic effects in its communication with satellites (see Audoin and Guinot (2001) for more tests). Experiments in biology, economics, and geology implicitly confirm the theory of time they employ too, but none test theories of time as severely as physics does.

Beginning with physical time is a large assumption. It's certainly possible to try the other direction, to go from "here" to "there," contrary to R.E.M.'s advice. Whether it is fruitful I can't say. Henri Bergson (and many phenomenologists) have pursued that course. In his doctoral dissertation of 1889, Bergson distinguished between what he called "lived time" (*durée réelle*) and the "mechanistic" time of science. (Note that this distinction was devised even prior to relativity.) He argued that the time of science is a distortion based on a mistaken extrapolation from lived time. Instead of lived time portraying an "illusory" time, as we commonly find today, here physical time and its "spatialization" is judged "illusory," or at least, distorted. Whether mistaken or not, the direction of explanation is opposite that here. Later, he and other phenomenologists seek to derive physical time from temporal experience, much like explaining how one arrives at a physical model of a table from the experiences of the manifest one. Given this option, and conceding that R.E.M. weren't referring to time, why not go from "here" to "there"?

My choice doesn't arise from a generally reductionist or scientistic attitude.[10] Instead I have three specific motivations:

1. I want to see if we *can* stick with physical time in the face of manifest time. We enjoy no guarantee that physics gets time essentially right. But it might. Suppose it does. Occam's Razor then counsels us to posit no more entities or structure than needed. Wanting the explanations physics offers, we know that we *already* need to attribute to time *at least* the structure that physics needs. Will that suffice for all purposes? While an answer will necessarily involve supplementing physics with higher-level structure, i.e., brains, macroscopic environments, and so on, can we make do without adding extra physical or metaphysical temporal structure? If possible, then one has escaped needlessly positing some surplus structure to one's world picture.

2. I have more faith in the methods by which physics has discovered temporal features than those by which we arrive at manifest time. Physics generally proceeds by looking for the best explanation of observable physical facts. Its explanations are subjected to withering criticism by rival theories and severe empirical testing, resulting in theories with many theoretical and empirical virtues. Manifest time, by contrast, is a somewhat vague set of theses that we arrive at through development. We're born into a world with all kinds of temporal and causal patterns: day, night, feeding, sleeping, etc. To thrive, it's crucially important to develop temporal representations of the world to make sense of all this. Amazingly, we do and they by and large *work*. Whether these representations are based on a *wholly accurate* picture of time is irrelevant for the purposes that drive its origin. That beliefs are conducive to fitness doesn't make them true. Skeptics might say the same of physics, that success in physics doesn't imply truth; but in this case we at least have an argument on our hands. Anyway, the point need not get so deep: no one would suggest that we stick with folk physics, and for these same reasons we should not rest content with the deliverances of manifest time.

3. So far replacing physical time with a metaphysically "beefed up" time hasn't worked. Impressed with manifest time, many researchers in physics and analytic

---

[10] I admit that, by nature, I have both.

THE PROBLEM OF TIME 29

metaphysics have claimed that physical time is either incomplete or inaccurate because it fails to respect what's manifest.[11] They may well be right. Still, the replacements for physical time are disappointing. Analytic metaphysicians have devised all manner of modifications or replacements for physical time in the past hundred years (see Dainton (2001) for some references). These theorists are sometimes called *tensers*, as they propose models of time that would provide objective counterparts for our tensed claims, i.e., they propose physical models that truly distinguish an objective and observer-independent past, present, and future. These models of time are typically sophisticated products and shouldn't be confused with manifest time. Instead they are models that adorn the time of physics with all manner of fancy temporal dress: primitive flows, tensed presents, transient presents, ersatz presents, Meinongian times, existent presents, priority presents, thick and skipping presents, moving spotlights, becoming, and at least half a dozen different types of branching! What unites this otherwise motley class is that each model has features that allegedly vindicate core aspects of manifest time. However, these tricked out times have not met with much success. Manifest time gets explained by fiat, in general, and why physics misses their metaphysical accessories is left a mystery (see Chapter 13). Before claiming a science is incomplete or inaccurate, we ought to at least give a more conservative explanation a fair shake.

For these reasons we begin with physics. But we will not end with the mind. That path is too long. I shall not try to close the gap between temporal *experience* and physical time. That would be to try (in part) to solve the mind–body problem, yet my aspirations are far less grand. Natural science has never explained a single mental experience directly. Deducing temporal experience from physics is a fool's errand. We can provide a passable answer to Eddington's two tables problem without solving the problem of the mind's relationship to natural science. We can hope for nothing more bold when answering the "two times" problem.

The way forward, I claim, is marked by some shifts in the way we study time. Taking our cue from Eddington's tables, notice that to understand the desk's solidity, one needs to know an awful lot of atomic physics, condensed matter physics, statistical mechanics, and more. Without understanding the physics, one will not recognize the microphysical counterpart of solidity when one sees it. But physics isn't enough. Explaining the desk's solidity will require appeal to boundary conditions, statistics, scale differences, and other supplements to physical theory. Explaining texture and color conceived as secondary qualities requires the addition of human beings and

---

[11] Here, for example, is Arthur Prior, one of philosophy of time's most eminent and founding figures: "[T]he theory of relativity isn't about real space and time" (Prior, 1996, 51). And here is the noted physicist Lee Smolin, "Logic and mathematics capture aspects of nature, but never the whole of nature. There are aspects of the real universe that will never be representable in mathematics. One of them is that in the real world it is always some particular moment" (Smolin, 2013).

hence the sciences of their perceptual systems. Features of the manifest table emerge as interaction effects between human agents and physical tables.

Two lessons result from this reflection.

First, we must not shy away from the science of time. We shouldn't be forced to use only the blunt tools philosophers of time offer us. A-properties and B-properties are important categories, and logical analysis is a powerful tool. But these distinctions and tools need supplementation with actual physics. You can't see the second law of thermodynamics with A- and B-properties, nor can you learn that the relativistic signature singles time out, nor will you see the connection between this signature and the laws of nature, and so on. Again and again we will find that the features useful in answering the two times problem are invisible to philosophy uninformed by science.

Second, when answering Einstein, Carnap switches from talk of "physics" to that of "science." We will follow suit, for it seems clear that more than physics is required to tackle the "two times" problem. Perhaps ultimately everything in some sense reduces to physics, but in our current state of knowledge we crucially require other sciences as well. Physics may be unable to explain the now, but that doesn't mean that physics and philosophy aided by other natural and social sciences cannot. If explaining tables calls for more than physics and philosophy, we can reasonably expect that to be true of the larger topic of time.

Reconciling manifest and scientific time is our goal. If I'm right, success requires an all-out interdisciplinary attack on the problem. The natural and life sciences are both crucial, but so too is philosophy. Philosophers bring to bear a distinguished history of dealing with time, a host of conceptual tools for clarifying problems and answers, and the freedom to look past a single discipline. Especially with a topic as puzzling as time, we need all the angles and tools we can get. The story described here is only the beginning. Whereas some may find the present fragmentary and uncertain state of research an obstacle, I consider it an exciting opportunity. In what follows we'll encounter a man who hears you speak before he sees your lips move, experiments on children's sense of self, a theory linking time to laws of nature, determinism from "east" to "west," no-go theorems for types of relativistic presents, time bias in judgments of responsibility, the latest speculation about time in quantum gravity, the evolution of the psychological present, and much more. With so many diverse elements, I haven't been able to sew together all the strands as tightly as I would like. Nevertheless, like a scene that comes into focus when you stand back, I trust that a coherent image emerges from this book's incomplete and scattered parts.

2

Lost Time
Relativity Theory

Does physical time have a now, a flow, or a direction? In Chapter 1 I asserted that it does not. It's time to make good on this claim. What I will argue in this chapter and later ones is that physics lacks many of the core features we associate with manifest time. Lacking a grand unified theory, our project is complicated by the fact that we have many physical theories—and worse, some of these aren't necessarily compatible with each other. We'll return to this last point in Chapter 4. Here we'll focus on a centerpiece of modern physics, relativity, and I'll argue that relativity lacks features of manifest time. That claim is easy to establish, as none of our physical theories employ a distinguished now. What takes a bit more work is the further claim that buried within relativity there isn't structure suitable for "animation" in the sense described in the previous chapter; that is, there are not good relativistic candidates for a flowing now. If one wants to "animate" a relativistic structure, it turns out that there are no good structures to choose. Indeed, an informal dilemma seems to hold:

the better a structure represents manifest time, the "less" relativistic it is; the "more" relativistic it is, the worse it represents manifest time.

To be clear, I think that we can explain why manifest time arises for us in a world governed by our physical laws. But doing so, if I am right, will require embedding a subject like us in a world like ours, and not simply finding some structure in physics that plays the "manifest time" role.

2.1 Classical Physics

Replaced as a fundamental theory by relativity and quantum theory, classical physics is still interesting to study with respect to time. It is historically important and large portions of its structure survive in contemporary theories. It is also widely employed in a vast amount of modeling, ranging from dynamical systems theory to statistical mechanics to engineering. The reason it is still used is that, in a certain physical regime, classical physics more or less gets everything right. Consequently, in that same regime it more or less gets time right too. It proves a good warm up to our later discussion.

Classical physics is not one theory. Newton, Hamilton, Euler, Lagrange, Jacobi, and others have all formulated the core ideas in different—and often strictly inequivalent— ways. However, all these formulations posit essentially the same temporal structure to the world. For the sake of expediency we can pretend that we have one theory in mind. Even here we have choices. We can describe the theory as Newton understood it, as a theory of three-dimensional space *and* one-dimensional time, or the way Cartan and Friedrichs did much later (Cartan, 1923, 1924; Friedrichs, 1927) using the modern coordinate-free geometric language of relativity, as a four-dimensional *spacetime* theory. For ease of comparison with relativity, we'll go the latter route.

Take as primitive the notion of an event. Birthday parties, King Philip's Great Swamp War, firecrackers going off, and just sitting still are all examples of events. We represent the set of all events with a four-dimensional smooth and connected manifold M. A manifold is a special type of set that is especially agreeable to doing physics: it allows a system of n coordinates near each point p of M and insists that these coordinate systems play nicely with each other. The manifold will have topological structure. In classical physics M is typically assumed to be topologically Euclidean, but other possibilities are allowed so long as M is smooth and connected.

Newtonian physics, stripped to its core, is a set of differential equations that tell us *how far* and in *what direction* matter must go for *how long*. Consider Newton's famous second law, $F = ma$. This law associates an acceleration with a body having a certain mass and impressed with certain forces. Acceleration picks out a measure of time and a measure of space. The acceleration states *how fast* instantaneous velocity is changing. Velocity is of course spatial distance divided by temporal distance, and instantaneous velocity is just the limit of this quantity as temporal distance goes to zero. The essence of the theory is about these distances and directions. They are verified experimentally. When classical physics says that with some assumptions about air resistance, drag, mass, and more, bodies falling toward Earth will reach a terminal velocity of approximately 32ft/sec^2, the physical laws are in effect telling us how long a second is. A second is the amount of time through which a perfect body obeying all the relevant assumptions travels that far toward Earth. You can change the units, but at bottom it can only go *so far* in *so much time*.

A topological manifold can't see such facts. The distance between my first and second birthdays, for instance, is not preserved by topological transformations. Metric structure is required to speak of going so far in so long. Metrics provide a notion of "distance" from which all manner of spatiotemporal geometry is constructed (curvature, angles, geodesics, and much more). Newton's original theory posited a notion of absolute rest. Using this notion of rest, one can define a single metric over the entire spacetime M, as we will do in relativity. Since classical physics doesn't require rest, however, it's common to instead represent classical spacetime with what is called either Galilean or neo-Newtonian spacetime. The Cartan version of this spacetime comes equipped with two metrics on it, a "spatial" metric h_{ab} and a "temporal" metric t_{ab}. I'll explain how these metrics work in a moment. Right now, note that these metrics, like all metrics, provide a map, where defined, from pairs of (tangent) vectors at a point p in M to real numbers. Using this map one can then define all the distances—the *how fars* and *how much times*—necessary for classical physics.[1]

Two more structures are needed to represent a classical world. Vectors live in tangent spaces attached to manifolds. If we wish to compare vectors in tangent spaces at different points, then we need some way to accomplish this feat. Enter the affine connection. Represented by the "derivative operator" ∇, the affine connection is a mathematical structure on manifolds that allows us to make sense of the parallel transport of vectors about a surface and hence "straightness." Unless explicitly stated to the contrary, we'll assume that ∇ represents a flat connection; that is, one such that vectors transported about a closed loop always come back pointing in their original direction. Of course we also need to add the matter content of a classical world, which can be represented by its mass-density ρ. Differential equations then describe how this matter evolves with time. When all of these pieces are put together, the following structure

$$\left\langle M^4, t_{ab}, h^{ab}, \nabla, \rho \right\rangle$$

characterizes a Newton–Cartan world.

Focus first on the temporal metric t_{ab}. If ξ^a is a vector at a point, then the temporal metric provides ξ^a a "temporal" length. Importantly, these lengths can only be positive or zero. If the temporal metric assigns a vector a positive length we'll call that vector *timelike* and if zero we'll call that vector *spacelike*. Since t_{ab} is well-defined for all vectors, we can use it to exhaustively sort vectors into the timelike and spacelike classes. The temporal metric does more than sort vectors into two classes. Its primary job is that it allows us to define functions that provide distances along curves on M. The details needn't detain us here (see Malament (2012) and references therein). Suffice it to say, with the help of the derivative operator, t_{ab} can assign "temporal lengths" to curves on M. Take a curve representing your travels from your last birthday to the current one. Then the temporal metric will assign it a length of one year.

The metric h^{ab} likewise assigns vectors "spatial" lengths. However, h^{ab} does not sort vectors exhaustively: it only assigns lengths to spacelike vectors, i.e., ones given zero length by t_{ab}. Otherwise it is undefined. Hence its main job is to define functions that provide distances along curves that are spacelike at every point. For events related by such curves one can then define "spatial" distances.

---

[1] Formally, the Cartan metrics are defined as smooth degenerate (i.e., vanishing determinant) tensor fields of type $(0, 2)$. They have signatures $(0, 1, 1, 1)$ and $(1, 0, 0, 0)$, respectively, and later we'll have the opportunity to focus on this fact. Because degenerate they represent an extension of the usual notion of metric. They warrant the appellation "metric" because they play all the roles associated with distances in physics. We also insist that the two metrics jointly satisfy various conditions. Of these, the only one I want to highlight is that the two metrics are orthogonal, i.e., $h^{ab} t_{bc} = 0$.

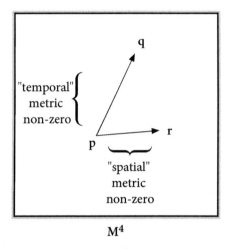

$$\mathbf{M^4}$$

Figure 2.1 Newton–Cartan spacetime

Together we have the following implications for distances between two events:

$$spatial - distance \neq 0 \Longrightarrow temporal - distance = 0$$
$$temporal - distance \neq 0 \Longrightarrow spatial - distance = \text{undefined}$$

In other words, when the spatial distance between two points is non-zero, the temporal distance between them vanishes, and when the temporal distance between two points is non-zero, the spatial distance between them is undefined.[2] See Fig. 2.1.

Points that have positive temporal distance from one another neither have nor don't have spatial distance between them. This consequence is an expression of the fact that absolute rest isn't defined in Galilean or Neo-Newtonian spacetime; for if it were defined, we could use it to provide a measure of spatial distance for non-simultaneous events, e.g., the spatial distance between a finger snap now and one later.

Using this structure, classical spacetime can provide the *how fars* and *how longs* necessary for classical physics.

2.1.1 Recovering space and time

Notice that in the above description we didn't need anything dubbed "space" or "time." The two metric fields and connection do all the work Newton's physics requires.

---

[2] In formulas the above discussion can be summarized as follows. The "temporal" length is given by $\left(t_{ab}\xi^a\xi^b\right)^{\frac{1}{2}}$. The "spatial" length is given by $\left(h_{ab}\xi^a\xi^b\right)^{\frac{1}{2}}$. The non-vanishing of one has the following effect on the other, respectively:

$$\left(h_{ab}\xi^a\xi^b\right)^{\frac{1}{2}} \neq 0 \Longrightarrow \left(t_{ab}\xi^a\xi^b\right)^{\frac{1}{2}} = 0$$
$$\left(t_{ab}\xi^a\xi^b\right)^{\frac{1}{2}} \neq 0 \Longrightarrow \left(h_{ab}\xi^a\xi^b\right)^{\frac{1}{2}} \text{ undefined.}$$

True, we dubbed one metric field "spatial" and another "temporal," but these four-dimensional fields taking vectors in tangent space as input and spitting out lengths as output aren't themselves space or time. Nothing in the physics demanded that we carve up the four-dimensional mosaic of events into anything resembling intuitive space or time. Surprisingly, not only does the physics lack any mention of a flowing now, but "space" and "time" aren't explicitly needed either!

Yet we know that there are Newtonian (and other) versions of essentially the same physics that do make use of structures that play more traditional space and time roles. Something like space and time must be recoverable from Newton–Cartan spacetime. Indeed that is so. Thanks to the special role t_{ab} plays, we can "find" a unique time and space that gives the spacetime a more familiar feel. Definitions don't cost us anything, so one can then insist that space and time were there all along. Alternatively, we could model classical physics beginning with a basic time and space split.

Let's recover time and space from Newton–Cartan spacetime. We can do so by using t_{ab}'s invariant and exhaustive division of vectors into the timelike and spacelike. Begin with "space." Call a hypersurface Σ in M spacelike if at every single point of Σ all vectors tangent to Σ are spacelike (zero temporal length). These hypersurfaces will be our "spaces."

Next, let's get a candidate for time. Making some assumptions about M, we can use the temporal metric to define a family of what are called *global time functions t*. These are mappings from M to the real numbers. The functions monotonically increase along every future directed timelike curve, so one can think of them as counting "ticks" along any timelike curve. There are infinitely many of these functions. Importantly, however, they all agree on the basic facts of duration (and therefore simultaneity). Take a smooth curve γ representing a particle, person, or planet existing from event p_1 to event p_2. Then the temporal length of γ is given simply by[3]

$$t(\gamma(p_2)) - t(\gamma(p_1))$$

The reader will note—for future reference—that the duration of γ depends only upon the endpoints of the path. It is a hypothesis of the theory—call it the *classical clock hypothesis*—that ideal clocks record the t_{ab}-length of worldlines. In any case, since

---

[3] More precisely, assuming M is simply connected, t_{ab} permits the introduction of a family of global time functions $t \colon M \to \mathbb{R}$ via a vector field t_a such that $t_a = \nabla_a t$ for a smooth time function t. Take a smooth timelike curve γ from event p_1 to event p_2. Let ξ^a be the tangent field associated with γ. Then the temporal length of γ is

$$\int_\gamma t_a \xi^a \, ds = t(\gamma(p_2)) - t(\gamma(p_1))$$

An important caveat here is that the actual real number, say $t(p_1, p_2) = 17$, has no physical significance. An alternative function $t' = t + b$, where b is an arbitrary constant, should describe the same durational facts, even if t' maps $t'(p_1, p_2)$ to $17 + b$, since it will give the same ratio of distances as t does. If t and t' agree that the duration between, say p_1 and p_2 is twice that between p_1 and p_3, and so on, that is good enough. One can bypass this annoyance by working directly with t_a instead of t (Friedman, 1983, pp. 75–6).

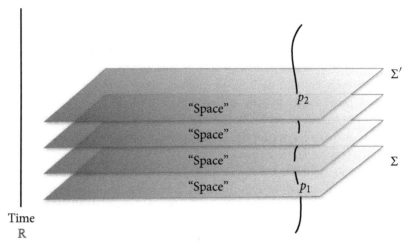

Figure 2.2 Foliation of M into time and space

these durations are the temporal facts needed for doing classical physics, any one of these global time functions can then act as our time.

Now let's link "time" and "space." There is a neat connection between any one of these global time functions and our above "spaces," namely, the condition that a hypersurface be spacelike turns out to be equivalent to the global time function t being constant across that hypersurface (Malament, 2012, p. 252). Therefore all the events comprising any spacelike hypersurface Σ read the same time, up to an arbitrary constant. A different time function t' will shift the meaningless origin point, but it will otherwise agree on what matters: if t is constant for all events on Σ, then t' is constant for all events on Σ too. So we are justified in regarding these spatial hypersurfaces as all of space at a moment of time, for all the events on Σ_t are simultaneous with one another, for all values t.

Simultaneity understood in this sense "foliates" M into a unique temporal sequence of spaces, each of which reads the same time.[4] As a result, classical spacetimes can be broken up into topological products of "time" and "space," i.e., $M = \Sigma \times \mathbb{R}$, as in Fig. 2.2. Moreover, since the duration defined above depends only on the endpoints, and the simultaneity slices agree on their value of t, we can also think of t as providing the duration between different slices Σ, Σ', and not merely along particular worldlines (so long as one endpoint is on Σ and the other on Σ'). Thus we can think of the real number line with the function induced by the temporal metric on it as representing "time," each tick of which is a unique slice that picks out a global subset of Euclidean

---

[4] A foliation of manifold M^n is a set of M^{n-1} submanifolds—the so-called *leaves* of the manifold—which depend on a parameter α, such that $M = \bigcup_\alpha M^{n-1}_\alpha$. In the present case $\Sigma = M^{n-1}$ and $\mathbb{R}$ indexes the leaves. Here I'm assuming that M^n is simply connected, so the time function is a global one; therefore the spacetime is decomposable into global simultaneity slices. If the time function is not a global one, then we get only the implication that the spacelike hypersurfaces are local simultaneity slices.

space (thanks to h_{ab} restricted to Σ). Space and time are back. Or since mathematically this M is not in any sense "prior" to $\Sigma \times \mathbb{R}$, perhaps it's better to state that they never left.

For later use, let's pause a moment to describe another way of looking at what was just done. The temporal metric t_{ab} in effect provides us with the resources to define an *earlier than* relation that totally orders the sets of simultaneous events on M. That is, for any two events, $p, q \in M$, the temporal metric determines whether p precedes q, q precedes p, or neither precedes the other. The last option, that neither p nor q precede one another, is an equivalence relation. It is better known, of course, as simultaneity. Simultaneity is thus an equivalence relation. Equivalence relations take sets and partition them into subsets known as equivalence classes. In the case at hand, M is the set and simultaneity partitions it (because equivalence classes are either disjoint or identical) into subsets of points that we can regard as spatial (vectors connecting any of these points have zero temporal duration). Here, then, is an invariant entity that we may regard as "space." Thanks to the spatial metric, any class of mutually simultaneous events is a subset of a three-dimensional Euclidean space.

What of "time"? Formally speaking, on this picture time is simply the quotient set of M associated with the simultaneity relation, i.e., M/sim. That means that if we "divide" through M by the equivalence classes, we get a new set. The points of this set are instants. Each instant is tied to an equivalence class of simultaneous events. Time is the set of all such instants. Due to M and t_{ab}, time has the structure of a one-dimensional real affine space, or more familiarly, the Euclidean line.

When asking where time is hiding in Newton–Cartan theory, I mentioned that one could begin with a separate time and space. Newton's own theory is one obvious way to do this. But those with the mathematical background can understand Newton–Cartan theory in way that highlights how special time is: represent spacetime as a fiber bundle. A fiber bundle with fiber F is a projective mapping π between a total space E and a base space B that is "locally trivial." Loosely put, "locally trivial" means that the total space looks like the product space $B \times F$ locally. In this picture, we can begin with a separate time represented by $\mathbb{R}$. Now think of time as the "base space" of a bundle with respect to which (assuming space is Euclidean) $\mathbb{R}^3$ are the attached "fibers." For any point in $\mathbb{R}$ (time), the set $\pi^{-1}(p)$ of all points in $\mathbb{R}^3$ is the fiber over p. If you like, to each instant of time we stitch a distinct fiber that represents space, as in Fig. 2.3.

In a sense, time is all of space collapsed to a point, for each fiber $\mathbb{R}^3$. Time is distinguished: one can't think of $\mathbb{R}$ as the fiber and any one of the $\mathbb{R}^3$s as the base.

2.1.2 Classical ideal clocks and manifest time

It will be a useful expository device to conceive of each physical theory we consider as positing one or more *ideal clocks*. We can imagine an ideal clock to be a point-sized test object that perfectly records the passage of time along any allowed worldline. Material bodies "listen" to the ticks of these clocks by obeying the laws of physics. Since

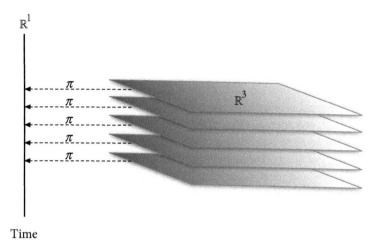

Figure 2.3 Time as a "base space"

there are no perfect bodies there are no physical instantiations of this perfect clock (unless you count the universe in its entirety). That's why it is ideal. But its ideality is no mark against claims about it being true. It has the same status as—is part of, really— the laws of nature. Empirical evidence for physics comes in the form of measuring bodies at particular places and times. To the extent that these measurements confirm the differential equations that we take to be the laws of physics, the measurements also confirm the ideal clocks posited by physics.

We can imagine each timelike curve as carrying an ideal clock measuring the duration of any path. However, as I mentioned, since duration is path-independent in a classical spacetime, one can also think of time as elapsing between any two simultaneity slices. If Jack and Jill leave work at noon and then meet up later for dinner, the amount of time that passes between work and dinner is the same for them both, as are the order of intervening events, even if Jack stayed home and Jill went to the gym (see Fig. 2.4). Time uniquely partitions spacetime, so classically it's needlessly cumbersome to picture each timelike curve with an ideal clock. One master clock serving the whole universe will do, as in our global time function above.

Here are a few features of this classical ideal clock.

- *Absolute.* The ideal clock is independent of the existence and arrangement of material bodies in the universe. The rate the ideal clock ticks doesn't hang on *what's where.* As a corollary, it is in particular independent of the systems that we happen to dub "clocks," i.e., Newton's famous "sensible measures" of time. Those physically instantiated clocks are just bodies that happen to approximate—some better, some worse—the ideal clock. (Whether a non-absolute reading of classical time is possible is an interesting and controversial matter that is not the subject of this book.)

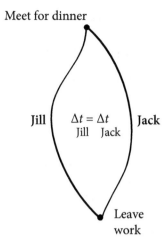

Meet for dinner

Jill $\Delta t = \Delta t$ Jack
        Jill Jack

Leave
work

Figure 2.4 Path-independence of classical duration

- *Time translation invariant.* We represented time with the topological space
$\mathbb{R}$ above. If we place a coordinate system on this set, as in the real number line,
it's important not to regard all of its features as also had by time. The real number
line has a unique '0' point on it, and this might tempt one to think an instant is
distinguished like the now. However, the physics is time translation invariant; that
is, the laws of the physical theory are invariant under shifts forward or backward
along any time t that parametrizes the moments:

$$t \mapsto t + k$$

for some constant $k \in \mathbb{R}$. The transformation leaves the form of the laws the same.
Empirically, what this means is that a Newtonian experiment done at, say, 2 p.m.
will have the same outcome as one at, say, 3 p.m., assuming everything else is the
same. Given this symmetry, the unique zero point could never show up physically.
- *Time reversal invariance.* The classical dynamics doesn't "care" about the direction
of time and will evolve a state the same way in the earlier and later directions of
time. Again, given a parametrization adapted to the moments:

$$t \mapsto -t$$

is a symmetry of the theory. This suggests that a privileged directionality is not
needed. Sometimes it is said that classical time requires an ordering. This claim
is half-right. A tertiary betweenness relation $B(a, b, c)$ among moments suffices
for classical physics. This relation says, of three moments, that one is between the
other two, but it does not distinguish a preferred directionality as *less than* does
over numbers or *earlier than* does over times. For our purposes, a betweenness
relation defined as

$$B(<; a, b, c) \iff a < b < c \text{ or } c < b < a,$$

where $<$ is a total order, is appropriate. In other words, there is a difference between the line and the directed line. It's important that the line be *capable* of holding a direction, not that a particular one be distinguished.[5]

How does the ideal clock of classical time compare with manifest time?

At first glance, not too badly. The global common nows can be regarded as the equivalence classes of simultaneous events, which are themselves global in nature. The temporal metric space also fits well: we don't normally think of manifest time as gappy, disconnected, etc. I think a good case can be made for thinking that manifest time is also unique and absolute in the senses specified above. The uniqueness of the foliation corresponds nicely with the idea that the flow is objective and unique. That the duration is independent of the particular matter arrangement is also a good fit.

On second glance, however, all three ingredients singled out in the Introduction as crucial to the manifest image are lacking. There is no distinguished now, nor any notion of a flow, and even past/future asymmetry is not needed. The time translation invariance described above seems at odds with any physically motivated distinguished moment (and hence one that flows) and the time reversal invariance appears to conflict with any physically motivated differences between past and future, at least as far as basic classical mechanics is concerned.

Classical physics doesn't provide us with natural counterparts of these three features. True, Newton asserts that time flows in the *Scholium*. Yet I don't read him as suggesting that it flows in the sense intended here. If he is, then he is positing more than his physics needs. Classical physics, I mentioned, is often said to be "friendly" to tensed time. However, if all it takes to be friends is to be ignored, that is not so impressive a partnership. At best, the equivalence classes of simultaneous events are good candidates for "animation," but they do not by themselves vindicate the animation at the heart of manifest time.

2.1.3 Trautman–Cartan theory

Earlier I mentioned that there are two ways in which physical theory may conflict with manifest theory, either by neglecting manifest theory or by outright conflict. What we witnessed above is a case of physics ignoring some features of manifest theory.

---

[5] It might be suspected that when we move to relativity we'll need to go back to order relations and leave behind betweenness. Robb's (1914) famous axiomatization of Minkowski spacetime, for instance, uses the primitive *after* relation, which would correspond to $\leq$ and not B. Many celebrated accounts of Minkowski spacetime—e.g., Reichenbach (1924), Grünbaum (1973), van Fraassen (1985)—have furthermore based the spacetime geometry on a basic notion of causation, another temporally directed notion. However, with Mundy (1986) I find this causation-obsession unnecessary. As he shows by example, there are axiomatizations of Minkowski geometry that don't use *before* or *after* or any "causal" notions among their primitives. John Schutz's (1997) elegant axiomatization also uses as primitives only events, paths, and a betweenness relation, for example. In that axiomatization no assumption picking a particular direction of time is needed.

Before leaving classical physics, it's worth pointing out that it may have aspects that are actively hostile to the manifest picture of time. Here is one possible example.

Standard Newtonian gravitation theory posits the existence of a gravitational force whose form is determined by the Poisson equation:

$$\nabla_a \nabla^a \phi = 4\pi G \rho$$

where ϕ is a scalar field that describes the Newtonian gravitational potential, ∇ is a flat derivative operator that determines how vectors behave under parallel transport, ρ is a scalar field representing the mass density, and G is Newton's constant. If this math is unfamiliar, know that the left-hand side of the equation describes the form of the gravitational force and the right-hand side describes the distribution of the massive "stuff." The equation constrains the relationship between the force and the stuff on a flat background spacetime. That is standard classical gravitation theory.

Trautman (1965) reformulated Cartan's theory with a curved derivative operator. The idea is to absorb gravity into the spacetime curvature, as in general relativity. Instead of being deflected by gravitational forces from geodesic paths in flat spacetime, free-falling particles now travel geodesics on a curved spacetime. Unlike in relativity, there is no spatial curvature in Trautman's theory, so the leaves of the foliation are spatially flat three-dimensional hypersurfaces. The major difference with the traditional theory is with respect to time. The connection ∇ now possesses non-trivial curvature along timelike directions. Move a vector in a closed loop on a three-dimensional spatial surface, and thanks to the spatial flatness, the vector will return pointing in the same direction as it did originally. But move a vector in a closed four-dimensional loop, and thanks to the curvature of time, it will *not* generally point in the same direction upon return to the original event. This temporal curvature affects the acceleration of massive bodies, and it does so in a way that makes it look like there is a gravitational force obeying Poisson's equation.

On Trautman's formulation, the counterpart of the Poisson equation is one relating the spacetime geometry R_{ab} to the matter density ρ:

$$R_{ab} = 4\pi \rho t_{ab}$$

where R_{ab} is the Ricci tensor. The important point for us is a simple one, namely, that on this formulation the matter distribution, described by ρ, affects the spacetime geometry, described by R_{ab}, and vice versa. Thus the particular and presumably contingent distribution of matter affects—among other things—what is a geodesic, or free fall "straight" path, in Newton–Cartan spacetime. This means that the rate at which even ideal clocks tick in Trautman spacetime is determined by the contingent matter distribution.

Why is this important to time? It's significant if we think that manifest time holds that the ticking of the world's ideal clock is independent of the contingent matter distribution. Presumably most think it is. Few believe, pre-theoretically, that the duration of events hangs on where I left my keys. Indeed, that *the matter distribution cannot determine the lapse of time* is a crucial premise in Gödel's (1949) famous argument for the ideality of time, a premise he draws from what he considers the common sense view of time. Strictly speaking, Trautman's formulation contradicts this common sense view: the rate at which Newton's time flows locally along a worldline is affected by the local matter distribution in Trautman's reformulation, and so, presumably imperceptibly, by where I left my keys. Given what was to come (general relativity), the Trautman version of gravitation theory is arguably the most natural formulation of classical gravitation. If so, one could read the history of classical physics as already possessing dormant threats to manifest time. Classical physics' relationship with manifest time may not have been as rosy as is sometimes thought.

2.2 Relativity

Relativity poses some of the most famous challenges to manifest time. No other theory's consequences strike as deeply into our core notion of time. This fact should be no surprise: Einstein explicitly said that reconceiving our usual notion of time was what enabled his discovery of special relativity in 1905. It took a while for these consequences to stand out. Initially many overlooked the radical consequences of Einstein's theory because it was not sufficiently distinguished from Lorentz's theory (which takes place in the above Galilean spacetime). Later, when general relativity was discovered (around 1915) and its high-profile predictions vindicated, the theory's structure came into sharp focus and the inevitable clash with manifest time became apparent to all in the 1920s. Some Kantians and phenomenologists then opposed relativity, often due to its departure from anything resembling manifest time. Now the idea that relativity challenges some aspects of manifest time is well-known, explored in works of physics, philosophy, and popular science. Despite the familiarity of these challenges, philosophy of time's take on it is still unsettled: some feel that relativity harbors what is needed for manifest time, others don't.

My own view is that in relativity there is very little worth dubbing time at all, at least if manifest time is one's standard. In classical physics we couldn't find "animation," but at least good candidates for animation exist. Here we can find good candidates only by distinguishing non-relativistic structures.

Before we get to that, I need to explain the theory. I make no attempt to describe relativity in all of its glory. Few areas of physics are covered so well from so many angles.[6] I aim only to describe a few core principles of the theory inasmuch as they bear on time. A word about terminology: by "relativity" I mean what many other authors call "general relativity." Since Minkowski spacetime, the spacetime associated with special relativity, is one solution to general relativity, there is no non-historical

---

[6] Some excellent resources include, in rough order of difficulty, Geroch (1981), Callahan (2000), Carroll (2003), Wald (1984), and Malament (2012).

reason to distinguish them as two theories. Philosophy of time, in my opinion, has been harmed by an unwarranted concentration on special relativity. By focusing on "relativity" I hope to underscore my insistence that we consider the full theory.

Relativity represents the arena of events with a four-dimensional manifold M, as we did above. Now, however, the all-important metric structure is characterized by a single *spacetime* metric, g_{ab}. The distances between events are not spatial or temporal but spatiotemporal.[7] This fact is slightly startling when first met. We're used to spatial distances, e.g., the coffee cup is 12 inches from the computer, and temporal distances, e.g., lunch is 4 hours after breakfast, but not spatiotemporal distances. Using this metric, non-trivial lengths exist between any two events. Suppose you were looking into the night sky from the Earth's southern hemisphere on February 23, 1987. Then you might have seen the supernova SN 1987A. That event happened a long time ago and far away in the Large Magellanic Cloud. Nonetheless, the metric provides a "distance" between the two events, the actual supernova and the event of seeing it on Earth, in addition to distances between the actual supernova and you reading this paragraph, the actual supernova and Benjamin Franklin's death, and so on.

I want to highlight two ways in which the relativistic metric differs from the classical ones. These differences are somewhat formal, but they have huge consequences and I'll use them again in Chapter 6—so I ask the reader to bear with me. Before describing the differences, let me adopt a parlance often used in relativity, namely, that of the line element. The spacetime metric g_{ab} works the same way as the classical temporal metric t_{ab} does: it takes pairs of tangent vectors defined at a point of M and returns a real number.[8] Suppose we imagine the tiniest of changes in position vectors, namely, infinitesimal coordinate displacements ds. Metrics can be expressed in terms of their action on these displacements. Doing so defines a line element via the following schema

$$ds^2 = g_{ab}\, dx^a dx^b$$

To quickly get a feel for this, note that in three-dimensional Euclidean space, the line element between two objects—written in Cartesian coordinates—is:

$$ds^2 = dx^2 + dy^2 + dz^2$$

As one can see, the square root of the distances squared in each of the three spatial directions gives the familiar Pythagorean total distance, ds, between objects.

Now to the differences. The first is that the *type* of metric is quite different than before and the second is that the relativistic metric is *not fixed* once and for all for all points, but is allowed to vary point by point. The physical ramifications of these alterations turn out to be monumental.

---

[7] In practice, the speed of light is often used as a conversion factor from meters into seconds or seconds into meters. Hence the reader may find distances occasionally expressed solely in meters, but it must be remembered that in this case it's then correct to speak of a meter of time.

[8] Like the temporal metric, it is a smooth tensor field of type $(0, 2)$ defined at every point of M.

The difference in the type of metric can be expressed succinctly if we avail ourselves of some formal terminology. Classical metrics are symmetric positive semidefinite metrics of Euclidean signature, but relativistic metrics are symmetric pseudo-Riemannian metrics with Lorentzian signature. That statement won't mean much to my non-mathematical readers! Fortunately we can explain how this difference matters quite easily. One important and odd consequence is that relativistic distances (but not classical ones) can be negative. Look at the above Euclidean line element based on the Pythagorean metric. This distance, like the temporal one, can be positive (if distinct locations) or zero (if the same location), but it can't be negative. That corresponds nicely with spatial intuition. What would it even mean for my coffee cup to be a negative distance from my computer? That the metric is "positive semidefinite" encodes the fact that the distance is always nonnegative. The shift to a pseudo-Riemannian metric is a shift to an indefinite metric, one that allows negative line elements, i.e., $ds^2 < 0$. For instance,

$$ds^2 = -dx^2 + dy^2 + dz^2$$

is an indefinite metric and one can easily see that it can be negative. This counterintuitive feature means that we're stretching the notion of metric a bit, but otherwise these metrics deserve the denomination because they play the roles we expect metrics to play. If it helps one accept the idea of a negative distance, bear in mind that these "distances" are spatiotemporal, not spatial, so our spatial intuitions are not the best guide.

Don't mistake the above change, which is a deep one, with a trivial one nearby. The trivial change is one merely about the coordinates in which one expresses the metric. Coordinates are simply our language for describing what's real. As with any language, we can alter the word–world assignments as we like. We can call cats "dogs" and dogs "cats" and then say that cats bark. Similarly, we can call dx "$-dx$" and try to erase the difference between the two types of metric. But the difference I'm discussing is a deeper one than that. It's a real coordinate-invariant feature of the metric, in particular, a difference in the type of *signature*. We'll discuss the signature in more detail in Chapter 6. Suffice to say for now, there is a genuine difference between Euclidean and Lorentzian metric signatures.[9]

---

[9] If we consider the matrix that represents the Euclidean distance, when diagonalized it will have 1's down the diagonal and zero elsewhere, i.e., $h_{ii} = 1$ and 0 otherwise. If we keep track of the number of positive, negative, and zero eigenvalues associated with this matrix—which is how we compute the signature—we have a (1,1,1) or (+++) signature. Such metrics are positive definite. (The classical temporal and spatial metrics, being degenerate, have zero eigenvalues associated with them, and hence were (1,0,0,0) and (0,1,1,1), respectively.) When moving to relativity, in addition to being indefinite, we also insist that the metrics be Lorentzian, which states that when diagnolized at a point the signature of g_{ab} has $n-1$ positive eigenvalues and 1 negative eigenvalue for any n-dimensional metric. In four dimensions, that means the signature is (+++−) and that signatures such as (++−−) are ruled out. As I'll explain in Chapter 6, these metrics are equivalent to ones with the positive and negative eigenvalues swapped.

This difference from classical spacetimes cannot be underestimated. We will spend much of Chapter 6 discussing its ramifications. For now, note that this shift is at the heart of the familiar lightcone structure central to relativity. Classically t_{ab} exhaustively sorted vectors into the "timelike" (positive temporal distance) and the "spacelike" (zero temporal distance). The existence of negative distances now demands an extension of this categorization. Given an event $p \in M$, we can exhaustively categorize displacement vectors into one of three classes based on whether ds^2 is positive, negative, or zero. Given the signature convention we'll adopt, displacements for which $ds^2 < 0$ are "timelike," displacements for which $ds^2 = 0$ are "lightlike" or "null," and displacements for which $ds^2 > 0$ are "spacelike." Notice that the displacements dubbed "spacelike" are no longer vanishing, as they were classically.

To visualize the structure this division induces, go into the tangent space at some point p of M. (We can't think of this lightcone structure as given on M itself, for the lightcone structure defined at point q may be quite different than that at distinct point p.) See Fig. 2.5. Notice that the invariant set of events with $ds = 0$ from p defines two $(n-1)$-dimensional boundary sheets that divide the space into three regions, resulting in the famous "lightcone" structure of relativity. Displacements can either point into one of the two lobes carved out by these two sheets, on one of the sheets, or in neither direction. The result is another way to visualize the difference with the classical: classically, the boundary sheet of the "future" timelike lobe met the boundary sheet of the "past" timelike lobe at an $(n-1)$-dimensional hypersurface; now they meet only at point p itself.

Although the lightcone structure is defined in the tangent space, it nonetheless makes sense to speak of curves in M as traveling in timelike, lightlike, or spacelike directions. That is, we can ask of a curve whether at every single point it travels

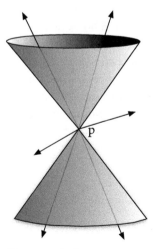

Figure 2.5 Lightcone structure

in the timelike, lightlike, or spacelike directions. "Most" curves aren't like this, as they shift direction. Relativistic physics doesn't care about these curves, for it insists that nothing physical is allowed to switch directions in this sense. Furthermore, it states that massive objects are represented by curves that are always timelike and light is represented by curves that are always lightlike. Relativity possibly permits the existence of particles traveling always spacelike paths—so-called tachyons—but no such particles are actually posited.

The second big difference between classical and relativistic metrics is that g_{ab} is allowed to vary point by point whereas the classical metrics are not. The centerpiece of relativity is the idea that the mass–energy distribution is related to the spacetime geometry via the Einstein field equations:

$$G_{ab}(g_{ab}) = 8\pi T_{ab} \tag{2.1}$$

Here G_{ab} is the Einstein tensor representing the spacetime curvature. It is a functional of the spacetime metric g_{ab}. T_{ab} is the so-called stress–energy tensor and represents the distribution of energy and matter at each point of M, encoding information from the pressure to the energy density. The Einstein field equations can be conceived as a constraint that must be satisfied between the spacetime geometry and the "stuff" at every point of M.[10] Written as I have above, the equations look sleek and compact. In fact it represents 256 coupled partial differential equations (until symmetry considerations are invoked), so it is rarely solved explicitly. The existence of all of these equations submerged in the above expression helps explain why it's wrong to understand Eq. (2.1) as saying that a given matter–energy distribution uniquely determines a spacetime geometry, or vice versa. In general that's not true. For instance, there are many distinct vacuum spacetimes, all sharing $T_{ab} = 0$ but having distinct spacetime geometries.

One important observation to make about Einstein's equation is that it is a local equation, that is, one holding at a point. Relativity demands little more than that the constraint embodied in this equation holds point by point. Tremendous freedom is therefore allowed. In particular, what this means is that the metric can vary point by point. In classical physics the two metrics were given once and for all. Here the metric at a point is sensitive to whether there is a planet there, where you placed your coffee cup, and more. That means that the lightcone structure can vary point by point, twisting, tilting, and expanding in countless ways. It also means the curvature can vary too. Although this massively underestimates the varieties of spacetime possible, as a hint of the possible riches, consider that the Gaussian curvature at a point can be positive (spherical), negative (hyperbolic, like a saddle), or zero (flat). The curvature may be positive at event p, negative at q, flat at r, negative at s, and so on, so long as certain continuity constraints are met. An infinite variety of shapes are therefore

---

[10] Although harmless here, this conception isn't strictly correct. Because T_{ab} depends on the metric, we can't think of the left-hand side as describing "geometry" and the right-hand side as representing the "stuff."

possible. Since all it takes to be a lawful model of the theory is that a triple $\langle M, g, T \rangle$ satisfy Einstein's equations at every point, relativity is very permissive.

In reaction, many physicists would like to impose additional constraints on the theory to "censor" these allegedly "unphysical" solutions. How and whether to do this is controversial. Although some regard "reasonableness" as an extra condition to be laid down on relativity, I regard this discussion as symptomatic of the fact that, in a sense, the theory is still under construction. The costs and benefits of different systematizations of relativity are being weighed, and we don't yet know where the balance will settle.

Let me highlight a major departure from classical physics that follows from what we've just said: duration becomes path-dependent in relativity. To put this point properly we need the notion of *proper time*. The proper time is the duration between events on a timelike worldline. Imagine drawing a line on a representation of a worldline, marking off "ticks" with little dashes and ordering these via $\mathbb{R}$. Any one you draw would parametrize the worldline. Proper time is a parametrization, but it is a special one that acquires physical significance through its definition via the metric. The proper time τ between two events separated by path P on a timelike worldline is

$$ \tau_P = \int_P \sqrt{-g_{ab}\, dx^a dx^b} $$

where g_{ab} is the spacetime metric. Since the proper time is simply an integration of a square of an invariant quantity, it too is an invariant of the spacetime. Among other things, what this means is that all good observers will in principle agree on this quantity. Proper time just *is* the spacetime metric evaluated along a particular type of path. In relativity as in classical physics it's a postulate—the *Clock Hypothesis*—that ideal clocks measure proper time.

It should be noted that proper time is not specific to relativity. Although proper time arises mostly in relativistic discussions, we can easily define a classical proper time by simply substituting $g_{ab} = -t_{ab}$ in the above definition. The resulting quantity will evaluate the temporal metric along a timelike path too, also yielding a duration along P. Nothing stops us from also defining a "proper length," either relativistically or classically. One need only evaluate the integral over curves that are spacelike.

Relativistic proper time, unlike (flat) classical proper time, is path-dependent. We're used to three-dimensional spatial distance being path-dependent between locations. Taking Interstate 5 to Los Angeles from San Diego is shorter than taking the route via Interstate 15. Temporal durations, by contrast, are path-independent between events and simultaneity slices classically. If you and I were at work at event p and later meet at a restaurant at event q, then the duration between p and q would be the same for each of us, no matter what we did or where we traveled. That's not true in relativity for two reasons, both connected to the above differences between classical and relativistic metrics (Fig. 2.6).

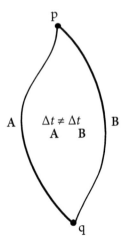

Figure 2.6 Path-dependence of duration

One, relativity uses a single Lorentzian metric, so when we calculate the proper time along a path, the duration depends on the non-timelike components of the metric as well as the timelike components. As a result, even when the metric is flat and the same at every point we obtain path-dependent duration. For example, in flat Minkowski spacetime the metric is everywhere

$$ds^2 = -dt^2 + dx^2 + dy^2 + dz^2$$

when expressed in "natural" Minkowskian coordinates (about which more later). Plug this metric into the above formula for proper time. Now consider two distinct paths, P and P', between distinct events p and q. Suppose that you are at rest along path P. Then all those spatial differences—dx, dy, dz—vanish, leaving one with the result that the proper time along P is just the amount of coordinate time t that passes for you, i.e., $\tau_P = \int_P dt$. Since any distinct path P' will have non-vanishing contributions from dx, dy, and dz, $\tau_{P'} \neq \tau_P$.

Two, because the metric can vary point by point and proper time takes the integral of these infinitesimal distances ds, different paths between p and q mean that one is adding up different contributions to lengths. Suppose that my route from p to q takes me through event r and that your route takes you through distinct event r'. In general the metric at r will not have the same form as the metric at r'. The metric at r might be approximately Minkowskian and therefore approximately flat, whereas the metric at r' might be approximately the Schwarzschild metric, the metric appropriate to the exterior of a black hole. Integrating over a path through r and over a path through r' will not generally yield the same value. (This kind of path-dependence is available to the Trautmanian classical mechanic, but not the first.)

The path-dependence of duration has recently been verified to a breathtaking accuracy. Using the Earth as the gravitational source, clocks only a meter higher than others have manifested measurable time dilation (Chou et al., 2010). That is, the clock at your waist "ticks" observably slower than the clock at your head! GPS devices correct for this and other types of dilation, so we can view most smart phones, tablets, and car navigators as implicitly testing this phenomenon all the time.

The path-dependence of duration leads to what I consider one of the *central differences* between classical and relativistic time. Classically there is a one "master" ideal clock. If we take arbitrary worldlines between two events or even between two simultaneity slices, the proper times mesh together to create a global clock. The physics of matter listens to the ticks of this global clock. In relativity, by contrast, one can often define a global time (see below), but matter-energy doesn't listen to it. Restricted to the clocks to which the physics listens, the ideal master clock of classical physics has *shattered into a huge plurality of miniature ideal clocks* in relativity, one for every timelike path.[11]

2.3 Where's Time?

The literature in philosophy of time is littered with attempts to find something like manifest time in relativity. Although the physics is usually quite straightforward, these debates are heated and opinions show no sign of convergence. My diagnosis of the never-ending quality of these disputes: an argument either that one has or hasn't found manifest time requires at least two big decisions, neither answered by logic or physics, and either leaving ample room for disagreement. The first decision is what counts as manifest time. The concept, we know, is rough around the edges. Different authors may precisify it in different ways. The second decision is what physical structure to identify with the features chosen in the first decision. Relativistic spacetimes possess many structures. Which one is the best expression of your understanding of manifest time? Here again there are many possibilities. The interplay between the two decisions can lead to otherwise clear-headed researchers talking past one another. When one adds diverging projects to the mix, confusion ensues. Some researchers are looking for candidates for tensed animation, whereas others are simply looking for structures to dub "the present." These different goals make different standards natural.

For the purposes of this book, our decisions are made for us. We are not looking to vindicate manifest time but only to explain why creatures like us might employ that notion. More locally, in this chapter, I am arguing that physical time doesn't provide those hoping to vindicate manifest time with any natural physical candidates for animation. Because "the now" is the foundation of the flow and the distinction

---

[11] The badly named "twin paradox" is simply an expression of relativity's path-dependence. Consider twins, one of whom remains home on Earth and another who departs in a rocket and returns later. Upon meeting they discover that the twin who left Earth is younger (i.e., fewer gray hairs, more optimistic, greater vertical leap) than the one who remained. The reason for the discrepancy is not because life on Earth is draining. The reason is that the twin who left traveled a path with shorter proper time than the one who stayed. Why is that path shorter? There may be an indefinite number of reasons due to the spacetime selected, but ultimately the answer is, as David Malament likes to say, simply a "hard fact of life" in a relativistic spacetime.

between past and future, we can simplify our discussion by focusing upon the manifest present. We are therefore interested in hewing as closely to the manifest present as we can and seeing whether any physical structures tolerably close to this are good applicants for this role.

Among the features often associated with the manifest present are

1. 'Present with' is an *equivalence relation*.[12] (The manifest present is typically identified with what is real, i.e., the real for relation, and that is considered an equivalence relation amongst all upright people.)
2. The present *bisects* the universe in two, i.e., it is global and non-intersecting. (We typically invest different ontological properties in the past and future, so we need the present to distinguish them cleanly.)
3. The present is *achronal*, i.e., no two events in the now are such that one is in the causal past of the other. (We don't think there are earlier and later parts of the present.)[13]
4. The present is composed of events *simultaneous* with one another.
5. Present events are *radar-synchronous* with one another. (This requirement is just an operationalized version of 4. For an explanation, see Box 1.)
6. The present is *productive*, i.e., present events influence the events on the "next" present.
7. It is *common*. Different observers can share the same present.

There is certainly much more to being present than listed above. Crucially, none of the "tensed" aspects discussed in Chapter 1 have been included. At best the above represents a small list of possibly necessary but far from sufficient conditions on the present. Depending on how one cashes out manifest time, some may not even be necessary. For example, surely it's a stretch to suggest that being radar-synchronous is necessary for present events.

There is no reason to be dogmatic about what's on the list. I ask only that we retain some reasonable constellation of features associated with manifest time so as not to trivialize the question. We must keep the proverbial eye on the ball: the postulated present should, ultimately, be a good candidate for "animation," i.e., something worthy of the importance manifest time attaches to it. Otherwise we are simply dubbing some events with the English word "present." Consider McEvent, the coarsely

---

[12] A binary relation is an equivalence relation if and only if it is reflexive, symmetric, and transitive. Such relations partition sets into cells such that every element of the set is a member of one and only one cell.

[13] Advocates of the so-called "specious present" may contest this requirement, insisting that the experienced present does extend over a duration. Fair enough. Readers with worries in this spirit can modify this requirement to make the manifest now have the duration they require. A modified achronality requirement might be that we don't regard large stretches of time, e.g., the life of Abraham Lincoln, as confined to the present, which is something everyone can accept.

Box 1   Einstein–Poincaré "Radar" Synchronization

When discovering their versions of special relativity, both Einstein and Poincaré provided operational definitions of simultaneity (see Galison (2004) and Jammer (2006) for historical discussions). Philosophers now generally regard it as a mistake to say that operations *define* concepts. Nonetheless, the operations do give us reasons to believe we've picked out the *right concept*, that a structure is up to playing the "simultaneity role." So it still might be important to our reasons for regarding some chosen structure as simultaneity to find out that it picks out the same structure as do some natural operations already associated with simultaneity.

Here is how the method works. Consider two timelike observers, A and B, tracing out worldlines in spacetime. We'll suppose the spacetime is otherwise a vacuum and that A and B are so light that their mass is negligible. A wants to know what event on her worldline is simultaneous with event b on B's worldline. One way to do this, at least in an idealized sense, is to shoot light from event a_1 at B so that it reaches B at exactly event b. B has a perfect mirror and immediately he reflects the light back to A. A receives the light at event a_2. Assuming the speed of light is constant, no delays in either path, etc. A reasons that b happened at the same time as point a on her worldline, where a is given by half the proper time between a_1 and a_2. In principle, given enough observers, mirrors, and photons, one could deduce an entire hyperplane of points in the universe simultaneous with a. The rule is simply to set b, the time of signal reflection, equal to:

$$a_1 + \tfrac{1}{2}(a_2 - a_1) = \tfrac{1}{2}(a_1 + a_2)$$

Simultaneity is usually thought to be an equivalence relation. Using this method, if inertial observer O deems a and b to be simultaneous, will all other observers also at rest *in that frame* agree with O? The answer is yes: O and any observer O' at rest with O will agree on the plane of simultaneity. After all, that is what makes them part of the same frame.

Indeed, a whole cluster of concepts come together in this case. The fact that clocks synchronized together in such a frame stay synchronized together, in addition to the fact that this synchronization is reflexive, symmetric, and transitive among observers in the frame, is what allows us to define a global time function whose "ticks" are the equivalence classes of synchronized events whose ideal clocks read the same time.

This method will not pick out an equivalence relation in all spacetimes. It's well-known (Reichenbach, 1957) that radar-synchrony is an equivalence relation only (for some special observers) when the spacetime metric is *static*. (A metric is static if it admits a timelike Killing vector K and possesses a surface of codimension one that is everywhere orthogonal to K. For those uninitiated with relativity, these two conditions might be conceived as counterparts of time translation invariance and time reversal invariance—the reader can then see that these are somewhat special spacetimes.) Since "most" spacetimes aren't symmetrical enough to count as static, radar synchrony won't work in most spacetimes.

defined but perfectly relativistic structure corresponding to the event of the first McDonald's opening. Suppose someone crowns McEvent the now and claims to have reconciled the present with relativity. I want to be able to respond by saying, fine, but McEvent lacks most of the seven features listed and therefore can't be used to recover much of manifest time. To name one shortfalling, McEvent won't yield a common settled past. Fast food lovers could insist on McEvent's significance, but it would be tolerably clear that we're talking past one another.

By precisely stating the physical conditions needed by the manifest now, we can in principle prove hundreds of theorems on the compatibility or incompatibility with relativity for different understandings of manifest now. I cannot summon the will to do this. Nonetheless, with existing knowledge we can peer into all this in-principle work and already make plausible the dilemma mentioned at this chapter's outset, namely, that one can find temporal structure worthy of manifest time or relativity—but not both. The dilemma is only loose due to the flexibility in the concepts. Looking at possible relativistic nows (below), however, convinces me that proposals always fall on one side or the other in a fairly straightforward way. The above-defined McEvent is relativistically acceptable, but it is obviously unworthy of grounding manifest time. The events I pick out as simultaneous with you reading this sentence fits many of the above conditions, but it's not relativistic. Proposals between these two exist, but—I submit—any compromise isn't very promising.

Classical physics was presented in a formulation devoid of any intuitive "time" or "space." But as in a *Where's Waldo?*[14] illustration, with enough work eventually Waldo pops out. In the classical case, we were able to find a "time" that in fact corresponds to all seven of the above features without much trouble. In relativity, by contrast, the work is in vain, for in this picture Waldo doesn't exist.

2.4 Minkowski Spacetime

Earlier I complained that philosophy of time concentrates too heavily upon Minkowski spacetime, the spacetime associated with special relativity. This is unfortunate because Minkowski spacetime is hardly a realistic spacetime (a flat vacuum spacetime) and it also contains many extremely special and unrepresentative symmetries. Since many of these arguments rely on these symmetries, the arguments are correspondingly parochial.[15] Nonetheless, given all the work devoted to this spacetime, I would be remiss to ignore it. So let me show how a very popular debate in Minkowski spacetime provides a nice instance of my informal dilemma. I will describe the infamous "Rietdijk–Putnam–Penrose" (RPP) (Putnam, 1967; Rietdijk,

---

[14] Or *Where's Wally* outside North America.

[15] Fine (2005), for instance, defends the option of relativizing existence to a global inertial frame in order to mesh relativity with tenses. But this option—even as unpalatable as it is—will only work in spacetimes possessing global inertial frames, something ours does not have. Fine provides no hint of how to generalize his idea in his discussion of relativity.

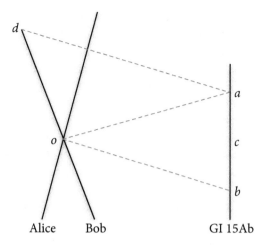

Figure 2.7 Relativity of simultaneity

1966) argument against an open future and Stein's (1968) response to it. The three authors in RPP in fact make distinct arguments, but the core ideas are similar enough to warrant unification. RPP has been controversial for over forty years. Yet with a few i's dotted, it is utterly convincing.

Two decisions need to be made: one, on the concept of the present, and two, on its implementation in relativistic structure. On the first decision, Putnam chooses what he calls "the man-in-the-street's" conception of time (a label itself showing the age of the debate). This conception is clearly meant to evoke the same target as my manifest time. In particular, since the person in the street wants to invest ontological differences based on this now, Putnam reasons that the events of the now must pick out a non-trivial equivalence class. Because he can't find any other equivalence relations, he chooses the constant t hypersurfaces picked out by coordinate time (in "natural" Minkowskian coordinates)[16] of an observer. In other words, he finds his nows in the planes of simultaneity that a relativistic observer would use.

Using this notion of the present, the threat posed by the relativity of simultaneity is obvious. Following Penrose (1989), consider two observers, Alice and Bob, who walk by each other on the street one day (see Fig. 2.7). Call this intersection event o. Alice

---

[16] Natural coordinates make use of Minkowski spacetime's global inertial structure. An inertial frame is a frame of reference picked out by a special set of massive bodies, namely, those free bodies that are mutually at rest. Classical physics defines them as frames wherein Newton's three laws hold (or hold in their simplest form). Relativity follows suit in defining them as frames wherein the relativistic laws of mechanics, electrodynamics, and the constant speed of light all hold in their simplest form. Because convenient, when a spacetime has global inertial structure it's desirable to adapt a coordinate system to an inertial observer who takes herself to be at rest. To do so, pick an always timelike inertial observer O who picks out orthogonal coordinates $\{e_0, e_1, e_2, e_3\}$ for M. If we then label events in terms of quadruples (x_0, x_1, x_2, x_3) then any given event in M is characterized by $x = x_0 e_0 + x_1 e_1 + x_2 e_2 + x_3 e_3$. If we further suppose that the axis coinciding with the direction of the trajectory of the observer is time, we have one example of coordinate time—say, x_0.

defines one frame, Bob another. Put coordinates on their frames such that o is the coordinate origin in each. Now draw the simultaneity planes that include o. Due to the relativity of simultaneity, Alice's simultaneity plane is distinct from Bob's simultaneity plane, despite both planes containing o. Extended all the way to Andromeda's exoplanet Gl 15Ab, o is simultaneous with event a for Alice and simultaneous with event b for Bob. Yet more than a few Earth days separates a and b.[17] Now focus on event c midway between a and b. When they cross in the street, c is unreal for Bob but real for Alice. Which is it? Assuming that it's more or less nonsense—or at the very least, desperate—to assert that *reality itself* is person-relative, we have a dilemma: either the original metaphysical view is wrong or we must arbitrarily distinguish some inertial frame.

Putnam's version of the argument seeks an outright contradiction. He begins by assuming that your here-now is real but that your future is not. Together with the assumptions already in place, this causes trouble. Pretend that you are Bob and your here-now is o. o is therefore real, as are the events simultaneous with o, including b. However, you're not special. Observers at Gl 15Ab count. We've established that event b on their worldline is real. Now we ask what events are simultaneous with b for the observers at Gl 15Ab? The answer is a set of events that include d on your worldline. Yet d is in your (Bob's) future but also real, contrary to the person-in-the-street's assumption that the future isn't real. To make the point salient, let's suppose that d represents your future death! We commonly suppose that to be open and unreal. However, if we choose the velocities of Bob's and Gl 15Ab's worldlines properly, your later death is as real as your here-now.

One might worry that RPP is framed in terms of coordinates or that it isn't very rigorous. Both concerns are misplaced. Addressing the first, replace Putnam's simultaneity slices with the set of events *orthogonal* to an inertial observer o at a given event. Long ago Robb (1914), in seeking a causal theory of time, defined a sense of orthogonality solely in terms of invariant Minkowski causal structure. Consider two

[17] Three points. First, the simultaneity slices for Alice and Bob drawn in the figure may strike you as counterintuitive. That is an artifact of the Minkowski diagram. The fact that all observers agree on the speed of light means that the lightcone bisects the angle between the temporal and spatial axes. So only one inertial frame is represented as having its spatial and temporal axes at right angles. As a result, Alice and Bob's simultaneity slices tilt in opposite directions. Second, Minkowski spacetime possesses Lorentz symmetry. A consequence of this symmetry is that global inertial frames are related by the famous *Lorentz transformations*. The transformation for time in natural coordinates in two dimensions is:

$$t \mapsto t' = \frac{t - vx/c^2}{\sqrt{1 - v^2/c^2}} \tag{2.2}$$

where c is the speed of light and v is the velocity of the unprimed frame of reference with respect to the primed frame. Using these equations it's immediately obvious that Alice and Bob must differ over what events are simultaneous. Suppose an event happens for Alice at $t = 0$. That means—for all the events they don't have in common—it won't happen for Bob at $t' = 0$ because Alice's present is drawn through the line $t' = -\gamma vxc^{-2}$, where $\gamma = (1 - v^2/c^2)^{-1/2}$. Third, readers new to this argument should be cautioned that neither Alice nor Bob will *notice* this disagreement when they intersect. It takes time for any signals from Gl 15Ab to arrive on Earth. They discover their disagreement theoretically, just as we did.

pairs of ordered points, p, q, and r, s, and the vectors connecting each pair, $\overrightarrow{pq}$, $\overrightarrow{rs}$. Then we can define orthogonality as $< \overrightarrow{pq}, \overrightarrow{rs} > = 0$ using the inner product provided by the Minkowski metric (which Robb ultimately derived from the earlier than relation). Robb-orthogonality has the nice feature that for any point on any inertial worldline, the set of all points Robb-orthogonal to that point is precisely the set of all points simultaneous with that point as determined via the Einstein–Poincaré synchronization procedure. We thus have a coordinate-invariant conception of simultaneity. If we now remove the restriction to Alice or any worldline ("no one is special"), the advocate of manifest time runs afoul of the following theorem (Malament, 1977a, Prop. 3):

Theorem 1. *Suppose R is a binary relation of $\mathbb{R}^4$ such that (i) R is implicitly definable from κ, the causal connectibility relation, (ii) R is an equivalence relation, and (iii) R is non-trivial, i.e., there are points p and q, such that Rpq and yet $p \neq q$. Then R is the universal relation.*

Treat the relation R as Putnam's equivalence relation, the one uniting the events in the person-in-the-street's now. This theorem is then a precise expression of the informal argument sketched above, answering the second worry.[18] If being related by R "colors" the events, then all the events in Minkowski spacetime are colored because R is the universal relation.

If it's a theorem, why has the argument repeatedly come under fire? Presumably it must be that one doesn't share the assumptions of the *reductio* on manifest time. This charge is hard to answer, as I've assumed that we share a pre-theoretical notion of the common now. I admitted that we may not outside the core of the concept. Here all I can say is that Putnam's choice is hardly idiosyncratic. It seems a very natural expression of some features of manifest time. In particular, within an inertial frame, it has *all of the first six* features listed above, 1–6. Compared to many other proposed presents, Putnam's now does remarkably well.

The most famous attack on RPP is a set of two papers by Stein claiming that what he calls "temporal becoming" is compatible with special relativity (1968, 1991). The centerpiece of these papers is a theorem proving that a unique "determinacy" relation R can be defined consistently in Minkowski spacetime in terms of only its geometric structure. Interpreting temporal becoming in terms of this determinacy relation, temporal becoming is shown to be compatible with special relativity.

Begin with the *definite for* relation R. R is a binary partial ordering relation between events, where 'Rxy' means the point y is definite as of x.[19] Like Putnam's *real for*

---

[18] Malament (1977a) is well-known for its relevance to conventionality. But it's also interesting to read in concert with RPP: Proposition 1 essentially demonstrates that Putnam's now is geometrically definable, Proposition 2 states that it is unique, Proposition 3 proves that it isn't generalizable across all distinct inertial observers.

[19] Stein doesn't tell us how to interpret determinacy. The idea is that certain events have become and are therefore determinate for other events. I'm not sure I understand what it is for an event to exist and yet be indeterminate (presumably, in some non-epistemic sense). Future me lives on M (and so exists) but has

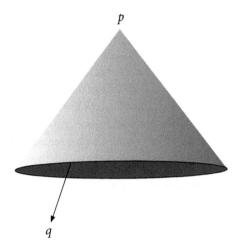

Figure 2.8 Stein's *definite for* relation

relation, R is supposed to be reflexive, transitive, and definable from the geometric structure of Minkowski spacetime. It differs from Putnam's relation by being antisymmetric and by the condition that if $\overrightarrow{xy}$ is a past-pointing vector (from x to y), then Rxy holds. Stein's theorem demonstrates that if Rxy obtains between at least one pair of events, then it can satisfy these conditions iff $\overrightarrow{xy}$ is a past-pointing vector. Since the set of all past-pointing vectors from x constitutes its backward lightcone, the proof shows that the region of definiteness for any event is precisely its past lightcone (Fig. 2.8).

Does this notion of becoming show that manifest time is definable solely in terms of Minkowskian structure? Perhaps in some sense, but that it falls short in many respects is made patently clear if we consider the present according to Stein. The present is defined as those events that are co-determinate. This demand shrinks the present down to a point. This result is a straightforward consequence of the antisymmetry of R: $Rxy \wedge Ryx \rightarrow x = y$ for all x, y. Antisymmetry is needed because Stein wants R to be an ordering relation, and ordering relations (partial or total) are by definition antisymmetric. We can call each of these tiny presents the "now," one for every event, but this now departs from the traditional notion in scores of ways. Except in trivial senses, it does not satisfy *a single one* of the above seven desiderata. Moreover, if we were to animate this "now," treat it as ontologically special, and imagine it flowing, the resulting metaphysics is more or less absurd.

no past-pointing vector Rab pointing to him (at b) from this event here now (a). Does future me have a freckle on his nose or not? All we really know is that *determinate for* had better not be *real for*. The reason is that few would accept that the *real for* relation satisfies the conditions laid down for the *definite for* relation, e.g., antisymmetry. As Callender (2000) points out, there is a difference between notions of becoming and flow that are observer- or event-dependent and those that are independent of observers or events. If one wants the *world* to become, as would-be vindicators of manifest time do, then one wants a more substantial perspective-independent sense of becoming.

As I discuss in Callender (2000), adding even the most meager requirement on the present, namely, that it have some spatial extent, quickly turns Stein's theorem into a no-go theorem for his becoming. Assume

$$\text{Non-uniqueness: } \exists x\, \exists y\, \exists R\, (Rxy \wedge Ryx \wedge x \neq y)$$

This condition merely insists that at least one event in the universe shares its present with another event. Think of it as a super-weak version of the idea that the present is global. As diluted as it is, this condition turns Stein's result into the following 'no-go' theorem:

Theorem 2. *Suppose R is a binary relation on time-oriented Minkowski spacetime such that it is (i) implicitly definable from time-oriented metrical relations, it obtains in at least one case, is (ii) transitive, is (iii) such that, if $y \in J^-(x)$, then Rxy,[20] and (iv) satisfies non-uniqueness, then R is the universal relation.*

Because it is a trivial modus ponens of Stein's modus tollens, I'll forego the proof. Intuitively, one sees that non-uniqueness stipulates that for at least one pair of events R is not antisymmetric, and this is enough to turn the existence proof into a no-go theorem.

If there is a problem with RPP's argument, it is only that it is limited to the Minkowski solution. Outside that context, we typically don't have global inertial frames at all. Or put geometrically, we don't have the symmetries necessary to show that the "space at a time" defined via Robb orthogonality is uniquely definable in terms of the causal structure of the spacetime, so we lack a good geometric candidate to use in a *reductio*. Minkowski spacetime isn't an attractive setting for manifest time. Here in the case of RPP and Stein we see an example of my informal dilemma. Putnam's present is congenial to manifest time but provably not relativistically invariant; Stein's present is provably invariant but not congenial to manifest time.

2.5 Lorentzian Time

Before leaving "special" relativity, let's pause to recall some history. As is well-known, both H. A. Lorentz and Henri Poincaré discovered the Lorentz transformations (Eq. (2.2) above), the equations with the defining symmetry of Minkowski spacetime, before Einstein did (see, e.g., Galison, 2004). Their paths to these transformations were different, as were their subsequent interpretations (for Lorentz's, see Brown, 2001; for Poincaré's, see Zahar, 2001). Like Einstein, both discovered these transformations while working in an entirely classical "3 + 1" space and time setting, not the "geometric" picture later suggested by Minkowski in 1908. This historical fact suggests that perhaps one could reinterpret the relativistic data in such a way as to preserve

---

[20] $J^-(x)$ represents the set of all points y in the causal past of x; that is, the set of all points connected to x by a timelike or null curve.

a classical time. After all, if Lorentz, Poincaré, and other great physicists could arrive at the correct transformations in a classical context, the Lorentz transformations *must* be compatible with classical time.

That is correct. Without going into details, remember that physicists Oliver Heaviside, Joseph Larmor, George FitzGerald, and others discovered features of electromagnetic theory that would lead to an inability to discover the correct inertial frame of reference in which classical electromagnetism was written. The classical theory models a magnet and conductor at rest with the ether very differently than it does a magnet and conductor moving with respect to the ether—yet no measurement can tell the difference. That this theory "leads to asymmetries which do not appear to be inherent in the phenomena" motivates Einstein in 1905 to seek a theory without such a feature. It must be emphasized, however, that the defect Einstein finds is not a failure of the theory to reproduce what is seen. It was a perceived theoretical defect, not an empirical flaw or problem of consistency.

One might reasonably view Einstein's motivating demand as negotiable. Lorentz clearly did, and indeed, it seems that historically one of his motivations was to maintain a traditional conception of time:

> My notion of time is so definite that I clearly distinguish in my picture what is simultaneous and what is not.   (1927, p. 221)

The theory Lorentz adopted can *in principle* be developed to handle the paradigmatically special relativistic effects. For instance, one can show that classical electromagnetism predicts that a complex charged system will contract in the direction of motion. This genuine "physical" contraction mimics what Einstein later considers a kinematical effect, what we now call Lorentz–FitzGerald length contraction. Lorentz's dynamical contraction is empirically indistinguishable from the kinematical effect. As Bell (1987b) shows nicely, the effects of Maxwell's equations on clocks and rods imply that we will never experimentally determine Lorentz's true states of motion. In this interpretation, which Lorentz continued to develop well after Einstein's discovery of relativity, the background spacetime is not Minkowskian but neo-Newtonian. Absolute simultaneity and all of the structure of classical time is restored. The classical spacetime can be foliated via equivalence classes of simultaneous events. We could then maintain the picture of equivalence classes of simultaneity planes successively flashing in and out of existence, while granting that we cannot experimentally determine which frame is the one that flows or becomes. The theory distinguishes a frame, but it also informs us that we can't tell which one.

Today such an interpretation of rod contraction and associated relativistic effects is regarded as introducing a conspiratorial interpretation of nature, albeit one that is not contradicted by observational facts. Why is the true "ether" frame systematically hidden from us? Why do the matter fields possess Lorentz symmetry but not the spacetime? One might reply—not unreasonably, I think—that the aspects of the Lorentzian picture that appear conspiratorial only do so when adopting the rival Minkowskian view. Conspiracies depend on background assumptions. Change those and the unexpected can become expected. From the Lorentzian perspective the shrinking of rods and so on are precisely the expected results of a massively successful theory, classical electrodynamics.

If the foregoing is correct, why can't the advocate of manifest time move to a Lorentzian picture and leave behind the inhospitable setting of relativity? Admittedly it is an unfashionable interpretation, but fashion shouldn't dictate what science we accept. While one can debate the methodological vices and virtues of Lorentzianism for a long time, two reasons stand out against such a move.

First, there is more to the world than bodies held together by electromagnetic forces. Atoms held together only by electromagnetic forces are unstable. So we need "Lorentzian" theories of strong forces and weak forces if we're really to have an adequate Lorentzian alternative. While this point stands, its dialectical power is perhaps weakened by the observation that the standard "kinematical" approach required such relativistic extensions too. Absent a reason to think such extensions aren't possible, perhaps this observation really just points to a new Lorentzian research program (Brown, 2001).

Second, Minkowski spacetime is only one solution of a far richer theory, "general" relativity. The choice is not between Minkowski and Lorentz, but instead between full-blown relativity, which includes our best theory of gravitation, and Lorentz. Yet there is no Lorentzian interpretation of general relativity. Lorentz (1900) did try to incorporate (classical) gravitation into his picture, as did Poincaré, but neither met with much success. Today there are attempts, most notably by Broekaert (e.g., 2007), to provide a Lorentzian interpretation of general relativity, but these pale in comparison to the real thing. One could say that this too points to the need for a new Lorenztian research program. Readers can decide for themselves how much credit they want to extend the Lorentzian gravitation program, but it seems clear that one research program is thriving while the other is not.

Although one *might* argue that historically one could have preferred Lorentz to Minkowski without "bending" the usual norms of science, one cannot now argue similarly for a choice of Lorentz over full-blown relativity. Relativity simply overwhelms its Lorentzian rivals in terms of both theoretical and empirical virtues. Perhaps the future will change this assessment. Today, however, going Lorentzian solely in order to save some aspect of manifest time is requiring science to bend in ways that don't optimize its virtues.

2.6 Outside Minkowski: Domes, Donuts, and Diamonds

Looking past Minkowski spacetime, many philosophers and physicists claim that "general" relativity is more hospitable to manifest time than "special" relativity.[21] One

---

[21] See the papers by Tooley (2008), Swinburne (2008), Crisp (2008), Lucas (2008, 1999, 1986), and Ellis (2006).

misguided reason for this assessment is the non-existence of a general RPP result outside Minkowski spacetime. Malament's theorem (see Theorem 1) is inapplicable in this more general context because most solutions don't have the special symmetries that Minkowski enjoys. However, the lack of a counterpart of Malament's theorem in the more general setting is often due to this setting being *so* inhospitable to manifest time that one can't even pose the question asked by RPP. To take a dramatic example, the infamous solution to Einstein's field equations known as Gödel spacetime doesn't contain a *single* everywhere-spacelike global hypersurface. Assuming the present is global and spacelike, one cannot show that Alice's present conflicts with Bob's because neither has anything worth dubbing a present at all! An RPP-like argument is at hand here, but it is so short that no one bothers to make it.

We can get RPP-like results outside Minkowski spacetime. We just can't get anything general, but must instead go spacetime by spacetime, proposed present by proposed present. Not only would this project be time-consuming and boring, but it also wouldn't be very illuminating. Better to find some "natural" candidates suggested by our experience with classical presents and see how they fare.

Classically we were able to "discover" time and space buried in $\langle M, t_{ab}, h_{ab} \rangle$. Can we do the same in relativity? Not if by "same" we mean formally the same. Any mechanical attempt to mimic the classical strategy is immediately stymied by the fact that the counterparts of two conditions that coincide for our classical instants of time come apart in relativity. Classical temporal instants, recall, are hypersurfaces that are (i) everywhere spacelike and (ii) orthogonal to time. If we look past the interpretation, what this meant formally was that our classical instants included events that happened at zero temporal distance apart, i.e., $ds^2 = t_{ab}\xi^a\xi^b = 0$, and that the two metrics were orthogonal, i.e., $t^{ab}h_{ab} = 0$. Since we don't have two metrics relativistically, there is no exact counterpart of classical orthogonality; just as bad, the condition $ds^2 = 0$ picks out null hypersurfaces, not spacelike hypersurfaces. If we try to use a natural replacement for orthogonality, namely, the vanishing of the relativistic norm, i.e., $g(u, v) = 0$, then it turns out that we can seek orthogonality or the classical counterpart of spacelike ($ds^2 = 0$), but not both.[22] So, to begin, unlike in the classical case, we must choose between orthogonality to the timelike and $ds^2 = 0$ for our presents.

The $ds^2 = 0$ option isn't very attractive. Identifying the now for event p with all those points zero distance away picks out the entire surface of the lightcone emanating from p. If M is temporally orientable, then we can make a consistent assignment of past or future lightcone, enabling us to restrict the now to, say, the surface of the backward

---

[22] If $ds^2 = 0$, then the hypersurfaces picked out are not orthogonal to the timelike. That is because vanishing distance picks out null hypersurfaces relativistically, and null vectors are orthogonal only to null vectors. If the counterpart of orthogonality holds, i.e., $g(u, v) = 0$ with v timelike and u non-zero, then the surface that is picked out is one for which $ds^2 > 0$. That is because u must be relativistically spacelike, yet the spacelike in relativity, unlike classically, is one for which distance is non-zero. The notions of orthogonality and zero distance come apart in relativity: the first picks out spacelike directions and the second null directions. $ds^2 = 0$ hypersurfaces that are orthogonal to the timelike don't exist.

p

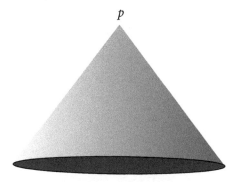

Figure 2.9 Lightcone presents

lightcone (Fig. 2.9). Still, if manifest time is the standard, this option is not compelling (although it has been proposed). Savitt (2000) explains in detail the reasons why not. Let me simply note that, aside from possibly including the Big Bang in your current present, this identification sacrifices the present being an equivalence relation, the present being spacelike, the present being synchronous, the present being shared, and more. Nearly the only feature it has going for it is that the lightcone structure is genuinely relativistic. Conforming to our informal dilemma, this 'now' does great relativistically but performs abysmally as a candidate for manifest now.

More promising is the "orthogonality" option, as that will typically demand that presents are spacelike.

The most natural way to implement this idea is problematic. Although spacetime may be variably curved locally, in the neighborhood about any point p in M, to a first approximation spacetime becomes Minkowskian. That's not strictly true, for the curvature at a point doesn't vanish. What is true is that one can define a local infinitesimal simultaneity slice, given a timelike observer O, its four-velocity field, and point p. Take a vector at point p. Then we can decompose it into components which are parallel to and orthogonal to O's velocity field. The infinitesimal subspace spanned by the orthogonal components can be conceived of as an infinitesimal simultaneity slice for O at p. One can even break apart "space" and "time" by expressing this decomposition in terms of separate temporal and spatial metrics (Malament, 2007, 2.4), so we have a kind of counterpart of our classical presents. Call this a *puny present* (Fig. 2.10). One can think of every observer as carrying around a little local now in his or her neighborhood. These puny nows won't serve our purposes, however, as they lack virtually all of our desired properties. Not only don't they divide the world into a past and future, they don't even divide finite regions into past and future. Relativistically impeccable, puny presents depart dramatically from the manifest present.

Here is another idea taking its cue from the classical case. Classically the set of points spacelike to p picks out an attractive candidate "present." If one regards classical spacetime as the limit wherein the relativistic lightcones tilt all the way until they

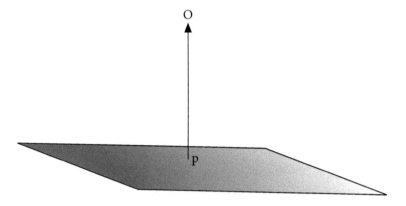

infinitesimal simultaneity slice

Figure 2.10 Puny presents

present   present

Figure 2.11 Donut presents

"touch," then perhaps the natural relativistic counterpart is not *one* particular spacelike hyperplane intersecting p, as Σ above, but rather the set of *all points spacelike related to p*. In relativity this will pick out a four-dimensional volume, a kind of "donut" around p, not a three-dimensional sheet. Call these *donut presents* (Fig. 2.11). Although natural in some respects, this "now" violates many of our suggested desiderata. For instance, the points in the absolute elsewhere for p are all spacelike to p, but that doesn't mean that they are all spacelike to each other. The set is achronal. Whole lives of distant observers might fit into p's now! Worse, with the assumption that this present is shared and non-trivial, we can quickly get an RPP-like "no-go" theorem in Minkowski spacetime

(Weingard, 1972). Color all the spacelike events from p red, where "red events" are real or definite. Pick one of those red events, q, not identical to p. Color red the events that are spacelike related to q. The result will be some red events *within p's lightcone*. Repeat *ad infinitum* and one colors red the whole M if the manifold is Minkowskian. An argument in this spirit holds in many non-Minkowskian spacetimes too; one may not be able to color the whole spacetime, but there will be choices of q that suffice for a *reductio*.

Squash the now back down to a hyperplane in M. Wüthrich (2013) considers the cosmologist E. A. Milne's notion of *private space* (referred to in Rindler, 1981). Consider not merely the hypersurfaces orthogonal to observer O at p, but the hypersurface picked out by all the inextendible *geodesics* ("straightest paths") running orthogonal to O at p. This *geodesic hypersurface*, which often looks like a "dome" in cosmological spacetimes, is dubbed *private space* (Fig. 2.12). Like our notion of Robb orthogonality in Minkowski spacetime, private spaces have the attractive feature that they are unique for a given O at p. Locally they are spacelike and even flat. Nonetheless, private space lacks many desirable features for a now. Apart from problems the reader will quickly discover, I note some of the less obvious consequences. (1) There is no guarantee that as we extend away from p that the submanifold remain entirely spacelike. By definition it will be spacelike in the direction pointed toward p, but geodesics defining the submanifold can become timelike or null separated from each other. (2) In some spacetimes (e.g., FLRW, see Page, 1983) the present will intersect the Big Bang, which is as counterintuitive as the $ds^2 = 0$ nows. (3) This now will cut the spacetime in two, as desired, yet in some spacetimes that division will not divide each timelike worldline in two. Some inextendible worldlines might never have a past relative to a given now! Because private spaces may not intersect every timelike observer, we cannot run an

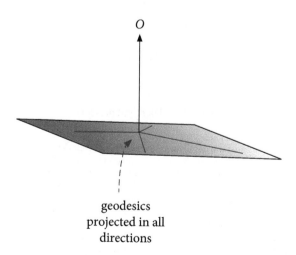

geodesics
projected in all
directions

Figure 2.12 Private presents

RPP–style argument for arbitrary observers. Yet we can make a similar "Alice and Bob" type argument based on the distinct private spaces of two or more distinct observers. Private spaces . . . are too private.[23]

Finally, taking a lesson from Goldilocks, perhaps a present not-too-big and not-too-small is our best bet? In this spirit, consider the structure known as the *Alexandroff interval*. This proposal is not inspired by classical reasoning, but rather by thinking about what we really need explained, and it's different from the others since it's relativized to a pair of instants on a timelike curve rather than an instant. Take two events p and q on a timelike curve λ. Suppose p and q are a certain finite proper time apart from one another, with $p \ll q$. Then the Alexandroff present $\mathrm{Alex}(\lambda; p, q)$ is defined by the set that forms the intersection between the interior of the future lightcone of p with the interior of the past lightcone of q, i.e., the region $I^+(p) \cap I^-(q)$. Interestingly, one can prove that these regions are globally hyperbolic (see the next chapter) if certain conditions are met (Ribeiro, 2010). The Alexandroff present looks like a nice "diamond" in the Minkowski representation—and hence we can call them *diamond presents* (Fig. 2.13)—but they are well-defined in all "well behaved" (i.e., causally stable—see Box 2) spacetimes.

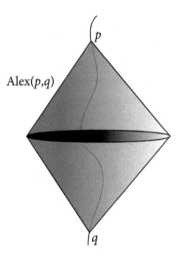

Figure 2.13 Diamond presents

---

[23] Here is a way to fix *that*. We want our presents to be shared, so Milne's notion of a *public space* beckons. Suppose we have a family of timelike geodesic observers emerging from point p. Then at some later arbitrary event q on one of these worldlines, the public space is the spacelike hypersurface that is orthogonal at every point to these timelike geodesic observers. So long as the spacetime is not rotating, the notion of public space should be well-defined and even unique. The nows generated in this way will be shared by all the timelike geodesics in the family. Do we have something that is unique, global, and shared? It depends who you ask. Everything looks good if you're a geodesic timelike observer. If not—and I doubt that *you* are a geodesic observer—then these nows don't have many of the features we desire. We bought a certain amount of commonality by distinguishing some observers over others, and so this time shares many of the problems we've already encountered and will meet again in the next chapter.

By themselves, diamonds aren't very promising manifest presents. They are local, achronal, fail to bissect the world, and typically aren't shared. One doesn't find the global tripartite division of reality that is so central to the manifest image of time here.

Still, perhaps what one finds is *good enough*. That is Savitt's (2009) thought (see also Savitt, 2014). As here, Savitt wishes to explain why human beings might come up with manifest time, not vindicate every aspect of it or find an exact physical counterpart. Much can be said for diamond presents in this respect. Savitt notes, for instance, that if one limits the proper time between q and p to a (roughly) psychologically realistic 1 second, then the Alexandroff interval is $300,000$ km across where it is widest. He calculates that if two observers walk past one another at 4 km/h, then their two diamonds include the same events to one half of one millionth of one percent—enough overlap so that no one would notice! Going back to RPP, Alice and Bob strictly will still disagree about what's present, but probably they will agree on every *noticeable terrestrial* event. With this in mind, Savitt writes:

> In the course of human conceptual development, it would be no more surprising that we developed the idea that this brief, fat structure was unbounded than that we developed the idea that the surface of the earth was flat.   (357)

The idea is that diamonds can plausibly *explain* the idea of a common global now arising in creatures like us.

Although I'm sympathetic to this enterprise, and will do much the same myself later, what is doing the explaining here? Diamond presents seem to be a kind of unstable attempt to satisfy two masters, relativity and psychology. Like those seeking compatibility with relativity, Savitt insists on finding the present in some relativistic invariant structure. However, it's the injection of psychology that allows Savitt to make any contact with the ordinary now. Psychology is suggesting a temporal length of about 1 second (if it does), not relativity. Relativity allows diamonds so big that the entire universe (or most of it) is present and diamonds so small that we won't find common overlap over noticeable terrestrial events. What's carrying much of the explanatory burden are the psychological facts. Of course we can fit these facts into a "diamond," but that was never the question. Once we let in psychology, I don't see why the diamonds should matter.[24]

If we're engaged in showing why real creatures like us develop a notion of manifest time like we do, we don't need to identify the present with some relativistically invariant structure. Creatures live in a relativistic spacetime, to be sure, but we don't suppose that what these creatures devise and dub "the present" must *itself* be an invariant structure. There need be a Lorentz invariant structure corresponding to the "present"

---

[24] Furthermore, as Carl Hoefer has remarked (p.c.), perhaps any psychologically realistic present requires *more* of the backward lightcone than that which makes it into the Alexandroff present. Why should the future lightcone of q determine how much of the backward lightcone of p is relevant to the psychologically realistic present? See also Dorato (2011).

no more than there need be one corresponding to the sandwich chain *Subway*'s famous "foot long" sandwiches. Being a foot long is not relativistically invariant, but we possess a good story of why we can successfully speak that way (involving the low relative velocities among those in line, those making the sandwiches, and so on). If we're instead engaged in the metaphysical project of finding a relativistic present suitable for tensed "animation," by contrast, then the present we need must be a good candidate for animation and ontological priority. These diamonds are not such structures. Apart from the departures from manifest time already listed, they are worldline dependent. One preferably doesn't want to swallow the pill of "saving" animated time by making reality observer dependent. So the diamonds aren't helpful in this project either. Causal diamonds are useful tools in mathematical physics. Yet if one keeps one's goals straight, they don't provide a metaphysical vindication of manifest time.

2.7 Conclusion

Does relativity imply that there is no special now? Let's be clear: the physical facts alone won't tell us whether we ought to believe in manifest time's temporal features. Physics by itself doesn't rule in or out much. The idea that it does is silly. Physics doesn't claim that basketballs exist (yet they do) nor that ghosts don't (they don't). Neither kind of entity is mentioned by physics. What makes it irrational to believe in ghosts is that *no good theory* requires them and *good methodology* exhorts us to posit no more than *good theory* needs. What makes it rational to believe in basketballs, by contrast, is that *good theory* (the theory of the external macroscopic world) requires them and *good methodology* exhorts us to posit what *good theory* needs. That's a lot of *good*s, and hence, many promissory notes in epistemology and methodology. What does the work in any argument for or against some piece of ontology is therefore more than whether physics employs said piece of ontology.[25] All we've seen is that relativity doesn't employ a special now nor is it easy to add one without doing some violence to the theory. That is only one theory, and we haven't yet approached the question of whether good theory as a whole requires a now. In the next chapter, we'll look at whether slight modifications of relativity are more congenial to the now.

---

[25] If one has an inkling that basketballs "reduce" in some sense to physics (as I do), then physics implicitly mentions basketballs. That would be of great comfort to those whose "good theory" includes a background commitment to physicalism. But that doesn't impugn what is said here, for the belief in physicalism is more than the belief in physics.

3

Tearing Spacetime Asunder

When first learning relativity we are taught that no one should tear asunder what Nature has joined together. Minkowski famously claimed that space and time were doomed to fade away into mere shadows. We ought not tear apart spacetime into space and time. It therefore may come as a surprise to some readers that physicists are constantly tearing spacetime asunder. Relativistic spacetime is commonly decomposed into "time" and "space" in so-called "3 + 1" formulations of relativity. First invented by Darmois (1927), these formulations slice spacetime into a set of three-dimensional spatial hypersurfaces ordered by a one-dimensional time parameter. They are used in solving the Cauchy problem (see Chapter 8), in seeking a theory of quantum gravity (see Chapter 5), and in numerical relativity, where computer simulations take data on an initial "spatial" surface and evolve it forward or backward to model cosmological regimes. How should we regard the time functions employed? Are they hospitable to manifest time? Outside the context of Minkowski spacetime there are often geometrical and material reasons to prefer some time functions over others. As we'll see, the question is how to understand these "preferences."

3.1 Cauchy Time

My daughter's kindergarten categorized students' behavior with a clip on a behavior chart: green for model behavior, yellow for slight infractions such as talking during nap time, and red for tantrums and the ultimate crime, biting. Relativists have invented a similar categorization for the behavior of spacetimes. Respect for causality is the behavioral goal. The hierarchy is known as the Causal Ladder, but among those who fear causality violation it plays the same role as Jacob's Ladder. The higher a spacetime makes it up the ladder, the closer it is to heaven. See Box 2 for a chart that characterizes some of the more common classifications used in relativity.

Focus on one particularly nice type of spacetime, namely, those that are globally hyperbolic. These spacetimes permit an especially attractive way of decomposing $\langle M, g \rangle$ into "space" and "time." The definition of global hyperbolicity is included in Box 2. Globally hyperbolic spacetimes are known as "predictable" spacetimes. We can appreciate why this is the case by noticing that global hyperbolicity is equivalent to both of the following statements:

- spacetime contains a closed achronal connected submanifold Σ such that Σ is intersected by every timelike curve in M exactly once
- spacetime contains a closed achronal connected hypersurface Σ such that $D(\Sigma) = M$, where $D(\Sigma)$ is the domain of dependence of Σ.

These statements are perhaps more illuminating than the official definition. Take the first characterization. Grab some point $p \in M$ on a timelike curve. That curve, if extended far enough, must intersect Σ at some point. Turning this around, that means that if we know the complete conditions across Σ, then in principle we can predict what happens at p, for any p. The dynamical equations are deterministic, so p will have a trace on Σ no matter where p is. Meanwhile, the second condition says that the domain of dependence of a particular type of slice through spacetime is the entire manifold. Consider some subset S of M. The domain of dependence of S, $D(S)$, is the set of all points p such that every timelike or null inextendible curve through p intersects S (Fig. 3.1). That $D(\Sigma) = M$ means that there are no bits of the world that are causally isolated from Σ. Intuitively, there is nowhere for p to hide from Σ. Like Big Brother, Σ sees all.

With these features in mind, one sees the connection between global hyperbolicity and in principle predictability of the entire manifold. One of the centerpieces of relativity is the proof that the so-called Cauchy problem (see Chapter 8, but for now, a kind of determinism) is solved for globally hyperbolic spacetimes (Choquet-Bruhat and Geroch, 1969). This predictability is obviously desirable for computer simulations of relativity, but it is also theoretically important.

The features that make such spacetimes so congenial to prediction also allow for a natural decomposition of M into "space" and "time." Assume that M is globally

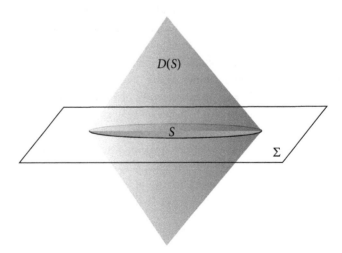

Figure 3.1 Domain of dependence

hyperbolic. Then we know at least two further facts thanks to what is sometimes called Geroch's *splitting theorem* (Geroch, 1970). We know that a particular kind of time function exists and that M can be topologically split into $\Sigma \times \mathbb{R}$.

Regarding the time function, global hyperbolicity implies that there exists a continuous onto map $t: M \to \mathbb{R}$ such that (i) $\Sigma_a \equiv t^{-1}(a)$ is a Cauchy hypersurface for all $a \in R$ and (ii) t is strictly increasing along any causal curve. Property (ii) means that t is a *cosmic time function*. Hawking (1969) proved that any spacetime stably causal or higher in the causal hierarchy allows such a time function. However, the surfaces of level t aren't just any hypersurfaces, but by Property (i), Cauchy hypersurfaces. A Cauchy hypersurface is a submanifold of spacetime that is intersected exactly once by every inextendible non-spacelike curve. The data on these slices are in principle sufficient to reproduce the rest of the spacetime. To mark this distinction, let's dub this species of cosmic time function a *Cauchy time function*. Regarding the split, Geroch shows that global hyperbolicity is equivalent to the claim that M is homeomorphic to (i.e., topologically equivalent to) the product $\Sigma \times \mathbb{R}$, where Σ are the Cauchy surfaces. Σ can be of arbitrary topology, so long as basic relativistic constraints still hold.[1]

The upshot of the splitting theorem is that we can chop up, or tear asunder, M into "spaces" Σ strung together by a "time" R (Fig. 3.2).

So far the splitting of M into "space" and "time" is entirely topological. The definition of a Cauchy hypersurface is topological ("submanifold intersected exactly once...") and the decomposition into space and time is the fracturing of M into a topological

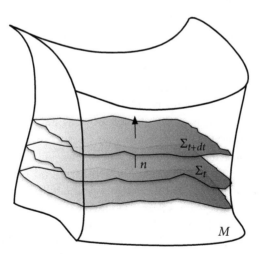

Figure 3.2 Tearing spacetime into space and time

---

[1] Two other interesting features follow, too. One is that all the level t surfaces Σ_t are homeomorphic with one another. Hence there is no spatial topology change (see Callender, 2000, for discussion). Another is the absence of closed timelike curves, so time travel is prohibited (but see Friedman, 1991). The second is an immediate consequence of the monotonicity condition on cosmic time and also an immediate consequence of the fact that timelike curves intersect Σ exactly once in globally hyperbolic spacetimes.

product manifold. There is more to the world than topology. A relativistic spacetime is given by $\langle M, g, T \rangle$ satisfying the Einstein field equations. We've broken up M into $\Sigma \times \mathbb{R}$, but we haven't yet split g or T, nor have we found $3 + 1$ expressions for Einstein's equations.

Our previous observation helps us break up g. Suppose $\langle M, g \rangle$ has a Cauchy time function. Then not only does M split up orthogonally but so does g as follows:

$$g = -N\,dt^2 + g_t$$

Here g_t is a nice Riemannian $(+++)$ metric on each slice of constant t and N is the so-called *lapse function*.[2] The lapse function $N(x^i, t)$ gives the amount of proper time that passes per unit t in the direction normal to Σ. For a clock moving a spatial distance $d\ell$ the proper time elapsed will be

$$d\tau = N^2(x^i, t)\,dt^2 - d\ell^2$$

where τ is the clock's proper time and t is the Cauchy time. We see that the path-dependence of duration survives in this formulation, as the clock's ticks are affected by $d\ell$. In fact, even if the clock is still ($d\ell = 0$), its ticking will be affected by the local lapse field. The lapse therefore acts as a kind of disturbing field: synchronize two clocks on Σ_t and let them evolve freely to $\Sigma_{t'}$; unless the lapse field is the same along both paths, they will not tick off the same amount of proper time. Notice that the choice of lapse field in effect chooses the foliation. Suppose $d\ell = 0$ in the above equation. The proper time is invariant; therefore fixing N determines dt, where t is our Cauchy time function parametrizing the foliation.

So far we've decomposed the spacetime geometry $\langle M, g \rangle$ into a pseudo-Riemannian product of $\langle \mathbb{R}, -dt^2 \rangle$ and $\langle \Sigma, g_t \rangle$. We still need to decompose T, decide on the initial data to include on Σ, and find the laws that will evolve this data such that it creates a spacetime that satisfies Einstein's field equations. At this point matters get very complicated and there are many different approaches. I won't ask the reader to wade through these methods as the differences won't matter to us. In the popular Arnowitt–Deser–Misner (ADM) formalism, for instance, the result is that relativity looks like a classical Hamiltonian system, albeit one with heavy constraints. The three-dimensional spatial metric is like the "position" and the extrinsic curvature field is like the "momentum," and the two act as the initial data placed on spatial hypersurfaces. Einstein's equations become four constraint equations, conditions that must be satisfied on each Σ, plus twelve evolution equations that push the initial data forward or backward in time, in addition to four arbitrary conditions for the lapse and its spatial counterpart, the shift vector. There are other options based on different decisions one can make, e.g., whether one obtains the evolution laws by projecting the

---

[2] I have the set the shift vector N^i to zero, as its value is freely chosen and we won't be focusing on it. The shift vector moves coordinate systems from slice to slice.

Einstein tensor (ADM) or Ricci tensor (numerical relativity) down onto Σ. See the excellent Gourgoulhon (2012) for options.

Here I just want to draw attention to the fact that this kind of decomposition can be done. I also wish to draw attention to the fact that N makes an appearance in the *evolution laws* in these $3 + 1$ formalisms. The laws that push the data on Σ forward in time are N-dependent, and therefore, they in effect demand a choice of foliation.

Return to the main thread. Cauchy surfaces are a natural extension of Putnam's choice of "space" to globally hyperbolic spacetimes. They enjoy a number of attractive features, including:

- Σ_t bisect the universe, i.e., they are nonintersecting global slices ($\Sigma_t \cap \Sigma_{t'} = \emptyset$ for $t \neq t'$) that each intersect once every inextendible timelike observer
- Σ_t are achronal
- Σ_t are productive in the sense that $D(\Sigma_t) = M$, where $D()$ is the so-called domain of dependence
- Cauchy time t is an equivalence relation
- For observers whose worldlines are orthogonal to Σ_t the events on Σ_t are *locally* radar-synchronous.

Apart from radar-synchrony being local and not global, we otherwise have reasonable counterparts of many items from last chapter's presentness wish list.

As expected, with the good comes the bad. We've only changed the presentation of the theory, not the models of the theory itself. We can't expect an invariant now to suddenly emerge via a change of language. Now that we have a candidate relativistic "now" we are in a position to run an RPP-like argument. The problem is that the choice of Cauchy foliation $\{\Sigma_t\}_{t \in \mathbb{R}}$ is arbitrary, or put differently, we have complete freedom in our choice of lapse function. The choice is a gauge condition corresponding to the infinity of possible foliations we can use. The decision is crucial, as the resulting $3 + 1$ *laws* hang on these arbitrary choices. This should have been expected. The general covariance of the four-dimensional theory doesn't disappear when moving to $3 + 1$. Rather, it manifests itself in the ability to recover M from an infinite variety of initial slices.

Hence we can easily run "Alice and Bob" stories like Penrose's to cause problems for tensers. Suppose Alice and Bob intersect at some event e. Let Alice use so-called *geodesic slicing* ($N = 1$) and let Bob use so-called *maximal slicing* (hypersurfaces where the mean curvature vanishes). Then I assert that for all but the most contrived spacetimes there are events now or real for Alice and not for Bob, and vice versa. Beware: some of these foliations can be very badly behaved. For instance, geodesic slicing ($N = 1$) fails to be well-defined when the fiducial observers (those orthogonal to Σ_t) collide with one another, which will happen when in the presence of a non-uniform gravitational field. This complication, among others, make the prospects of a *general* counterpart of Theorem 1 (which showed that a notion of the present is

---

Box 2   Spacetime Behavior Chart

For events $p, q \in M$, we'll write that $p \ll q$ if there is a future directed timelike curve from p to q and we'll write that $p < q$ if there is a causal curve from p to q. The former is the relation of *chronological precedence* and the latter is the relation of *causal precedence* (which extends the former to timelike and lightlike curves from p to q). The chronological past of p is defined as $I^-(p) := \{q \in M : q \ll p\}$ and the causal past of p is defined as $J^-(p) := \{q \in M : q < p\}$. The chronological future and causal future is defined analogously. Going down the list, spacetimes are classified as more and more causally ordered:

- Chronologically vicious: $\forall p \in M, p << p$. There is a closed timelike curve at each point.
- Non-totally vicious: $\exists p \in M, p \not\ll p$. Some points don't allow closed timelike curves.
- Chronological: $\forall p \in M, p \not\ll p$. No closed timelike curves.
- Causal: $\forall p, q \in M$, if $p < q$ and $q < p$ then $p = q$. No closed causal curves
- Past Distinguishing: $\forall p, q \in M, (I^-(p) = I^-(q)) \implies p = q$. Two points that share the same chronological past are the same point. An analogous definition exists for Future Distinguishing.
- Strongly Causal: $\forall p \in M$, there exists a neighborhood U of p such that no timelike curve passes through U more than once. No causal curves are almost closed.
- Stably Causal: no arbitrarily small perturbation of the metric can make the spacetime contain a closed timelike curve.
- Globally Hyperbolic: the spacetime is strongly causal and $\forall p, q \in M$, such that $p \ll q$, $J^+(p) \cap J^-(q)$ is compact or empty.

---

not invariant in Minkowski spacetime) impossible. Nonetheless, a *reductio* can still be made on the proposition that the "future" is unreal. The present understood as level surfaces of Cauchy time suffers essentially the same fate as simultaneity slices in Minkowski spacetime.

3.2 "Unique" Time Functions

The $3 + 1$ formalism provides a natural decomposition of $\langle M, g \rangle$ into "time" and "space." However, the decomposition relies on a gauge condition, and the reliance on this condition allows us to make an RPP-type argument. Can we eliminate this arbitrariness? We arrived at Cauchy time by insisting that spacetime be globally hyperbolic. If we permit further restrictions, different time functions—subspecies of Cauchy time—emerge with the amazing property that they are *unique*, or at least, physically preferred. The literature in relativity is replete with times: not only cosmic times and Cauchy times, but Misner time, Maupertuis time, WKB time, Kodama time, Chitre–Misner time, Kijowski and Tulczyyev's "material time," and the "cosmological time function" of Andersson, Galloway, and Howard. In virtually every case, the time defined will exist only in a proper subset of Einstein's field equations. These times may be preferred by restrictions on the matter–energy content of the spacetime or the geometrical features of the spacetime, or both. I'll discuss two examples and show that, contrary to initial impressions, they conform to my informal dilemma of Chapter 2. Later I will discuss the nature of these "preferences."

The most famous case is the cosmological time of relativistic cosmology. This time is often celebrated as returning us to a kind of classical time. When we peer into the night skies, if we're willing to "squint" enough the distribution of matter looks spatially isotropic. Locally that is not so. Some directions have desks or planets in them, others do not. Pan out, however, and at large enough scales the universe looks roughly the same in every direction: radio sources, X-ray and γ-ray radiation, and the 3K thermal radiation that bathes the universe all seem evenly distributed in all directions. These facts strongly support the claim of large-scale spatial isotropy. Cosmologists furthermore assume that we aren't special. Hence this isotropy should be the case at every point, i.e., the universe is spatially homogeneous. (Alternatively, making certain additional assumptions about spacetime's shape, spatial isotropy *implies* spatial homogeneity.) Together these assumptions—that on large scales the universe is spatially isotropic and homogeneous—are known as the *Cosmological Principle*. This principle suggests that we can approximate space with a model that has the same curvature at each point. We make contact with the history of our universe if we additionally assume what is sometimes called *Weyl's Principle*. This principle states that the worldlines of galaxies form a non-intersecting bundle of timelike geodesics orthogonal to the above spacelike hypersurfaces (Rugh and Zinkernagel, 2011; Narlikar, 2002).

If these assumptions are correct, relativity simplifies tremendously. Weyl's principle allows us to describe a so-called comoving frame of reference wherein the fundamental observers, the galaxies, are always at rest with one another. Hence for these observers the spatial coordinates x_i are constant. This means that $g_{0i} = 0$, which implies that one can impose a time–space split. Pick a time coordinate t that is identical to the proper time of these galaxies. In these coordinates the two assumptions restrict the spacetime metric to a very special form:

$$ds^2 = dt^2 - a(t)^2 \, d\ell^2 \tag{3.1}$$

known as a Friedman–Lemaître–Robertson–Walker (FLRW) metric, named after the physicists who independently devised it. Here $d\ell^2$ describes the spatial metric of a spatially homogeneous and isotropic geometry, either the three-sphere S^3 (or elliptic space P^3), an open Euclidean universe R^3, or a three-dimensional hyperboloid H^3. I haven't specified the exact spatial coordinates because the best choice depends on which spatial geometry $d\ell^2$ describes, e.g., spherical for the first, Cartesian for the second, and hyperbolic for the third. Note that $d\ell^2$ doesn't depend on time, reflecting

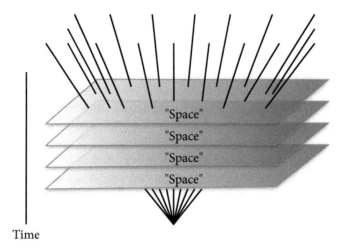

Time

Figure 3.3 Friedman–Lemaître–Robertson–Walker time

the time–space split in the line element. The only spatial feature that varies with time is a function known as a universal scale factor, $a(t)$, which is a field that is loosely conceived as the radius of the universe.

The simplifying assumptions we made to arrive at the FLRW metric also imply that matter has the form of a uniform fluid with density of matter $\rho(t)$. Use this to describe the matter and plug the above metric into Einstein's field equations. The result is a massive simplification. The 256 coupled partial differential equations of Einstein reduce to 2 simple relationships between the scale factor and the density of matter known as the *Friedman equations*. Solutions of these equations are known as FLRW spacetimes. These solutions and their perturbations are the workhorses of cosmology.

Call the t in Eq. (3.1) the *FLRW time* (Fig. 3.3). Because FLRW universes are globally hyperbolic, FLRW time is a species of Cauchy time; therefore it inherits Cauchy time's nice features. Better than that:

- The hypersurfaces with constant spatial curvature are *unique*.
- For so-called fundamental observers, the coordinate time is identified with the proper time.[3] Physics "listens" to the ticking of Friedman–Robertson–Walker–Lemaître (FLRW) time if you're a fundamental observer.
- We have a new "synchronization procedure": via Friedman's equations one can pick out moments of t not by radar synchronization (except perhaps locally), but via the density ρ. That is, one can use $\rho(t)$ as the basic clock. Every fundamental observer at a given t-moment will see the clock at the same hand, i.e., will read the same value of ρ from experiment.

---

[3] Fundamental observers are observers who stay constant in co-moving coordinates; these observers, but not others, will see the universe as isotropic, so they are also sometimes called "isotropic" observers.

Essentially what we have done is pick special observers—the galaxies—and coordinates that keep these observers at rest in space despite space expanding or shrinking. In a sense, this "straightens" all these distinguished worldlines to create what looks like a global inertial frame, but now the level surfaces with respect to which they are orthogonal are picked out by the matter density clock and not by radar synchrony. One can understand the excitement generated by this set of spacetimes. We can carve the universe into distinct spatial geometries, letting the "ticks" of a common proper time index the leaves of a unique, shared, and observable foliation.

Before evaluating FLRW time, note that some other time functions exist with similar benefits. One known as CMC time, or York time, seems to have been proposed by Crisp (2008) and Monton (2006) as a useful candidate for tensed animation. "CMC" stands for constant mean curvature. Given by the trace of the extrinsic curvature, it is the average of the principal curvatures of a space at a point. A time function counts as CMC time if the $t^{-1}(a)$ leaves are CMC Cauchy hypersurfaces whose mean curvature is equal to a, for every $a \in I$ (Andersson et al., 2012). FLRW time is a subspecies of CMC time since the constant curvature of FLRW universes implies that they have constant mean curvature. And both times are subspecies of Cauchy time. CMC time is not defined for a host of solutions of Einstein's field equations, even including some globally hyperbolic spacetimes. But when it is, and when other conditions hold (e.g., certain energy conditions are met, "space" is closed), it is sometimes provably unique (Rendall, 1997). Like FLRW time, CMC apparently rebuts the charge of arbitrariness applied to Cauchy time.

Are either of these times the answer to the tenser's dreams? Some have thought so, e.g., Swinburne (2008), Crisp (2008), Monton (2006). Yet several considerations speak against such a conclusion (see also Callender, 2010; Bourne, 2004).

First, FLRW time depends on elaborate averaging, and this point is no less true applied to CMC time (the "M" is for mean, after all). At most spatial scales the universe is not even close to being isotropic and homogeneous. Look at all the sharp density differences everywhere. No matter how messy your office is, it's not true that its matter density is isotropic and homogeneous. It's not even true at the cosmological scale, as anisotropies in the cosmic microwave background are well-known. The standard model irons out these differences, as is only proper in a model. Yet why on earth should fundamental time, if it exists, march to that particular averaged scale? Averaging is a way of losing information about the world. Isn't it odd that something purportedly metaphysically fundamental crucially hangs on information loss?

Second, we know that the actual world is not one described by either FLRW or CMC solutions. Regarding the former, we know that we're not fundamental observers. Fundamental observers see an isotropic world. Yet we're moving on the Earth, the Earth is moving with respect to the Milky Way, and the Milky Way is headed toward Virgo. We travel at a few hundred kilometers per second with respect to the cosmic rest frame. As a result of these so-called peculiar motions, we can see a dipole anisotropy in the cosmic background radiation. Nor is it true, as assumed, that the fundamental observers don't intersect one another. Galaxies collide. When they do, they are not co-moving and the whole cosmic rest frame defined above breaks down (two values of ξ^a correspond to the same point in the coordinate system). CMC time is no less idealized, as it too requires that the spacetime has a locally trivial geometry. No interesting local physics can be modeled in spacetimes with constant mean curvature, assuming these spacetimes include all matter in them. So we know that the actual world is not a CMC world nor an FLRW world.[4]

Third, the time functions defined are unique, but still the whiff of arbitrariness remains. Relativity, after all, doesn't distinguish the isotropic observers of FLRW space-times as special in any way. (Good thing, for we're not strictly isotropic ourselves.) What makes fundamental observers first-class citizens of the universe and non-fundamental observers second-class citizens? Nor does relativity single out the CMC foliation as any more special than the one on which the number of McDonald's is monotonically increasing (as appears to be the case) or other more complicated functions than the mean of the curvature. Even restricted to such spacetimes, there remain perfectly legitimate observers and foliations that do not march to the beat of these times. Fundamental observers aren't really fundamental.

Fourth, what is the precise connection between manifest time and the notion of the present found in the density or mean curvature clock? Is there any reason to think that what we regard as present are all the spacelike related events having the same ρ-value, for instance? All I can see is that these foliations lead to certain judgments about duration that are more intuitive than some others. For instance, when we claim that the universe is 13.6 billion years old, as opposed to 0 years old (based on the proper time of a photon emerging from the Big Bang), we are using something like FLRW time. For many purposes this clock is pragmatically very useful, but these reasons don't add up to grounds for believing that either of these times is metaphysically preferred.

3.3 Time, Stuff, and Laws

The above discussion raises an interesting set of questions. All of the special time functions impose some kind of restriction upon the solutions of Einstein's field equations. Hawking's cosmic time is defined only for stably causal spacetimes, Cauchy time only for globally hyperbolic spacetimes, CMC time only for spatially compact globally hyperbolic spacetimes, FLRW time only for constant curvature spacetimes, and so on. Call Ξ the set of *all* solutions to Einstein's field equations. In every case, if we restrict the theory to the solutions with a particular type of time, we reduce the set of

---

[4] One can reply that these spacetimes can be viewed as approximations to more realistic spacetimes. Unquestionably our peculiar motions are negligible when compared to the cosmological redshift resulting from the expansion of space and that this small departure from isotropy wouldn't be noticed by creatures like us (even if, granted, it's noticed by creatures like us with huge telescopes). But the question isn't whether the approximation is a good one—of course it is—but whether the realistic model has the temporal virtues the less realistic one enjoys. My suspicion is that the more realistic the spacetime, the less likely one will be able to show its time function, if it has one, is unique.

solutions from Ξ to a proper subset—call it Π (whose content depends on the type of time desired). Ultimately the basic question for all of these approaches is the following: is the nature of the restriction to Π *lawlike* or *factlike*? And if lawlike, will the set of possible worlds in Π suffice for the needs of physics? Are those who think a kind of cosmological time supplies us with something like manifest time asking us to adopt a *new theory*, one only with solutions Π, or are they simply making the observation that the actual world contingently seems to meet the conditions needed to define their favorite time?

I take it that in the project of *vindicating* tensed time as the *correct* view of time one is after the more ambitious goal. No tenser I've come across thinks that time is only contingently tensed. Typically they hold that time is metaphysically necessarily tensed—from which it follows that it is physically necessarily tensed. As such, their time then needs to hold in all physically possible worlds, and hence one is limiting relativity to the set of models Π.[5] This is a big step, for one is proposing essentially a new theory of physics.

As far back as 1936 the physicist Sir James Jeans seems to endorse such a view:

Einstein tried to extend the theory of relativity so that it should cover the facts of astronomy and of gravitation in particular. The simplest explanation of the phenomena seemed to lie in supposing space to be curved . . . It was natural to try in the first instance to retain the symmetry between space and time which had figured so prominently in [special relativity], but this was soon found to be impossible. If the theory of relativity was to be enlarged so as to cover the facts of astronomy, then the symmetry between space and time which had hitherto prevailed must be discarded. Thus time regained a real objective existence, although only on the astronomical scale, and with reference to astronomical phenomena.

(Jeans, 1936, p. 21)

Now the second property [in addition to expansion] which all mathematical solutions have in common is that every one of them makes a real distinction between space and time. This gives us every justification for reverting to our old intuitional belief that past, present and future have real objective meanings, and are not mere hallucinations of our individual minds—in brief we are free to believe that time is real. ... we find a distinction between time and space, as soon as we abandon local physics and call the astronomy of the universe to our aid.   (Jeans, 1936, pp. 23–4)[6]

What Jeans is actually endorsing isn't so clear. He immediately qualifies the strong assertion at the end of the first paragraph quoted with "although only on the astronom-ical scale, and with reference to astronomical phenomena." It's hardly obvious what

[5] Earman (1995, p. 198), discussing Gödel's famous argument for the ideality of time, considers the idea that tensed time is contingent. The question of whether time is tensed is then comparable to the question whether space is open or closed. While this position is logically available, I won't treat it here because I'm not aware of any tenser endorsing it.

[6] Suppose that Jeans had his way here and also that one formulates Newtonian spacetime à la Trautman (see above). Then the relativistic revolution would have simply shifted what is curved, space or time. Classically, à la Trautman, time is curved but space is not, but relativistically, à la Jeans, space is curved but time is not!

"real objective existence" means when so modified. To my ears, that sounds like time *emerges* at the astronomical scale. But an emergent time won't do for most tensers, for they believe that the tensed theory is fundamentally correct.

Assuming advocates of some kind of cosmological time have the ambitious goal in mind, there is a natural way to appraise their proposals. We simply need to see whether the restriction to Π leaves us with a good physical theory.

What is a good physical theory? Obviously this is a big question, but for our purposes we can take our inspiration from the idea that scientific theories strive to simultaneously optimize simplicity and theoretical power. They want theoretical elegance, unification, and compactness, but they also prize informational power, possible worlds ruled out. These twin virtues often come at the expense of one another.

Drop the pretense that relativity is a finished theory. True, historically Einstein had a more or less complete version by 1917, but this historical fact doesn't mean that the above balancing act stops. More of the universe and theory is understood to us, so it's natural to see if modifications to relativity can strike a better balance that achieves more strength. From a methodological perspective there is nothing unusual about this desire for increased simplicity or strength. Many quark models in the early 1970s were empirically adequate, yet some were eliminated as contenders because they didn't narrow down the range of phenomena as tightly as others did (Massimi, 2007). Indeed, relativity already places all kinds of restrictions on the theory to arrive at Ξ. We insist that metrics be non-degenerate and continuous, that hypersurfaces be perfectly smooth, and so on, yet we have no direct observational evidence for these claims—even though these demands remove as lawful perfectly conceivable behavior, e.g., signature changing spacetimes. From this perspective, our question is whether a system of laws corresponding to Π is on balance simpler and stronger than a system of laws corresponding to Ξ.

Put like this, it's clear that a restriction to worlds with FLRW time isn't plausible. Gravitational science needs way more than FLRW and the Friedman equations. Too many other solutions to Einstein's equations are used to model the real world, e.g., Schwarzschild, de Sitter, perturbations of FLRW, and these all get smoothed over in FLRW. All the strength of relativity in terms of "local physics" gets eliminated by a restriction to FLRW.

Truncating relativity to the Friedman equations would clearly be mistaken. Are less severe restrictions plausible? Many physicists feel that relativity is too permissive, that some of its solutions need clipping. To the delight of philosophers and the horror of physicists, relativity permits many "ill-behaved" spacetimes. These unruly spacetimes are genuine solutions to Einstein's field equations, yet they permit one or more type of "unreasonable" behavior, e.g., closed timelike curves, spatial topology change, supertasks, naked singularities. Active programs exist whose goal is to stamp out bad behavior. The physicist Roger Penrose's program of Cosmic Censorship, for instance, is a program designed to expunge relativity of naked singularities. Penrose's (1979) original proposal for cosmic censorship is equivalent to the constraint that the world be extremely well-behaved, namely, globally hyperbolic. Note the connection to time: as we saw, this demand would therefore insist on the existence of a Cauchy time function.

One can imagine several arguments in favor of such a restriction. First, with a small hit to simplicity we potentially get a large payback in strength. That is, strength is increased by ruling out solutions with features we don't see or believe actual: naked singularities, naked points at infinity, spatial topology change, multiply connected topologies, closed timelike curves, and gravitationally indeterministic regions. Why posit all of this—arguably—surplus structure to the solution space? Second, the Einstein field equations are *local*. They hold at a point. That means the metric in relativity is in principle allowed to vary point by point. Yet amazingly, the metric of our world forsakes most of its freedom. At any given point the metric *could* be such that a lightcone structure is imposed that would ruin global hyperbolicity. In an amazing coincidence, so far as we're aware that never happens. It seems a miracle, from this perspective, that the actual spacetime never exploits its freedom to go causally bad. The parent in me asks: surely there is a reason for this conspiracy of good behavior?

Yet most physicists seem uneasy about a blanket nomic restriction to globally hyperbolic spacetimes. Hawking writes:

> [M]y viewpoint is that one shouldn't assume [global hyperbolicity] because that may be ruling out something that gravity is trying to tell us. Rather, one should deduce that certain regions of spacetime are globally hyperbolic from other physically reasonable assumptions.
>
> (Hawking and Penrose, 1996, p. 10)

The hope is that relativity possesses some kind of mechanism whereby it protects itself from unreasonableness. Loosely put, the aim is to formulate and prove an assertion to the effect that *unreasonable* physics does not occur in *generic* spacetimes if *reasonable* states of matter are used in Einstein's field equations. The worry with large-scale criminal sweeps is that some of the good get caught up with the bad. Same here. Hence the hope is to be able to find some mechanism that would tend to eliminate the bad without actually doing so by fiat and thereby eliminating potentially useful spacetimes.

Unfortunately, we just don't know whether this restriction or something like it rules out what is really surplus structure. It's scarcely clear that our spacetime is *actually* globally hyperbolic, never mind *not possibly* non-globally hyperbolic. A search on the physics archive (arXiv.org/gr-qc) reveals scores of non-globally hyperbolic spacetimes being entertained for modeling our world. It's important to note that the existence of Cauchy surfaces may not be something we could ever know observationally.[7]

---

[7] In Malament's (1977c) classic treatment of observationally indistinguishable spacetimes, one pair he considers is Minkowski spacetime M, which is globally hyperbolic, and anti-deSitter spacetime glued to Minkowski spacetime, which together composes a spacetime M' that is not globally hyperbolic. Yet M and M' are observationally indistinguishable, as he shows. In such a world one could never learn (empirically) whether the world had Cauchy time or not.

At present, limiting physics to globally hyperbolic solutions seems risky, so few endorse such a move as more than pragmatic.

3.4 Conclusion

Are there structures in relativity that satisfy the twin demands of (a) corresponding to manifest time and (b) being relativistically invariant? Although one lacks a general proof, the answer seems to be No. All the candidates surveyed seem to lack one or the other feature. There are no shortage of interesting relativistically invariant structures—points, infinitesimal simultaneity slices, causal diamonds, and more—but none of these are close to manifest time, none are good candidates for animation. The theory has the resources to mimic a classical now—level surfaces of Cauchy time, instants of FLRW time, Milne's public space, and more—but these counterparts aren't fully relativistic, i.e., either the structures described aren't unique across observers or they are unique only within a serious truncation of the theory. If we idealize enough, a distinguished time emerges; but unless one wants to change the laws of relativity, this selection or idealization is always pragmatic and rarely a promising modification of the laws. Picking out a particular set of observers or concentrating on one highly idealized set of solutions will therefore always seem arbitrary from the perspective of basic physics. At heart, this was RPP's complaint in Minkowski spacetime. We hear its echo throughout all of relativity. Without seeing a clear physical need for a preferred time, relativity doesn't posit one.

4

Quantum Becoming?

There is more to basic physics than classical relativity. There is also quantum theory, our best theory of the non-gravitational forces and the "particle zoo" discovered in the twentieth century. Quantum theory introduces into the discussion of time many new topics, such as the famous energy–time uncertainty relations, puzzles involving quantum clocks, and new wrinkles on the problem of the direction of time. Given the focus of this book, however, I'll refer the reader to the relevant literature on these topics and instead concentrate on quantum theory's relationship with manifest time (see Butterfield, 2013). We will not launch a proper investigation of quantum time comparable to what we did for relativity last chapter. We can afford to neglect this investigation because—at least as regards our previous concerns—quantum mechanics will merely force us to repeat ourselves. For all its weirdness, quantum theory doesn't demand a different background spacetime arena than those described above. Non-relativistic quantum mechanics operates on a background classical Galilean spacetime; "special" relativistic quantum theory assumes a background Minkowski spacetime; and "general" relativistic quantum theory, where it exists, lives on globally hyperbolic spacetimes (Wald, 1994, 2009). We have already studied time in all of these cases. Despite what you may have heard, time in quantum theory is not discrete, ill-defined, or non-classical.[1]

If that were the end of the story, the shift to the quantum might not be very interesting for time. Quite the opposite is true. First, two of quantum mechanics' more famed and spooky features have been invoked in defending the idea that quantum time is congenial to manifest time. Quantum non-locality is said by some to make a preferred foliation of spacetime necessary, and the collapse of the quantum wavefunction is held to vindicate temporal becoming. We'll tackle these topics in the present chapter. Second, as we'll see in the next chapter, the need to extend quantum theory to gravity mandates stark choices about time, some of which suggest a friendly environment for manifest time and some of which suggest even greater hostility.

---

[1] That impression arises from mistaking quantum clocks with time itself. But we shouldn't confuse these two any more than we should confuse the coordinates of space with the positions of particles (Hilgevoord and Atkinson, 2011).

4.1 Quantum Mechanics

At bottom a physical theory typically posits some *stuff* and explains how it *changes* with time. Unfortunately quantum theory is beset with massive interpretational questions—none more pressing than the notorious measurement problem—so we don't really know what the "stuff" is until the theory is developed. We can begin our treatment, nonetheless, by briefly describing the common ground that is used to get the theory's amazingly accurate predictions.[2]

At an abstract enough level, this common ground is very simple. In Schrödinger's representation of the non-relativistic theory, systems of the world are represented by complex-valued quantum states $\psi(q, t)$, where $q = (q_1, \ldots, q_n) \in \mathbb{R}^{3N}$. Configuration space $\mathbb{R}^{3N}$ represents all the possible positions N-particle systems can have. Wavefunctions evolve through time via Schrödinger's equation

$$i\hbar\frac{\partial\psi(q, t)}{\partial t} = H\psi(q, t) \tag{4.1}$$

where $\hbar$ is Planck's constant and H is the quantum Hamiltonian, a function of the system's kinetic and potential energies. The equation takes an input quantum state, $\psi(q, t_0)$, and transforms it into an output state quantum state, $\psi(q, t)$, thereby determining the infinitesimal evolution in time. However, we don't get the predictions directly from the output state $\psi(x, t)$ but instead from something called Born's Rule. Named after its discoverer, Max Born, this rule tells us that the probability density $\rho(q, t)$ of particles in configuration q at time t is:

$$\rho(q, t) = |\psi(x, t)|^2$$

At this basic level the theory couldn't be simpler.[3] The quantum state evolves with time via Eq. (4.1). When you want a prediction, take the absolute value squared of the appropriate quantum state for that time. That's it.

Putting these pieces together into a consistent picture of the world turns out not to be so easy. The Schrödinger evolution is linear and unitary.[4] Getting definite measurement outcomes, by contrast, seems to demand an instantaneous nonlinear

---

[2] Quantum theory comes in many forms and raises many foundational questions. Here I will not describe the theory in all generality, nor will I attack the many foundational questions except insofar as they impinge on the current project. For an introduction to the physics there are many good textbooks, and for philosophical commentary, I recommend Albert (1992), Bell (1987b), Home (1997), Hughes (1992), Lewis (2016), Maudlin (1996), and Redhead (1989) as a start.

[3] More generally—since there are observable physical properties apart from position—for a system prepared in state Ψ, Born's Rule tells us that the probability $P(a = \lambda_i \mid \psi)$ that a particular value λ_i of the observable A is found when measured is simply

$$P(a = \lambda_i \mid \psi) = |c_i|^2$$

where $\psi = \sum_i c_i e_i$, where e_i are the eigenbasis vectors of the observable and c_i are scalars.

[4] An equation being linear implies that if f and g are both solutions, then $f+g$ is also a solution. When it is said that the evolution is unitary, what that means is that the Hamiltonian evolving the state can be expressed

and non-unitary transition. Moreover, the former rule is deterministic but the latter is probabilistic. The difficulty of reconciling these two equations into a happy marriage is one way of framing the measurement problem (see below).

The relativistic extension of this theory, known as quantum field theory, replaces the quantum state indexed by particles with states indexed by quantum fields. Quantum fields are the analogues of classical fields, e.g., the electromagnetic field. Despite the differences with the non-relativistic theory, the above outline is shared. The wavefunction, which may have components representing properties such as spin, still is represented by a vector in a high-dimensional state space, still evolved by a Schrödinger equation (or generalization), and still provides empirical probabilities via Born's Rule. The main difference, for time, is whether the quantum state is attached to a classical or relativistic time. In relativistic quantum theory it is obviously attached to the latter.

Even with this spare explication, we can appreciate an important point about quantum time (one that reappears in Chapter 6). Recall the adage that says time is that which makes contradictions possible. This is true. The rolled die can't *simultaneously* have both '5' and '3' facing up. Only at different times can '5' and '3' both face up. This deep feature of time makes itself conspicuous in a probabilistic theory. If I roll a fair die, the probability of any given face appearing is 16.67%. For this probability to make sense (or *be* a probability, according to the axioms of Kolmogorov) the sum of the six exclusive (contradictory) outcomes must sum to one. Here is an important difference between time and space, for the probabilities of each outcome must sum to one *at a time*, not necessarily *at a place*. The definition of probability, as given by the usual axioms, doesn't distinguish time from space; but in quantum theory, it does. For example, suppose we know a particle exists but don't know where it is. We demand that the total probability of finding the particle anywhere is one:

$$\int_{\text{everywhere}} dP(\mathbf{r}, t) = 1$$

where $P(\mathbf{r}, t)$ is the probability of finding a particle in some minuscule region $d^3\mathbf{r}$ surrounding $\mathbf{r}$. The insistence that the sum of all possible outcomes for any event sum to one entails a profound restriction on the permissible types of temporal evolution allowed, namely, that the evolution always be *unitary*.

What does this mean? Recall from linear algebra that operators can rotate vectors into other vectors in a vector space. These operators may actually represent rotation, as when one rotates arrows around the two-dimensional plane, but there are many other possibilities. In quantum mechanics, we can regard the initial wavefunction as one vector and the final wavefunction as another, both living in a complex vector space known as Hilbert space H. The Schrödinger evolution through time

as a unitary matrix. Unitary matrices preserve norms, and in quantum mechanics, the importance of this is that such evolution preserves the probability amplitudes.

can be represented in this formalism as corresponding to an operator, one that "rotates" the initial vector representing the system into the later one. A unitary operator restricts the time evolution operator, $U(t_1, t_2): H \to H$, to ones that satisfy $U^*U = UU^* = I$, where U^* is the adjoint and I is the identity operator. The reason this is important is that such operators preserve inner products on vectors in H, just as the rotation operator in the plane leaves distances and angles alone when acting on arrows.[5] And that matters because probability amplitudes in quantum mechanics are represented by inner products, and probabilities, as we just saw, are represented by squares of probability amplitudes. The Schrödinger equation guarantees that the temporal evolution is unitary. This restriction for time evolution is responsible for the probabilities making sense, and in particular, the probabilities being conserved through time, both locally and globally.

What may initially appear to be an abstract constraint in effect says that the flow of time shouldn't interfere with the probabilistic structure of quantum theory. The time–space split thus runs very deeply in the theory. The whole operator formalism is, in a sense, specially tuned to it.[6]

In sum, quantum mechanics requires a kind of global ideal clock that has at least the structure of the Cauchy time function we encountered earlier. Both the physical evolution and probabilistic apparatus "listens" to the ticks of this quantum clock.

With this exceptionally brief primer in place, let's now turn to a special challenge posed by the famous non-locality of quantum mechanics. Advocates of this challenge hold that these non-local correlations demand the reintroduction of a preferred frame in physics. In this way quantum mechanics is said to be friendly to manifest time.

4.2 Popper's *Experimentis Crucis*

Sir Karl Popper, reflecting on contemporary experiments suggesting quantum non-locality or action-at-a-distance, writes:

> It is only now, in the light of the new experiments stemming from Bell's work, that the suggestion of replacing Einstein's interpretation by Lorentz's can be made. If there is action at a distance, then there is something like absolute space. If we now have theoretical reasons from quantum theory for introducing absolute simultaneity, then we would have to go back to Lorentz's interpretation.   (1982, p. 30)

---

[5] Probability amplitudes can be represented as inner products of vectors on Hilbert space: $\langle X \mid Y \rangle$. Under a transformation U, $|Y\rangle$ transforms as $|Y\rangle \to |Y'\rangle = U|Y\rangle$ and $|X\rangle$ as $\langle X| \to \langle X'| = \langle X|U^\dagger$. If U is unitary, then $\langle X'|Y'\rangle = \langle X|Y\rangle$ because $U^\dagger = U^{-1}$.

[6] This fact is related to why quantum mechanics uses complex wavefunctions. To get something that will play the role of probability, we need to be able to extract from the state another phase-matched state, at a time. The doubling of states due to complexity accomplishes this without relying on time evolution. See Barbour (1993). It is possible to try to overcome this special tuning, as in the *general boundary formulation* of quantum field theory by Oeckl, e.g., (2006; 2007). This approach attempts to generalize quantum field theory to transition amplitudes between spacetime regions, not hypersurfaces at times.

Recall the Lorentzian interpretation of relativistic effects discussed in Section 2.5. According to Popper, the underdetermination of theory by data between Lorentz and Einstein, which he believed had persisted for more than sixty years, finally met its *experimentis crucis*. The experiment was Aspect's 1981 demonstration of the violation of Bell's inequality. This, Popper thought, proved that quantum mechanics is non-local and that absolute simultaneity is needed.

For those desiring vindication of manifest time in physics, Popper's reasoning to a physically preferred foliation of spacetime is a welcome respite from the news of the previous chapter. There we witnessed many ways in which physics might "prefer" one foliation over another, but it was at best a stretch to declare any of these preferences lawlike. Here, by contrast, theory suggests and experiments confirm the existence of robust, *nomic* correlations between spacelike separated events. Arguably we have a much better candidate for singling out a foliation than we had previously. Unlike cosmology, quantum theory is thought to be a fundamental theory of physics (or if one is worried about quantum gravity, at least as fundamental as we have now).

Is Popper right?

To answer this question we must take a detour through the philosophical foundations of quantum mechanics. In what may seem an odd decision, I'll stick to nonrelativistic quantum theory instead of the relativistic quantum field theory. This choice is justified by convenience and by the fact that it sacrifices little as far as the modest points I wish to make. It is convenient because it will make the following discussion more familiar to many readers. It sacrifices little because the relativistic theory predicts the same non-local correlations (Smith and Weingard, 1987) and suffers the same measurement problem as does the non-relativistic theory (Maudlin, 1994).

Quantum mechanics describes correlations between many observable physical properties. Ignoring other degrees of freedom, consider the correlations between two spin-1/2 particles created in the infamous singlet state. In this reformulation of the Einstein–Podolsky–Rosen (EPR) argument, two electrons emerge from a common source in a superposition of eigenstates of the x-spin operator, S_x. For a pair of electrons labeled 1 and 2, the singlet state is represented as

$$\psi = \frac{1}{\sqrt{2}} (|\uparrow_x\rangle_1 |\downarrow_x\rangle_2 - |\downarrow_x\rangle_1 |\uparrow_x\rangle_2)$$

where ψ is a ray in two-dimensional complex Hilbert space and the two eigenstates are up or down in the x-direction. Note that, according to the usual conventions, ψ is not in any particular state of x-spin because it is not an eigenstate of S_x. Suppose, however, that we measure the two electrons for x-spin via the canonical arrangement, namely, with particle 1 shooting to the left from the source and particle 2 shooting to the right of the source, whereupon they each meet an x-spin sorter such as a Stern–Gerlach device. What happens? Thanks to spin conservation, for an ideal experiment the probability of the measurements on systems 1 and 2 disagreeing is one. Indeed, due to the spherical symmetry of the singlet state, that's true for any measurement orientation. Therefore, upon measuring electron 1 and finding a definitely spin up or down state, we know with certainty the result on electron 2. Assuming locality, that is, that the measurement of 1 didn't affect the state of distant particle 2, EPR reason that particle 2 *already* had a definite spin state—even when it was in the singlet state, which doesn't have a definite spin state in the quantum formalism. Hence EPR's dilemma: either quantum mechanics is non-local (the measurement at one wing instantaneously affected the other result, ensuring 100% compliance) or it is incomplete (2 already had a definite spin).

Later, in 1964, John Bell derived one of the results for which he is most famous, Bell's theorem. This theorem produced an inequality based on the assumption that the world is local. For various situations quantum theory predicts the *violation* of this inequality; and in a host of experiments, most notably Aspect's in 1981, systems have vindicated this prediction. Since Aspect's seminal work, many experimental loopholes have been closed or shrunk. Now there exists a wide consensus that theory and experiment have discovered spacelike correlations not attributable to any local "hidden variable" theory. Complete or not, quantum mechanics seems to be non-local.

Does quantum non-locality conflict with the relativity of simultaneity?

The first thing to say is that it doesn't in any experimentally detectable way. The randomness of quantum measurements means that the above correlations are uncontrollable in a rather deep way. One can't use them to send superluminal signals. From looking at the frequencies of spins on the two wings over large ensembles, one can deduce that no locally hatched plan could have been responsible for the correlations holding. But that is all. One can't use these correlations to signal or to detect a privileged foliation. The so-called 'no-signaling theorem' (e.g., Redhead, 1989) guarantees this fact. If Popper thought that Aspect's experiments *empirically* picked out a particular frame when he spoke about an *experimentis crucis*, he badly misinterpreted them.

More likely is that Popper thought that the *best theory* of what is going on behind the experiments requires a distinguished but hidden frame. That would be the more charitable reading of what he says and better fits the mention of Lorentz. Does the best theory of quantum non-locality suggest a conflict with relativity? Here Popper is in good company:

> For me then this is the real problem with quantum theory: the apparently essential conflict between any sharp formulation [of quantum theory] and fundamental relativity. That is to say, we have an apparent incompatibility, at the deepest level, between the two fundamental pillars of contemporary theory ... It may be that a real synthesis of quantum and relativity theories requires not just technical developments but radical conceptual renewal.
>
> (Bell, 1987c, p. 172)

Why does Bell feel that quantum mechanics and relativity are at odds, despite the "peaceful co-existence" guaranteed by the no-signaling theorem?

The problem is that when expressed clearly—in a "sharp formulation"—quantum mechanics commits one to trouble. Quantum mechanics, as the reader no doubt is aware, has many interpretations. Which one is correct is a topic of ongoing research and often heated debate. The interpretations provide physical models of the quantum phenomena. In so doing they must explain what happens when states like ψ above are measured. The problem—the famous measurement problem—is the fact that, if we measure a state like ψ, the superpositional aspect of the state doesn't go away. Instead, the superposition just gets entangled with more objects, in this case the measuring device and even you, the observer. Let D represent the state of the measurement device, which we can imagine as a pointer reading either "up" or "down," and O the observer reading the measurement device. Then the post-measurement state of the system at time t, assuming it always evolves according to Schrödinger's equation and setting the interaction Hamiltonian to zero, is:

$$\psi_t = \frac{1}{\sqrt{2}} \left(|\text{sees "up"}\rangle_O |\text{"up"}\rangle_D |\uparrow_x\rangle + |\text{sees "down"}\rangle_O |\text{"down"}\rangle_D |\downarrow_x\rangle \right)$$

According to the usual interpretative rules, this superposition of pointer states and observers does not represent definite states of pointing or seeing because it not an eigenstate of the operators corresponding to these properties. Hence the measurement problem. Assuming we are in a definite state of seeing something, it seems that, in the words of Bell (1987a), "[e]ither the wavefunction as given by the Schrödinger equation is not everything or it is not right" (p. 22). Or less dramatically but more generally, either ψ_t isn't the resulting state, there is more than that state, or the conventional interpretational rules of the theory must be changed, i.e., the association of properties only with eigenstates.

Now adopt the standard "Copenhagen" view that the Schrödinger equation is not always right. This may not qualify as a "sharp" interpretation according to Bell, but it is sharp enough to cause trouble. According to this interpretation, the above wavefunction spontaneously "collapses" in a nonlinear, indeterministic fashion to a definite eigenstate upon measurement. The probabilities of collapse to a particular state are given by Born's Rule. The idea is that when we measure ψ_t, the state will instantly transition from ψ to either

$$\psi_t' = |\text{sees "up"}\rangle_O |\text{"up"}\rangle_M |\uparrow_x\rangle$$

or

$$\psi_t'' = |\text{sees "down"}\rangle_O |\text{"down"}\rangle_M |\downarrow_x\rangle$$

with (in this case) a half chance of either outcome. Definite outcomes are thereby secured.

To get a sense of the trouble, place these real collapses in Minkowski spacetime (for more see Aharonov and Albert, 1981). Consider two spin-1/2 particles in the singlet state. Both particles emerge from a common source, with particle 1 traveling to the

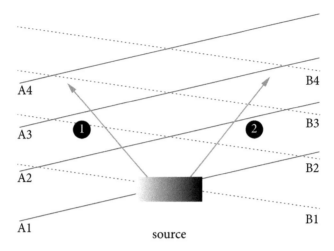

Figure 4.1 EPR–Bell experiment

left in the diagram and particle 2 traveling right. At event L particle 1 is measured for its value of x-spin; at event R, which is spacelike related to L, particle 2 is measured for its value of z-spin. Now consider two foliations of spacetime, Alice's t_A, wherein R happens first, and Bob's t_B, wherein L happens first. See Fig. 4.1.

Bob and his fellow inertial travelers tell themselves the following narrative. At time $t_B = 1$ the system is described by ψ. By $t_B = 3$ an x-spin measurement happens at L. Suppose the result is x-spin up. Then the quantum state instantaneously reduces to:

$$\psi_{B;3} = |\uparrow_x\rangle_1 |\downarrow_x\rangle_2$$

Later, by time $t_B = 4$, event R has occurred and Alice finds particle 2 to be z-spin up:

$$\psi_{B;4} = |\uparrow_x\rangle_1 |\uparrow_z\rangle_2$$

Alice and her fellow inertial travelers, by contrast, tell themselves the following narrative. At time $t_A = 1$ the system is again described by ψ. But by parallel reasoning, Alice gets the following sequence:

$$\psi_{A;3} = |\downarrow_z\rangle_1 |\uparrow_z\rangle_2$$
$$\psi_{A;4} = |\uparrow_x\rangle_1 |\uparrow_z\rangle_2$$

Look at how different the two histories are at $t_{A/B} = 3$! Alice and Bob disagree about whether event R or event L collapse the singlet state into a factorizable state. They therefore disagree about what each side measured, a definite spin state or the singlet state. That means they disagree about which events were determined and which ones were chancy. Even more, they disagree about what states occur in the world.

If we take the wavefunction at all seriously disagreements like this will not do. If real wavefunctions really collapse, then it seems that either Alice's story is right or wrong.

Ditto for Bob and Carl and . . . If we insist that one story is correct, then we have picked out a preferred set of timelike inertial worldlines, contrary to relativity. Whether this invites a return to Galilean spacetime or a preferred foliation of Minkowski spacetime is debatable (Maudlin, 1996). In either case, Bell's worry isn't idle.

4.2.1 Quantum preferred frames and time

To one defending an RPP–like argument against manifest time being physically fundamental, the above situation is discomfiting. By adopting the principle of relativity RPP claims that there can't be anything in the world that doesn't commute with the symmetries of Minkowski spacetime. The above speculation suggests to the contrary that one of our best scientific theories violates RPP's reasoning!

Here is Lucas (1998) savoring the irony:

> But physics goes further. It not only defeats the would-be defeaters of the tense theory, but offers positive support. Quantum mechanics, if it is to be interpreted realistically, distinguishes a probabilistic future of superimposed eigen-states from a definite past in which each dynamical variable is in one definite eigen-state, with the present being the moment at which—to change the metaphor—the indeterminate ripple of multitudinous wave-functions collapses into a single definite wave. . . . there is a unique hyperplane advancing throughout the whole universe of collapse into eigen-ness.

Lucas finds at least two elements of quantum mechanics attractive. One is the unique global hyperplane presumably singled out by collapses. Another is the suggestion in the theory that what is to the future of this hyperplane is genuinely open. Here let's continue our examination of this preferred frame. In Section 4.3 we'll tackle the second idea.

4.2.2 Caveats and alternatives

Bell's worry is a serious one, one deserving greater attention within physics. The problem is easy to ignore because (a) empirically a peaceful co-existence between relativity and quantum mechanics is ensured due to the no-signaling theorem, a theorem demonstrating that the non-locality can't be used for signals or any other information transfer across spacelike hypersurfaces, and (b) one only sees the problem if one tries to interpret the theory. Concern with empirical results and a desire to stay away from the vexed issue of measurement together lead many to ignore the worry. Sticking one's head in the sand, however, will not help. Given the pressing desire for a theory bringing together relativity and the quantum, it is potentially harmful to ignore a spot where the two theories may conflict.

That Bell's concern is a serious one doesn't mean it can't be answered. Tension between relativity and quantum non-locality was revealed above by using the standard Copenhagen interpretation of quantum mechanics. There are abundant reasons not to use that model (see Albert, 1992, and Bell, 1987a, for compelling criticisms) and there exist plenty of better interpretations. So we can ask, of any given *sharp formulation*, whether it invites the sorts of trouble that Popper saw.

Unfortunately this isn't the place to engage in that huge and controversial question. Here I can only state that the answer is *perhaps not*. Everything hangs on what counts as a sharp formulation and what one means by compatibility with relativity. There are plenty of interpretations that do not clash with relativity in the manner described above and there are tricky ways of understanding relativity and the wavefunction whereby hyperplane dependent stories like the above EPR case pictured in Fig. 4.1 count as relativistic (Fleming, 1996). Whether any of these paths escape Bell's concern is a question we'll leave to others.

We can afford to punt on these questions because our present interests demand the opposite: interpretations that *do* privilege a preferred frame. I want to argue that *even if* we adopt such a theory, matters are hardly rosy for tensers. What I want to do is consider the preferred foliation from the perspective of Bohmian mechanics, the best worked out "realistic" no-collapse interpretation of quantum mechanics. The Bohmian non-relativistic theory and its extensions to relativistic equations (see Bohm and Hiley, 1993, and Dürr et al., 1999, for extension to the Dirac equation, Berndl et al., 1996, and Nikolić, 2005, for extension to Klein–Gordon) demand a preferred frame. Not all Bohmians are satisfied with this state of affairs and the question is considered a matter of ongoing research. For the sake of argument, however, we'll assume that the preferred foliation is here to stay for the Bohmian. I want to concentrate on the Bohm theory because, for the tenser seeking to escape relativity via quantum mechanics, things will get no better than if an interpretation with this consequence is true. Yet we'll see that, because Bohmian mechanics is a realistic theory in principle describing everything, it allows us to formulate a complaint against the tensed theory that echoes later themes in this book.

4.2.3 The coordination problem

The basic idea of non-relativistic Bohmian mechanics is that there are, in addition to the wavefunction, particles. Relativistic versions of Bohmian mechanics have been developed for some fields, but the basic structure is the same. There is the wavefunction and there are the so-called "beables" (in the words of Bell, 1987b). Beables here are the elements of ontology proposed in addition to the wavefunction. The wavefunction evolves according to the relevant linear dynamical equation (Schrödinger equation, Dirac equation, Klein–Gordon equation, etc.) and the beables have a velocity that takes the wavefunction as input. In the non-relativistic particle version, the particles' configuration $Q = (Q_1, \ldots, Q_N)$ evolves according to:

$$\frac{dQ_k}{dt} = \frac{\hbar}{m_k} \operatorname{Im} \frac{(\psi^* \nabla_k \psi)}{\psi^* \psi} (Q_1, \ldots, Q_N)$$

where ψ is the wavefunction of the system, m_k the mass of the kth particle, and $\nabla_k = \frac{\partial}{\partial q_k}$ the gradient with respect to the coordinates q_k. Essentially, the velocity formula is the quantum probability current divided by the probability density. In the non-relativistic case, the Schrödinger equation and this guidance equation are the two fundamental laws of nature according to the Bohmian. One often makes another assumption, namely, that the initial probability density of particles is given by the absolute value of the initial wavefunction squared, $|\Psi|^2$. Call this assumption the *distribution postulate*; some Bohmians view it as a law of nature (Callender, 2007). The amazing thing is that if you assume the distribution postulate and that the beables and wavefunction evolve according to the above equations, you will find that the particles' predicted statistics precisely match those of "standard" quantum mechanics.

Consider our earlier experiment with two spin-1/2 particles in the singlet state. Stern–Gerlach devices essentially split the regions of positive wavefunction support into two disjoint sections, an upper and lower section as measured along the vertical axis of the magnet. In the version of Bohm's theory under consideration, spin is not fundamental. The value of spin is contextual, meaning that it depends on the initial location of the particle in the wavepacket and the kind of device it meets. If the Bohm particle starts off in the upper half of the wavepacket (along the ith dimension, where $i = x, y,$ or z), then the dynamics will evolve it to the region we call spin-up in the ith dimension; if the lower, then to the position we call spin-down in the ith dimension. There is no possibility of a transition from low to high, or vice versa. This impossibility is due to the fact that the Bohmian dynamics is first-order and deterministic; the combination means that trajectories can't cross in configuration space.

Suppose the wavefunction is in our singlet state above and that the configuration point representing the two-particle system is located in the upper left half of the wavepacket before the system is measured. (See Fig. 4.2, fashioned after Barrett, 1999, p. 142.) Suppose also that the measurements occur at spacetime separation, so that

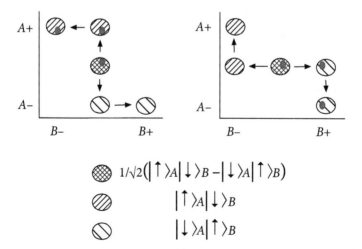

Figure 4.2 Bohmian configuration space

the measurement events occur in different orders in at least two different foliations of spacetime. According to one foliation, A measures first; according to another foliation, B measures first. Suppose A measures first. Since the particle is in the upper half of the wavepacket, A will find the particle to be x-spin-up; when the observer corresponding to B's foliation subsequently measures, the observer must find the particle to be x-spin-down. That is, A's measurement will have effectively collapsed the state to $|\uparrow_x\rangle_A|\downarrow_x\rangle_B$. Now suppose that B measures first. Then B will find the particle to be x-spin-up and therefore A will find the particle to be x-spin-down, i.e., B's measurement would have effectively collapsed the state to $|\downarrow_x\rangle_A|\uparrow_x\rangle_B$. In Bohm's theory, the actual *outcome* depends on who measured first!

But 'first' is not a relativistic invariant for spacelike separated events. There shouldn't be a fact of the matter about who measured first, yet on Bohm's theory there is. Hence one foliation must be preferred over the others.[7]

We now have enough background to ask a simple question. If the tenser is correct, there is a metaphysically preferred foliation according to which the world unfolds; and if the Bohmian is correct, there is a physically preferred foliation to which the Bohmian dynamics "listens." Are the two foliations the same?

If they are not, then the preferred frame of quantum theory is of no help to the tenser. Far be it from quantum mechanics saving tenses, the tenser merely trades one conflict with fundamental physics for another (see Fig. 4.3). Here the conflict would be between the two preferred frames. That would replace the conflict between the preferred frame of becoming and the lack of such in relativity. No progress is made.

So the tenser intent on using quantum theory must devise some rationale for thinking that the two frames coincide. Dean Zimmerman proposes the following principle as a generalization from experience:

> For any events e^1 and e^2, e^2 is causally dependent upon e^1 only if, when e^2 was happening, e^1 was happening or had already happened.   (2011, p. 236)

In other words, when reflecting on causation, it always seems that the events that are happening or are in your history have been caused by events that have already happened. Zimmerman holds that this principle captures one aspect of the core thesis of presentism, the view that only the present exists. Presentism is a "tensed" theory of time, one posited to vindicate manifest time. Elevated to a principle of presentism, this

---

[7] Why doesn't this demonstrate that Bohm's theory is false? Recall the distribution postulate. We don't know where the point representing all the particle positions is in configuration space. If we get spin-down in our experiment, we know the point corresponds to a particle being in the lower half of the initial wavepacket; but there is no way to know this beforehand. Moreover, it can be shown that if the distribution postulate is satisfied, then we can in principle never find out (see Dürr et al., 1992). Intuitively, discovering that the point is in the upper half of the wavepacket in the x-direction doesn't allow a reliable inference about whether it is in the upper or lower half in the y-direction. In any case, with the distribution postulate the so-called "no signaling theorem" holds in Bohmian mechanics too. It's therefore impossible in Bohmian mechanics to exploit these spacelike correlations for communication. At the statistical level, special relativistic symmetries hold. It is only at the sub-quantum level that the outcomes pick out a preferred frame.

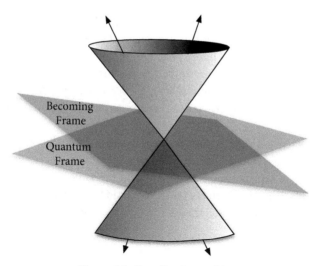

Figure 4.3 Coordination problem

connection serves to tie together the Bohm frame with the tensed frame. The Bohm guidance equation evolves according to one frame; assuming that the state (ψ_0, Q_0) on some slice Σ_0 can be said to *cause* the state (ψ_t, Q_t) on some slice Σ_t, then we can take this as a principled means of picking out the foliation of distinguished presents. (Alternatively, one might use the more controversial superluminal causation from one wing of the EPR experiment to the other as what determines the distinguished frame.) Fair enough. Zimmerman is correct that one may posit additional metaphysical principles that link the Bohm foliation with the tense foliation. Here he links causation with the unfolding present. One might also simply appeal to theoretical simplicity to identify the two frames.[8]

The problem is that this answer seems to undermine a principal motivation of the tenser—that one *experiences* becoming. Because the theory is complete, assuming mind–body physicalism, *you* are a Bohmian system corresponding to a piece of the universal wavefunction and a bunch of Bohmian particles. *Whenever you experience anything or even introspect you are making a Bohmian "measurement."* The same general limitations on Bohmian measurements therefore hold for you. If the distribution postulate is a law of nature, then the laws of nature in a Bohmian world prevent you from having any reliable feeling or impression or introspective reflection that could indicate which frame is the becoming frame. Zimmerman advertises the above

---

[8] Interestingly, Valentini (2008) argues that cosmic time in the 3 + 1 formulation of general relativity should be adapted to the Bohmian preferred frame. This proposal is more satisfying than the idea of adapting the cosmic time to hypersurfaces of constant mean curvature (on which, see Wüthrich, 2010), it seems to me, because here what is picking out the distinguished foliation is something putatively fundamental and not merely averaged at some special scales; and it is more satisfying than the adaptation to tensed time because theoretical virtues like simplicity might suggest squashing the two distinguished foliations together.

principle as a generalization from his experience (2011, p. 193). Assuming the world is Bohmian—a presupposition of this argument—he can't so generalize. For all he knows, his "EPR" experiences haven't been telling him the above principle is true. The only exception is if brains are somehow performing an *experimentis crucis* of their own, acting like a Michelson–Morley device but with a non-null result. But brains can only smell the ether wind if the laws of Bohm's theory are suspended or incomplete.[9]

Since tensers regularly appeal to experience to support their theory, this conclusion cannot be congenial to them. Hence the tenser faces a dilemma: either the becoming frame and preferred quantum frame are one and the same, in which case Bohmian mechanics implies that no physical experience could be a reliable guide to this frame, or they differ, and then the tensed theory conflicts with physics over the order of some events.[10]

4.3 Quantum Becoming via Collapses?

Defenders of manifest time sometimes claim inspiration or vindication for their views from objective wavefunction collapse. We met this view in Lucas (1998) above and it is found in Lucas (1999, 2008); Popper (1982); Shimony (1993, 1998); Stapp (1977); Tooley (2008); Whitrow (1961) and elsewhere. Quantum mechanics is claimed to require wavefunction collapse, and wavefunction collapse is supposed not only to pick out a preferred frame but also to allow the future to be open, indeterminate, or mutable. Here is Lucas:

> There is a worldwide tide of actualization—collapse into eigenness—constituting a pre-ferred foliation by hyperplanes (not necessarily flat) of co-presentness sweeping through the universe—a tide which determines an absolute present . . . Quantum mechanics . . . not only insists on the arrow being kept in time, but distinguishes a present as the boundary between an alterable future and an unalterable past.   (1999, p. 10)

The collapse of the wavefunction, interpreted realistically, suggests a picture of a fixed past (wavefunctions collapsed to the eigenstates of the relevant observable) and an open future (wavefunctions as superpositions of such eigenstates). In fact, the path from objective collapses to tenses is a two-way street. While some reason to tenses from collapses, others reason to collapse interpretations from a prior commitment

---

[9] Naturally, one can avoid the argument by denying physicalism and adopting some form of mind–body dualism. One's nonphysical mind could then "sense" the becoming frame and see that it is identical to the frame in which superluminal causation occurs, even though one can't sense superluminal causation physically. I suppose this is possible, but at this point my imagined opponent and I lack enough shared ground to make much progress.

[10] Monton (2006) claims that our experience of flow entails the *existence* of a preferred frame, but not *which one*. That I don't know how fast the wind is blowing doesn't mean that I don't genuinely feel wind. Same here, says Monton. But there there is an important disanalogy with the point about the wind. The Bohm frame is not *felt* at all, so long as feeling is a physical interaction.

to tenses. Here is Joy Christian on his motivation for pursuing Penrose's collapse theory:

> [It] implicitly takes temporal transience in the world—the incessant fading away of the specious present into the indubitable past—not as a merely phenomenological appearance, but as a bona fide ontological attribute of the world . . . For, clearly, any gravity-induced or other intrinsic mechanism, which purports to actualize—as a real physical process—a genuine multiplicity of quantum mechanical potentialities to a specific choice among them, evidently captures transiency, and thereby not only goes beyond the symmetric temporality of quantum theory, but also acknowledges the temporal transience as a fundamental and objective attribute of the physical world. (2001, p. 308)

Many who crave a tensed time find just what they need in objective wavefunction collapse—and vice versa.

Does quantum mechanics, on a collapse interpretation, support one's pre-theoretical views of the openness or mutability of the future, as Lucas suggests? It's not so clear. We can approach this question by asking whether the open/fixed distinction maps neatly into the superposition/eigenstate distinction. The answer is "no." To begin, the symmetry of Hilbert space implies that we can write out our wavefunction in any of an indefinite number of bases, e.g., position, momentum, spin. A wavefunction that is a superposition in one basis may not be a superposition in another; for instance, the wavefunction of x-spin down is a superposition of up and down spins in the z-spin direction. Here a collapse to fixity in x-spin buys openness in z-spin. Furthermore, how should we view the future measurement of systems already in eigenstates of the relevant observable (such as measuring a system definite in x-spin in x-spin)? These measurements aren't open in the sense of a superposition collapsing to the eigenstate of the relevant observable. Are the outcomes nonetheless open because future or are they fixed because eigenstates? If the former, then quantum mechanics has little to do with openness; if the latter, then again we have a drastic departure from our pre-theoretic intuitions of openness—for unlike claims such as "tomorrow $2 + 2 = 4$" the quantum claims that would be fixed are regarded as contingent ones.

To the problem based on Hilbert space's egalitarianism of bases, one may respond that the collapses are real physical mechanisms taking place in a distinguished basis. A natural choice might be the position basis, as suggested by perhaps the best realistic collapse theory, GRW (Ghirardi et al., 1986). A preferred position basis will entail (if it solves the measurement problem) the absence of superpositions of distinct *macroscopic* properties in other bases too. Eigenstates in position space still correspond to non-eigenstates in momentum space, but the proponent of this move won't care about non-positional superpositions: only the positions—or other chosen beable—are real. Yet note that in any realistic collapse theory such as GRW one doesn't get collapses onto eigenstates, but only near eigenstates. One can eschew a realistic trigger mechanism or consider states as definite even if only "near" eigenstates or pick different bases to distinguish. It won't matter. Once squashed down to an eigenstate, or near one, the Schrödinger evolution will quickly spread the wavefunction again. The spread of actualization Lucas gets from quantum mechanics is more a series of partial drips and splashes than a worldwide surge.

Perhaps the link with openness and transience arises instead from the single-case objective probabilities needed for a collapse theory? Shimony (1993; 1998) and Popper (1982) stress throughout their work the benefits of a "truly" probabilistic process. In it they find an open future, the flow of time, and even freedom. The intuitions underlying these links are clear enough. Suppose at time t there is an objective chance of 0.5 that a radium atom will decay tomorrow. For this to be true (it is natural to think), there must not "already" be at t a unique determinate future with (say) a decayed radium atom in it tomorrow. That would entail, on one way of understanding objective chances, an objective probability of 1, not 0.5. Since the tenseless theory of time entails that there is a unique determinate future—in a sense—the existence of non-trivial objective probabilities requires the tensed theory of time (see Shanks, 1991).

All of the inferences in this reasoning are more tenuous than is usually acknowledged. Here are two worries.

First, if the reasoning goes through at all, it does so only for some interpretations of chance and not others. The inferences are perhaps most naturally understood via a Popperian propensity interpretation, but there exist other (and I would argue, better) interpretations of objective single-case chances that won't yield the desired conclusion. On Lewis' (1994) theory of chance, for instance, non-trivial chances are compatible with a tenseless theory of time. Lewis views chance as a theoretical entity that increases the theoretical virtues of the best systematization of nature. Crucially, on this theory information about whether or not a radium atom decayed after t is "inadmissible" at t and therefore doesn't affect the value of the chance at t. See Loewer (2001) and especially Hoefer (2011) for more on this topic.

Second, and at least as important, the justification for the line that a "fixed" future implies trivial values of objective chance is similar if not identical to the famous argument for fatalism. The sea battle tomorrow spoils freedom today just as the radium atom's decay tomorrow spoils non-trivial values of chance today. If one believes, as I do, that the argument for fatalism is flawed (see Sobel, 1998, for a penetrating critique) then the existence of the sea battle tomorrow doesn't undermine freedom today; one therefore needn't see the tenseless view as implying any threat to freedom. Nor need it be a threat to non-trivial chances. The actual world may contain our radium atom in it decayed tomorrow, yet today it still may have a one-half chance of decaying. This possibility is obviously allowed on a "frequentist" conception of probability, but it is also permitted on many other conceptions.

4.4 Conclusion

Although many philosophers and physicists seek relief from relativity's assault on time in quantum theory, assistance is not so easily found. Whether quantum non-locality demands a preferred frame—and if it does, whether this frame is helpful to tensers—is far from obvious. If I'm right, such a frame is probably of no help to the tenser whose evidence for the frame is experiential. Others find an open future vindicated by quantum collapses. However, if there are collapses, the future they leave open is quite different from that found in manifest time, and the argument getting that conclusion is fraught with obstacles and alternative paths.

5

Intimations of Quantum Gravitational Time

The history of natural philosophy is characterized by the interplay of two rival philosophies of time—one aiming at its "elimination" and the other based on the belief that it is fundamental and irreducible.

Whitrow (1961)

"Quantum gravity" is the term bestowed upon research programs seeking to consistently mesh the quantum and gravitational aspects of matter. Relativity successfully handles gravity. Quantum theory successfully describes all the forces (strong, weak, and electromagnetic). Yet physical objects are not neatly separated into the quantum and the gravitational. A quantum object like an electron possesses mass and surely produces a gravitational field, but we have no fundamental theory that incorporates the gravitational effects on its motion. Alternatively, gravity can clump matter together so tightly that quantum effects should be relevant; but again, we have no fundamental theory capable of dealing with this. We need a theory that consistently handles both the quantum and gravitational aspects of matter. Obtaining one has been a top priority in theoretical physics for many decades.

Quantum gravity is not so much a developed theory as a set of research programs that embody commitments to specific basic physical assumptions and mathematical techniques. Superstring theory, loop quantum gravity, canonical quantum gravity, causal sets, causal triangulation, non-commutative geometry, and more are examples of different approaches. Each is based on an attractive "core idea" or "clue" that its advocates hope will pay off if developed.

The project inevitably demands hard and deep decisions about time. Even if quantum theory doesn't require a preferred foliation, as threatened in the previous chapter, it is still committed to two temporal features that cause trouble for any possible meshing with relativity. First, the quantum states are tied to global inertial frames, but relativity doesn't in general possess such frames. Second, time is absolute in quantum theory, not dynamical. The arrangement of matter doesn't affect the temporal ticking to which quantum states listen. Viewed from quantum mechanics' perspective, the problem is that relativity doesn't seem to have a time candidate up to the job of pushing quantum states forward. Bodies in motion don't care about the coordinate time, and the proper times they do care about are too local to characterize global quantum states. From a relativistic perspective, the problem is not so much that there isn't any time in quantum gravity (as is often said), but that *there is one* in quantum mechanics. This cluster of tensions, and the particular technical problems they engender in each program, go by the name "the problem of time" in quantum gravity.

Strategies for pursuing quantum gravity split in two, as do corresponding attitudes about time. If one feels that the quantum revolution provides the firmer or more promising foundation, then—as in superstring theory—one begins there and hence starts with a full-blooded time. If by contrast one believes that relativity provides the better starting point, then—as in loop quantum gravity—one begins with a theory wherein time is already demoted in some sense. In both cases one knows that the final product will be significantly different from relativity or quantum mechanics, and one needn't end where one begins. Nonetheless, these initial decisions about time tend to have serious ramifications later.

Because the field is so split, one can find confirmation of almost anything one thinks about time. There are approaches that hew to a more classical conception of time and proclaim that quantum gravity takes relativity off their backs. There are also approaches that claim to eliminate time altogether. While there is a sense in which anything goes on the frontiers of physics, I do wish to show that the time of relativity—such as it is—is quite resilient. It is both harder to kill off and harder to improve upon than is usually thought. To this end, I will give a fascinating example wherein "becoming" is possibly restored, followed by an elegant example of the opposite, wherein time "disappears" altogether. I ask some questions about the latter case suggesting that time is not yet purged of the theory. Regarding the former, I show that relativity has a way of causing trouble for becoming, even in quantum gravity. My "verdicts" in each case are hardly the final word. Rather, they represent notes of caution when it comes to the consequences for time from quantum gravity. If nothing else, these discussions show that philosophy of time and quantum gravity can interact in a mutually profitable way.

5.1 The Best of Times: "Asynchronous Becoming" in Causal Sets

Physicists developing the approach known as causal set theory (CST) have argued that their framework is consistent with a fundamental notion of temporal "becoming."[1] As such, it is advertised as providing a physical justification of what I've been calling manifest time. A causal set, or causet, is a discrete set of events partially ordered by a relation of causality. The idea behind the theory is that these sets "grow" as new events

---

[1] This section is based on work co-authored with Christian Wüthrich.

are added one by one to the future of already existing ones; furthermore, this "birthing" process is said to unfold in a "generally covariant" manner. Here is the founder of causal set thoery (CST), Rafael Sorkin, advertising the philosophical pay-off:

> One often hears that the principle of general covariance... forces us to abandon "becoming"... To this claim, the CSG dynamics provides a counterexample. It refutes the claim because it offers us an active process of growth in which "things really happen," but at the same time it honors general covariance. In doing so, it shows how the 'Now' might be restored to physics without paying the price of a return to the absolute simultaneity of pre-relativistic days.   (2006)

The claim is that the CSG dynamics (see below) rescues temporal becoming and our intuitive conception of time from relativity. Does it? Although the new physics definitely brings new opportunities, to a large extent the dialectic is, as Aerosmith put it, the same old song and dance.

5.1.1 The basic kinematics of CST

The guiding idea of CST is that the fundamental structure of the world consists of a discrete set of elementary events partially ordered by a relation that is essentially causal. The theory finds its inspiration in a theorem by Malament (1977b) that shows the precise sense in which the causal structure of a sufficiently well-behaved relativistic spacetime determines its geometry, up to a conformal factor. Paraphrased roughly, the causal order of spacetime contains all the information we care about except for the scale or "size" of spacetime. Motivated by this result, Rafael Sorkin and others formulated an approach to quantum gravity wherein discrete events supply the scale information and causal relations supply the rest (see, e.g., Sorkin, 2006). Quantum versions are being developed, but we'll focus on the classical theory (i.e., an approach to classical gravity that is supposed to be a stepping stone to quantum gravity).

The basic structure of the theory is the *causet* C, i.e., an ordered pair $\langle C, \preceq \rangle$ of a set C of otherwise featureless events and a relation "$\preceq$" defined on C which satisfies the following conditions:

1. $\preceq$ induces a partial order on C, i.e., it is a reflexive, antisymmetric, and transitive relation;
2. $\forall x, z \in C, \mathrm{card}(\{y \in C | x \preceq y \preceq z\}) < \infty$, i.e., the sets are locally finite.

These simple conditions constitute the basic kinematic assumptions of CST. The local finitarity of causets means that they are discrete structures, and this discreteness leads to some relevant differences concerning the issue of becoming in relativity.

How do relativistic spacetimes emerge from causal sets? A classical spacetime $\langle \mathcal{M}, g_{ab} \rangle$ is said to *faithfully approximate* a causal set $\langle C, \preceq \rangle$ just in case there is an injective function from C to M that preserves the causal relations, on average maps one element into each Planck-sized volume of the spacetime (Smolin, 2006a), and the result is that the spacetime doesn't have "structure" below the mean point spacing.

The first condition demands that the causet's causal relations are preserved on the emergent level of the relativistic spacetime. The second condition fixes the local scale. The third condition captures the idea that a discrete structure should not give rise to an emerging spacetime with significant curvature at a scale finer than that of the fundamental structure. It turns out that such an injective function can easily be found if we simply identify its image by a random "Poisson sprinkling" of events onto $\mathcal{M}$. Given such an image in $\mathcal{M}$, we obtain a causal set satisfying the kinematic axioms by lifting the set of events together with all the relations of causal precedence obtaining among these events. It is thus straightforward to find a causal set that is approximated by a given globally hyperbolic spacetime with bounded curvature. This spacetime is not itself fundamental or "real," but rather an approximate way of of talking about the underlying causal set.[2]

Note that it is important that the selected events in $\mathcal{M}$ are picked randomly and do not form some regular pattern such as a lattice. If the set of events exhibits too much regularity, then Lorentz symmetry is broken because we would be able to distinguish, at an appropriately coarse-grained scale, such a lattice from its Lorentz boosted analogue. Since there can be no such observable differences, the selection of events must be sufficiently random and irregular (Dowker, 2004).

In Chapter 2, I advertised a dilemma for any reconciliation between relativity and becoming: if we stick to a notion of the present close to that found in manifest time, then the present won't be relativistically acceptable, and if we instead find something relativistically acceptable, then the identified structures will have radically different properties from those ordinarily attributed to the present by manifest time. Do we face a similar dilemma in the context of CST? Theories posit kinematic and dynamical structure, and CST differs from relativity in both ways. First, let's see if the kinematics helps. Then we'll turn to the dynamics.

Beginning with the event of the here-now, one very natural definition of the events co-present with the here-now are those events on a spacelike slice. In CST the counterpart of a spatial hypersurface is a *maximal antichain*, i.e., a maximal set of events such that any two events are incomparable (not related by causal precedence). The problem with maximal anti-chains is roughly the same as the problem with spacelike slices: they are not unique. For any given event here-now, there are in general many maximal antichains of which it is an element. Thus, in a loose analogy to the many ways in which Minkowski spacetime can be foliated, the present in the sense of the set of events co-present with the here-now would thus not be uniquely defined (see Fig. 5.1). These presents are the counterparts to the presents assumed in the RPP

---

[2] Two points. First, the notion of emergence here is distinctive and philosophically interesting. See Samuel Fletcher's "Reduction and Causal Set Theory's Hauptvermutung" (manuscript) for discussion. Second, most causal sets sanctioned merely by the kinematic axioms do not stand in a relation of faithful approximation to spacetimes with low-dimensional manifolds. This is the so-called "inverse problem" or "entropy crisis" of causal set theory (Smolin, 2006a). The hope is that a suitable choice of dynamics will solve this problem.

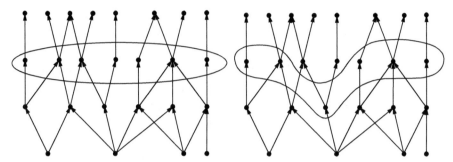

Figure 5.1 Two different maximal antichains

argument and they suffer the same fate. Neither are invariant structures. Putnam's present is not invariant under automorphisms of Minkowski structure and maximal antichains are not invariant under automorphisms of the causal set structure. Thus, it seems that a "spatially extended" present in a causal set would very much run into difficulties of the sort encountered in special relativity.

One interesting new twist is there is at least one other natural choice for the present in CST and that this choice *is invariant* under automorphisms of CST structure. Bouncing off of Stein's theorem (see Chapter 2), Wüthrich and Callender (2015) use what are known as *anti-Hegelian sets* to show that one can define in a relativistically impeccable way a relation of co-presentness in CST, one wherein distinct co-present events are always spacelike (unlike for Stein's becoming). It's surprising that this is possible, given that the only ingredient is the relativistic causal precedence relation. I won't go into the details of the construction here, as they require too much setup, but I can say that these sets require incredibly unlikely contrivances amongst event births to create the right kind of pattern.[3] As a result, they turn out to be unhelpful, and in fact, make our informal dilemma for relativity more rigorous. These anti-Hegelian sets of events are small and rare—hence unlike the manifest present (not global, etc.)—and in the unlikely cases where they grow large they exhibit the regular "lattice" structure that I mentioned breaks Lorentz invariance—and hence are not relativistic.[4] Therefore in this formalism it is strictly true that the closer one is to manifest time (i.e., the bigger

---

[3] For those interested, the relation is defined via the notion of a *non-Hegelian subset*. These are sets $H \subseteq C$ of events in a causal set $\langle C, \preceq \rangle$ consisting of distinct events $x_1, \ldots, x_k$ in C with the same relational profile, i.e., $\{x_1, \ldots, x_k \mid \forall x_i, x_j \forall z \in C$ such that $z \neq x_i$ and $z \neq x_j, \sim (x_i \preceq x_j)$ and $z \preceq x_i \leftrightarrow z \preceq x_j$ and $x_i \preceq z \leftrightarrow x_j \preceq z$, where $i, j = 1, \ldots, k\}$. (Singleton sets of events will satisfy this definition except that they do not contain *distinct* events. We insist on distinctness because we are only interested in non-Hegelian subsets of cardinality 2 or higher. Thus, $1 < k \leq n$ for a causal set of n events.)

[4] There are few analytical results to back this up at present, but the claim is not controversial. In conversation David Meyer points out that the expected number of non-Hegelian pairs for a causet obtained from N samples of a uniform process in an Alexandrov neighborhood of $(1 + 1)$-dimensional Minkowski spacetime is 1 in the limit as N goes to infinity. Since larger non-Hegelian subsets would contain more non-Hegelian pairs, their expected number would presumably be smaller and quickly tend to zero in the limit as the non-Hegelian subset grows.

the set of co-present events), the less relativistic the set is, and vice versa. The original dilemma (see the introduction for Chapter 2) for relativity haunts the kinematics of CST.

These new twists are interesting and even surprising. But as far as rescuing temporal flow from physics goes, the Aerosmith song should begin playing in your head about now.

5.1.2 Taking growth seriously

To find anything smacking of growth, one needs to turn to the dynamics. Here we meet some novel physics ripe with possibilities for time. The dynamics for a causal set is a *law of sequential growth*. What grows are the number of elements, and it is assumed that the "birthing" of new elements is stochastic. Suppose $\Omega(n)$ is the set of n-element causets. Then the dynamics specifies transition probabilities for moving from one $C \in \Omega(n)$ to another $C' \in \Omega(n+1)$.

Innumerable growth laws are possible. However, in a remarkable theorem David Rideout and Rafael Sorkin (Rideout, 2000) show that if the classical dynamics obeys just a few natural conditions then the dynamics is sharply constrained. In particular, it must come from a class of dynamics known as generalized percolation. Since the differences within this class won't matter for what follows, we can illustrate the idea with the simplest dynamics that satisfies the Rideout–Sorkin theorem, namely, *transitive percolation*, a dynamics familiar in random graph theory.

A simple way to understand this dynamics is to imagine an order of element births, labeling that order using integers $0, 1, 2, \ldots$ such that they are consistent with the causal order, i.e., if $x \preceq y \rightarrow \text{label}(x) < \text{label}(y)$. (The reverse implication doesn't hold because the dynamics at some label time may birth a spacelike event, not one for which $x \preceq y$.) We begin with the causet's "big bang," the singleton set. Now when element 2 is birthed, there are two possibilities: either it is causally related to 1 or not, i.e., $1 \preceq 2$ or $\neg(1 \preceq 2)$. Transitive percolation assigns a probability p to the two elements being causally linked and $1 - p$ to the two elements not being causally linked. Ditto now for element 3, which has probability p of being causally linked to 1 (2) and $1 - p$ of not being causally linked to 1 (2). The dynamics enforces transitive closure, so if $1 \preceq 2$ and $2 \preceq 3$, then $1 \preceq 3$. Another way to conceive of the dynamics is that when each new causet C' is born, it chooses a previously existing causet C to be its ancestor with a certain probability.

The heart of the idea that CST rescues becoming involves taking sequential growth seriously:

> The phenomenological passage of time is taken to be a manifestation of this continuing growth of the causet. Thus, we do not think of the process as happening "in time" but rather as 'constituting time'. (Rideout, 2000)

Becoming is embodied in the "birthing" of new elements.

Although interested in becoming, we should immediately remark that sequential growth is certainly compatible with a tenseless or block picture of time. In mathematics a stochastic process is defined as a triad of a sample space, a sigma algebra on that space, and a probability measure whose domain is the sigma algebra. Transition probabilities are viewed merely as the materials from which this triad is built. In the case at hand, the sample space is the set $\Omega = \Omega(\infty)$ of labeled causets that have been run to infinity. The "dynamics" is given by the probability measure constructed from the transition probabilities; for details, see Brightwell et al. (2003). On this picture, the theory consists simply of a space of tenseless histories with a probability measure over them.

However, let's take the growth seriously. There are different extents to which this can be done. At a more modest level, and consistent with explicit pronouncements by advocates of causet becoming, we can articulate a localized, observer-dependent form of becoming. Here, the idea is that becoming occurs not in an objective, global manner, but instead with respect to an observer situated within the world that becomes. The only facts of the matter concerning becoming are local, and are experienced by individual observers as they itch toward the future. In Sorkin's words, which are worth quoting in full,

> [o]ur "now" is (approximately) local and if we ask whether a distant event spacelike to us has or has not happened yet, this question lacks intuitive sense. But the "opponents of becoming" seem not to content themselves with the experience of a "situated observer". They want to imagine themselves as a "super observer," who would take in all of existence at a glance. The supposition of such an observer *would* lead to a distinguished "slicing" of the causet, contradiction the principle that such a slicing lacks objective meaning ("covariance").   (2007, p. 158)

According to Sorkin, instead of "super observers," we have an "asynchronous multiplicity of 'nows'." Yet it seems fairly straightforward that a perfectly analogous kind of becoming can be had in the context of Minkowski spacetime. Indeed, "past lightcone becoming," based on Stein's theorem, and "worldline becoming," as articulated by Clifton and Hogarth (1995), both satisfy the bill.[5] Here it seems we see the same old dance.

Although Sorkin himself remains uncommitted concerning whether the analogy holds, Dowker (2014) rejects it, arguing that "asynchronous becoming" is not compatible with general relativity, but only with CST with a dynamics like the one provided by the classical sequential growth (and hence also not with the purely kinematic CST). The reason for this seems to be ultimately metaphysical, because only with the dynamics do we get not just the events, but their "occurrence." Since in general relativity spacetime events do not "occur," goes the thought, there is no genuine form of becoming possible.

---

[5] See Arageorgis (2012) who makes a similar point.

Against this, observe firstly that (a large subsector of) general relativity certainly can be described in a "dynamical" manner via its many "3 + 1" formulations (see Wald, 1984, ch. 10). To make her objection, Dowker would first need to elaborate the reasons why a 3 + 1 dynamics do not provide the "occurrence" she desires. Furthermore, we note here a possible tension. If occurrence is simply a label for some events from the perspective of other events, then there is no problem—but then we note that such labels can be given consistently in general relativity too. But if occurrence implies something metaphysically meaty, such as existence or determinateness—then there is a possible tension between occurrence and the local becoming envisioned by Sorkin and Dowker. If spacetime events that are spacelike related do not exist for each other, for instance, then that is a radical fragmentation of reality.[6] Not only would that be a high cost to introduce becoming, but it is also one that, again, could be introduced in the ordinary theory.

What we are really interested in here is to determine whether a more ambitious, objective, global, observer-independent form of becoming is compatible with CST-cum-dynamics in a way that does not violate the strictures of relativity. Stein-like results have always had limited appeal. At best it defines a notion of becoming compatible with a spacetime. But if one desires that spacetime *itself* grows or changes, as many metaphysicians of time do, then Stein's project is simply seen as irrelevant. There is a difference between notions of becoming and flow that are observer- or event-dependent and those that are independent of observers or events (Callender, 2000). If one wants the world to become, as tensers do, then one wants a more substantial perspective-independent sense of becoming. Can CST provide us with this?

Even before getting into the details, one might be concerned that an analogue of Jack Smart's "how fast does time fly?" objection applies when we turn to the dynamics Smart (1966). Smart argued that if time changes and change is the having of different properties at different times, then it seems that at least two times are needed for any metaphysics wherein the present moves. That seems to be the case here too. Remember that the elements being created are spatiotemporal. What does a dynamics over variables that are spatiotemporal even mean? We have an external time given by the dynamics—the time in which growth happens—and an internal time given by the spatiotemporal metric the causet inherits from its embedding into a relativistic spacetime. The causal set counterpart of Smart's question beckons: how fast are elements born?

Is Smart's objection fatal to the idea of cosmological "growth"? Here philosophical opinion divides. Anticipating Smart's question, Broad (1938) argued that the kind of change that time undergoes is a *sui generis* kind of process. It is not to be analyzed as qualitative change, i.e., the change of properties with respect to time. It is its own thing.

---

[6] In Pooley's view (2013, p. 358n), dynamical CST should best be interpreted as a "non-standard *A* Theory" in Fine's (2005) sense, i.e., as giving up "the idea that there are absolute facts of the matter about the way the world is" (2013, p. 334).

We get a hint of that answer in Rideout and Sorkin's claim that birthing *constitutes* time and is not *in* time. The causet growth is time, in some sense, not something that happens in time. Like Broad, Skow (2012) believes that a second-order time is not required to make sense of a substantive sense of temporal passage. He regards this apparent second time dimension as a kind of metaphor to understand the action of primitive tense operators. For philosophers such as Broad and Skow, Smart's objection has no purchase. Others, however, might complain that appeal to *sui generis* processes and primitive logico-linguistic devices leaves a lot to be desired in terms of physical clarity.[7] No matter the reader's reaction, we'll bracket this worry since we're trying to give CST its best chance at a robust sense of becoming.

The problem with taking the primitive growth as vindicating becoming is that advocates of CST uniformly wish to treat the labeling time as "fictitious." The reason is that the choice of label is tantamount to picking a time coordinate x_0 in a relativistic spacetime. Any dynamics distinguishing a particular label order will be non-relativistic. Not wanting the dynamics to distinguish a particular label ("coordin-atization"), the authors impose *discrete general covariance* on the dynamics. This is a form of label invariance. The idea is that the probability of any particular causet arising should be independent of the path to get to that causet. In particular, if α is one path from the singleton causet to an n-element causet, and β is another path, then the product of the transition probabilities along the links of α is the same as that for β (and any other such path).

To get a feel for this, suppose that the singleton set births a timelike related element, Alice's birthday, at label time $l = 1$, and then this 2-element causet births a third element, Bob's birthday, spacelike related to the other two events at label time $l = 2$. That is path α. Path β instead births Bob's birthday spacelike related to the singleton set, and then births Alice's birthday timelike related only to the singleton set. Discrete general covariance implies that the product of the transition probabilities getting from the singleton to that 3-element causet is the same. Used as a condition to derive the dynamics, all sequential growth dynamics compatible with CST possess this symmetry. The further interpretation is that the probabilities respect this symmetry because the labels are pure gauge, that there is no fact of the matter about which path was taken.

With this simple example in mind, one can immediately see the trouble with regarding this growth as a real physical process (see Fig. 5.2). Suppose the event a timelike related to the singleton set is Alice's birthday party and suppose the event b spacelike related to both is Bob's birthday party. To enforce consistency with relativity, there is no fact of the matter about which one happened right after the singleton element event. To say which one happened "first" is to invoke non-relativistic concepts. It is therefore hard to understand how there can be growth happening in time. Seeing the difficulty here, Earman (2008) suggests a kind of philosophical addition to causal

---

[7] Or simply not evade the problem, as pressed by Pooley (2013).

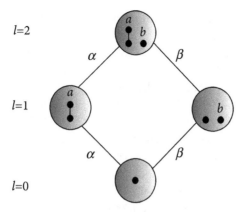

Figure 5.2 Alice and Bob's birthdays come into being

sets, one where we imagine that "actuality" does take one path or another. With such a hidden variable moving up the causet, we do regain a notion of becoming. But as Arageorgis (2012) rightly points out, such a move really flies in the face of the normal interpretation of these labels as pure gauge.

Perhaps the sensible reaction to this problem is to abandon the hope that CST does produce a novel sense of becoming. Hovering over this discussion, after all, is the fact that the above tenseless interpretation exists and is possibly superior because it does not ask us to imagine that one event came first.

Still, one might be tempted to press on. True, the dynamics is written in terms of a choice of label, a label that we need to interpret as gauge. That's unfortunate. Still, we know that a consistent gauge invariant dynamics exists "beneath" this labeled dynamics.[8] And one thing that we know is gauge invariant is the number of elements in any causet. Focusing just on these and ignoring any labeling, we do have transitions from C to C' and so on. There is gauge-invariant growth. Can we base becoming upon this growth?

The challenge we must face is that we are prohibited from saying exactly what elements exist at any stage of growth. Take the case of Alice and Bob above. The world grows from C_1 to C_2 to C_3. That's gauge invariant. We just can't say—not due to ignorance, but because there is no fact of the matter—whether C_2 consists of the singleton plus Alice's party or the singleton plus Bob's party. Causal set reality doesn't contain this information. There simply is no determinate fact as to whether C_2 contains a or b; but there is a determinate fact that it contains one of them. If it is coherent, therefore, to speak of a causet having a certain number of elements but without saying

---

[8] How do we know this? Ironically, if we rewriting the theory in terms of a "tenseless" probability measure space, as indicated above, one can quotient out under relabelings to arrive at a label-invariant measure space (for construction and details, see Brightwell et al., 2003).

what those elements are, then CST does permit a new kind of—admittedly radical and bizarre—temporal becoming.

Whether this notion of becoming is coherent depends on the identity conditions one has for events. If to be an event, one has to be a particular type of event with a certain character, then perhaps the idea is not coherent. After all, what is the C_2 world like? It does not have Alice *and* Bob in it (that's C_3), nor does it have *neither* Alice nor Bob in it (that's C_1). The world determinately has Alice or Bob in it, but it does not have determinately Alice or determinately Bob. "Determinately" cannot penetrate inside the disjunction. Notice that this feature is a hallmark of vagueness or of metaphysical indeterminacy more generally. Without going into any details of the vast literature on vagueness, let us note that there is a lively dispute over whether there can be ontological vagueness.[9] The causal set program, interpreted as we have here, supplies a possible model of a world that is ontologically vague yet which grows element by element.

One may be worried that on this notion of becoming in CST, no event in a future-infinite causet may ever be determinate until future infinity is reached, at which point everything snaps into determinate existence. This worry is particularly pressing as realistic causets are often taken to be future-infinite. So does any event ever get determinate at any finite stage of becoming? In general, yes. One way to see this is by way of example. As it turns out, causets based on transitive percolation in general have many "posts," where a *post* is an event that is comparable to every other event, i.e., an event that either is causally preceded or causally precedes every other event in the causet. Rideout and Sorkin (1999) interpret the resulting cosmological model as one in which "the universe cycles endlessly through phases of expansion, stasis, and contraction . . . back down to a single element" (024002–4).

Consider the situation as depicted in Fig. 5.3.

There is a post, p, such that N events causally precede p, while all the others—potentially infinitely many—are causally preceded by p. At stage $N - 1$, shown on the left, there exist $N - 1$ events. At this stage, all the "ancestors" of p except those three events which immediately precede p, must have determinately come to be and are shown in black. Of the three immediate predecessors, shown in gray to indicate their indeterminate status, two must exist; however, it is indeterminate which two of the three exist. At the prior stage $N - 2$, the gray set of events existing indeterminately would have extended one "generation" further back, as it could be that two comparable events are the last ones to come to be before the post becomes. At the next stage, stage N, N events exist and it is determinate that all ancestors of p exist. There is no ontological indeterminacy at this stage. Event p has not yet come to be at either stage and is thus shown in white. At stage $N + 1$, not shown in Fig. 5.3, event p determinately

⁹ Sider (2003) argues that existence cannot be vague. Interestingly, he asserts (2003, p. 135) that anyone who accepts the premise that existence cannot be vague is committed to four-dimensionalism, the thesis that objects persist by having temporal parts.

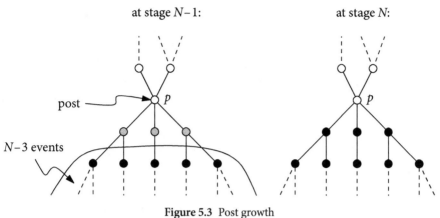

Figure 5.3 Post growth

comes into existence. At stage $N + 2$, one of the two immediate successors to p exists, but it is indeterminate which one. And so on.[10]

I close with a discussion of some of the strange features of this metaphysics.

First, note that many philosophers, from Aristotle to today, have thought that the future is indeterminate (see, e.g., Øhrstrøm and Hasle, 2011, and references therein). According to some versions of this view, it's determinately true that tomorrow's coin flip will result in either a head or a tail, but it is not determinate yet which result obtains. Vagueness infects the future. We note that the above causal set vagueness is quite similar, but with one big difference: on the causal set picture, the past too can be indeterminate! In our toy causal set, it is not true at C_3 that C_2 determinately is one way rather than the other.

Second, note that as a causet grows, events that were once spacelike to the causet might acquire timelike links to future events. If we regard the growth of a new timelike link to a spacelike event as making the spacelike event determinate, modulo the above

---

[10] One may object that this interpretation of the dynamics of a future-infinite causet presupposes a given final state toward which the causet evolves. Even though everything in the preceding paragraph is true under the supposition that the final causet is the one represented in Fig. 5.3, the objection goes, at stage N it is not yet determined *that p is a post*, as there could have been other events spacelike-related to p. Given that it is thus indeterminate whether p is indeed a post, and since this is the case for all events at finite stages, no events can thus snap into determinate existence at any finite stage of the dynamical growth process. However, even if this objection succeeds, it is still the case that it is objectively and determinately the case that at each stage, one event comes into being and that thus the cardinality of the sum total of existence grows. Although the ontological indeterminacy remains maximal, there is a weak sense in which there is objective, observer-independent becoming. In addition, if the causet does indeed not "tend" to some particular future-infinite causet, then all existence would always be altogether indeterminate (except for the cardinality). There would be no fact of the matter, ever, i.e., at any finite stage, of how the future will be, or indeed of how anything ever is. If this is the right way to think about the metaphysics of the dynamics of CST, we are left with a wildly indeterminate picture. Finally, it should be noted that the mathematics of the dynamics is only well-defined in the infinite limit; in particular, for there to be a well-defined probability measure on Ω, we must take $\Omega = \Omega(\infty)$ (Sorkin 2007, p. 160n; Arageorgis, 2012, section 3).

type of vagueness, then this is a way future becoming can make events past. That is, there is a literal sense in which one can say that "the past isn't what it used to be."

Finally, although we don't have space to discuss it here, note that despite appearances transitive percolation is perfectly time reversal invariant. This allows the construction of an even more exotic temporal metaphysics. If we relax the assumption that events can only be born to the future of existing events, then it is possible to have percolation—and hence becoming—going both to the future and past. Choose a here-now as the original point. Then it is possible to modify the theory so that the world becomes in both directions, future and past. Of course, similarly, we could have a causal set that is future-finite and only grows into the past, and thus is past-infinite.

5.2 The Worst of Times: Disappearing Time in Canonical Quantum Gravity

History is filled with religious zealots who cry that the end of the world is nigh. Similarly, it's also filled with philosophers who insist that the world is fundamentally timeless, e.g., Zeno, McTaggart, Gödel. The zealots' reasoning all suffer a common flaw: the evidence they use is too old, flawed, and coarse from which to derive any specific prediction, even if we assume their religious views were correct. The philosophers advocating timelessness all suffer a common flaw too: these arguments succeed in eliminating time only by raising the standards for what *counts as time* impossibly high. Zeno's argument for a changeless reality assumes that modest "Russelian" theories of motion aren't adequate. McTaggart insists that time be identified with an A-series (coupled to the little-discussed C-series—see on page 290) and that a B-series isn't good enough. Gödel desires a similarly A-theoretic lapse of time.[11] The natural response in all cases is simply to make do with less. It's only by adorning time with so much heavy jewelry that they are able to knock it over.

Philosophers familiar with this history might reasonably suspect that all cases of alleged timelessness share this same flaw. I want to show that this suspicion is not borne out in some recent claims of fundamental timelessness made in quantum gravity. The "semiclassical time" program in canonical quantum gravity provides a beautiful example of a fundamentally timeless world immune to this common response. The claim may have other problems (which we'll briefly discuss), but one of them is not the identification of time with a metaphysically rich batch of properties. Instead here the role time plays couldn't be more spare. Time is simply identified *as that parameter with respect to which the quantum matter fields evolve*. In other words, simply invert the Schrödinger equation for time. The Schrödinger equation encodes a mass of correlations amongst different variables, all with respect to the "t" parameter in the

---

[11] See Dainton (2001) for discussion of these cases and references.

equation. To be time is simply to play the functional role that this "t" plays. No high falutin' properties necessary. Let me now describe the result as non-technically as I can.

The "core idea" motivating canonical quantum gravity is at first blush very natural: express relativity in a suitable form, and then "quantize" it, just as is done in the non-gravitational case.[12] Quantization is a kind of formal algorithm for converting classical fields into operators that act on quantum states of the field. It is how one gets the theory of quantum electrodynamics from classical electromagnetism, for instance. To implement this idea relativity must be put in the right form. We met it earlier: canonical quantum gravity relies on the ADM "3 + 1" decomposition of spacetime mentioned in Chapter 3. When so expressed, the motivating idea is very intuitive. Classical mechanics describes three-dimensional particles evolving with time governed by a Hamiltonian; quantize, and then we get quantum theory evolving states of three-dimensional particles with time evolving according to a Hamiltonian. Follow the same recipe again, except this time replace three-dimensional particles with three-dimensional spatial geometries. Although technically *very* challenging, one can formally quantize Hamiltonian versions of relativity. The result is a set of partially interpreted equations. Prominent among these equations is the so-called Wheeler–DeWitt equation:

$$H\Psi = 0$$

Here H is the quantum gravitational super-Hamiltonian and Ψ is a wavefunction defined over three-dimensional spatial geometries. The original idea was that H would evolve the spatial geometry forward, stitching together spatial geometries into a quantized four-dimensional world. As in the Schrödinger equation, H would "update" the quantum state Ψ with time. A quick look at the equation, however, tells you that wasn't what was found. The right-hand side, the side where time should be, is zero. On its face, there is no time development of the quantum state. This result leads to a host of difficulties collectively referred to as *the problem of time*.

That alarming fact by itself doesn't mean the theory describes a still universe. As Peter Bergmann, a postdoctoral student of Einstein, made clear as early as 1959 (and published in Bergmann, 1961), time "disappears" even in classical relativity when it is put in Hamiltonian form. General covariance forces a kind of timelessness upon the theory—see Belot and Earman (2001). Yet we know that general relativity in its ordinary four-dimensional formulation works and ably handles change. Its successes include the motion of Mercury's perihelion, the deflection of starlight by the Sun, the gravitational redshift of spectral lines, and much more. The theory suffices to describe a very busy, dynamic universe. If the Hamiltonian formulation of general relativity is really equivalent to ordinary four-dimensional relativity (or at least a relevant set of solutions), and the latter has "time enough" in it, then so does the former if it's a

---

[12] On whether the idea is really so natural, see the discussion in the introduction of Callender and Huggett (2001).

viable formalism, albeit in a hidden way. There exist plenty of examples of constrained Hamiltonian systems that appear to be timeless and yet with a change of variables time pops out. Motivated by such considerations, advocates of canonical quantum gravity don't abandon the program but instead try various strategies of extracting a useful physics out of the Wheeler–deWitt equation.

Roughly put, there are essentially three types of approaches. The first is the hidden time program that tries to identify a fundamental time in the formalism. Because this is often done even prior to quantization, this is sometimes called a *tempus ante quantum* program. The hidden time might be some function of the spatial scale factor, or a special global time function such as York time, met in Chapter 3. Others try to identify time with some kind of matter clock or reference fluid.

The second more radical idea is to get by with no time at all—*tempus nihil est*—and see how much physics one can do nonetheless. Thinking that this is the message of relativity, Rovelli (2009), one of the founders of loop quantum gravity, tells us that the way forward is to "forget time." Similarly Barbour (1999) proposes that time is an illusion. While this more radical idea is exciting, in all of its manifestations it faces a basic challenge: even if the world is fundamentally timeless, still it seems like it has time in it. There is change, and change is at least the having of different properties in time. An urgent question for anyone espousing timeless quantum gravity therefore is explaining why it is that the world *seems* temporal. Relativity explains the appearance of a classical time by showing that locally deviations from classical theory get small, and this explains why classical mechanics works as a good approximation. Can we do something similar in the present case, that is, accept a basic timelessness and nonetheless find an emergent non-basic time?

This line of thinking leads to a kind of compromise between the hidden time program and the "forget time" program, namely, emergent time. Here, fundamentally, there is no time, but nonetheless time (and with it change, predictions, and so on) arises in certain regimes. Since time typically emerges in these schemes after quantization, the third group is sometimes dubbed a *tempus post quantum* set of programs. We'll focus on one known as *semi classical time*.

5.2.1 Semiclassical time

This idea, associated with Banks (1985), Kiefer (2012), and many others, is not as popular as it once was because canonical quantum gravity's stock has fallen. Nevertheless, we're interested in the basic idea of a fundamentally timeless world, and this example is still among the most rigorous and attractive, as well as relevant to newer programs' struggles with time.

As Kiefer (2012) notes, a deep paper from quantum mechanics' past can help motivate the idea. In 1931 Neville Mott described the collision between an alpha particle and an atom. To model the total system, Mott used the time-independent Schrödinger equation, which is analogous to our time-independent Wheeler–DeWitt equation. This equation lacks any time. What's interesting is what he did next: dividing the system into subsystems and using an approximation for the alpha particle (representing it as a plane wave, justified by its high velocities), he used the state of the alpha particle as a "clock" for the atom subsystem. In fact, it wasn't just any clock, for he showed that the atom, relative to the alpha particle, obeys the time-*dependent* Schrödinger equation. That is, if we treat the alpha particle as a clock, the atom "listens" to the time of quantum mechanics—even though the system as a whole is timeless! Hidden in the timeless equation for the total system is a time for the subsystem.

In more detail, the quantum state for the total system is

$$\Psi(r, R) = \psi(r, R)e^{ikR}$$

where small r refers to the atom and big R refers to the alpha particle. Suppose we *identify time* via a directional derivative as

$$i\frac{\partial}{\partial t} \propto ik \cdot \nabla_R \qquad (5.1)$$

That identification might strike you as arbitrary. Why should *that* be time? The answer is that if one makes this identification then one can derive the time-*dependent* form of the Schrödinger equation for the atom subsystem. The right-hand side of Eq. (5.1) plays the functional role of time for the atom. That is, the atom listens to the "ticks" of $ik \cdot \nabla_R$ just as it would $i\frac{\partial}{\partial t}$. Put loosely, if you were the atom, you would "evolve" to the fundamentally non-temporal directional derivative. This suggests that we may think of the directional derivative as time, that time—that parameter according to which the matter fields govern—is in fact this directional derivative. And since classical mechanics and classical time emerge from quantum mechanics, we know that this time will behave like classical time in the appropriate physical regimes. So there is every reason to think of the right-hand side of Eq. (5.1) as time *if* we can derive that the subsystem it governs obeys the time-dependent Schrödinger equation.

The italicized *if* hints that we cannot always make that derivation. That is indeed so, and part of what is interesting. Making the substitution of Eq. (5.1), we can't always derive the time-independent Schrödinger equation. We can only do so if certain conditions are met—essentially, with respect to systems and scales where it's legitimate to approximate the alpha particle with a plane wave. Recall our minimal definition of time: that parameter with respect to which the matter fields evolve à la Schrödinger. When the correlational structure of the system as a whole is such that the Schrödinger equation can't be derived using Eq. (5.1), then the directional derivative relinquishes its right to be dubbed time. Time "emerges" only when the subsystems enjoy certain types of correlations with the system as a whole.

Can the same ideas work for the timeless Wheeler–DeWitt equation? The Wheeler–DeWitt equation describes a universe with exactly zero energy. However, that doesn't imply that all the subsystems of the universe have no definable or zero energy. The subsystems can have definable energies. And where there is energy, there is the potential for time development. The Wheeler–DeWitt equation governs much more than Mott's system, so the equation is much more complicated and even more approximations must be invoked; but the basic idea is the same. Break the timeless system into subsystems, and treat the more classical subsystem (the gravitational part) as a clock with respect to which the other subsystem (the quantum field) evolves; then see if that subsystem obeys any of the "time-dependent" equations describing the matter fields. As DeWitt (1967) first showed, Banks (1985) elegantly proposed, and Kiefer (2012) continues developing and refining, this is indeed possible. Amazingly, if certain approximation conditions obtain, one can treat the more classical system as a clock with respect to which the other subsystem evolves via the time-dependent Schrödinger equation. The time-*dependent* Schrödinger equation for the subsystem can be derived from the time-*independent* Wheeler–DeWitt equation governing the total system. Time emerges from timelessness when the world is semiclassical and divided into subsystems.

To explain the derivation in all of its gory detail would be inappropriate for present purposes. For us, it's suitable if we sketch the basic idea.

The Wheeler–DeWitt equation describes the wavefunction of the universe Ψ, a functional of spatial three-geometries with three-metrics h and various matter fields. Take a simple case where we just have one matter field represented by a scalar field ϕ. Now the important step is the approximation. We note that the actual world is "semiclassical" in certain respects: gravity doesn't display its quantum aspects at large scales, changes in the quantum degrees of freedom happen rapidly with respect to changes in the gravitational degrees of freedom at many scales, spacetime is approximately flat at large scales, and so on. In particular, the relevant masses are much larger than the Planck mass. Assuming we're in such a semiclassical regime— which itself seems justified, given what we currently see—we appear to be warranted in using certain approximations.

Using an analogue of the Born–Oppenheimer approximation, we find that

$$\Psi \approx \exp\left(iS_0[h]/\hbar\right)\psi[h,\phi]$$

is an approximate solution of the Wheeler–DeWitt equation. Here $\psi[h,\phi]$ is a wavefunction over the spatial three-metrics, h, and the matter field, ϕ. S_0 obeys the Einstein–Hamilton–Jacobi equation for a gravitational field, so it describes a classical spacetime solution to Einstein's field equations. Together the approximate solution is taken to describe a classical spacetime and a quantum field. This solution is just the first-order part of the Born–Oppenheimer method. Subsequent orders can be expressed which get one closer to the genuine solution.

An approximate solution in hand, we proceed to the main event. The approximate solution still has no time in it. Yet it can be shown that $\psi[h,\phi]$ obeys the following (again, approximate) equation:

$$i\hbar\nabla S_0\nabla\psi \approx H_M\psi \qquad (5.2)$$

where H_M is the Hamiltonian for the non-gravitational fields ϕ and ∇ are functional derivatives with respect to h. A peek at the Schrödinger equation then suggests the following identification:

$$\frac{\partial}{\partial t} \propto \nabla S_0 \nabla \qquad (5.3)$$

When plugged into Eq. (5.2), we obtain a Schrödinger equation for ψ. The right-hand side of Eq. (5.3) is in some sense entirely spatial. But as in the Mott case, its identification as time is justified by the role it plays. If time is identified as that parameter according to which the matter fields evolve, then the functional derivatives of Eq. (5.3) are by all rights time. Once again, like a rabbit from a hat, we have pulled a time-dependent equation from a time-independent one (Fig. 5.4).

As in the Mott case, the derivation is awash in approximations. "Time" is only recovered "when" these approximations hold. In that sense the time is emergent and semiclassical. Fundamentally, there is no basic time, nothing worthy of the name (which means, in this context with our low standards for time, that nothing in the basic physics or full solutions obeys a time-dependent Schrödinger equation). Unlike the arguments of Zeno, McTaggart, and Gödel, not much is asked of time, and nonetheless time doesn't exist (fundamentally). Better yet, the above approximate derivation answers the most pressing challenge to such timelessness, namely, why time *seems to exist* if it doesn't. Here we have a precise answer to that question. Put in terms of the earlier discussion, the idea is that something answering quantum mechanics' job advertisement for time resides in inherently timeless physics. In the present case, instead of the alpha particle acting as time for the atom, the gravitational degrees of freedom are acting as a clock for the non-gravitational fields.

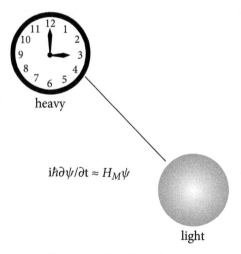

Figure 5.4 Semiclassical time

It's interesting to note that on this view the existence of time seems entirely contingent. Time didn't *have* to emerge. If the initial quantum state of the universe were a Gaussian, for instance, we cannot repeat the above procedure. There would be no time.[13]

I would be remiss if I didn't point out that, evaluated as a physical theory, canonical quantum gravity and "semiclassical time" leave us wanting more. Canonical quantum gravity suffers from a number of persistent technical difficulties. More than that, the theory as described seems incomplete. We want the theory to say something about domains that aren't semiclassical. That's what it's *for*. The universe can't always be broken up into a nearly classical "clock" subsystem and non-classical matter field. We can calculate perturbations from this situation, but outside this approximation we're in trouble. Without time we don't know how to interpret the theory.

Here is one reason why. Recall from Chapter 4 that the quantum probabilities are connected to time. Quantum mechanics is unitary, conserving probability with time, and this allows it to provide coherent probabilistic predictions of outcomes. Outside the semiclassical regime, we have no time. No time means no probability. No probability means no predictions. We can't answer: what frequencies should we expect for such and such value of a particular measurement? We can only ask that within the approximation. In a way, this limitation is payback for developing the theory tailored to an approximation justified by prevailing conditions; it means that outside those conditions, what the theory says is not always clear. What is the approximate solution *an approximation to*? There have been attempts to solve this problem (Unruh and Wald, 1989), but the theory is calling out for new physics, physics that can make predictions well outside the approximation scheme employed here.

5.2.2 Justifying the approximations

Suppose canonical quantum gravity overcame all of these physical and mathematical challenges. Can we really claim that time *disappears* in the theory? Although I trumpet this derivation as a great example of timelessness, I do not mean to imply that it is thereby free of philosophical worries or even successful in achieving true timelessness. To see if the emergence story really makes sense, ideally we would want the theory interpreted, see its ontology, and then check if the ontology's (timeless) behavior warrants the approximations invoked. Left as stated, the picture is rather impressionistic. I speak of "objects" and "subsystems" but all we really have are a mass of correlations in the super wavefunctional of the universe. We can rightly demand a solution to the measurement problem, an understanding where the probabilities come from, and much more. Answers to *some* of these questions are suggested in

---

[13] One might want to insist that the initial conditions are nomically determined, and perhaps in a way that demands an emergent time—in which case time would not be contingent. I'm simply pointing out that, on its face, without that discussion, the physics allows solutions lacking anything worth dubbing time.

Callender and Weingard (1996). Everettian interpretations of quantum theory may also see "objects" and "subsystems" "emerge" from this super wavefunctional.

Still, lurking behind this interpretational worry is the concern that we haven't discharged the "transcendentalist" worry that we *assumed a time to get a time*. That is, can we really state and justify all the approximations invoked in a vocabulary devoid of time? Or are the approximations only justified by assuming that we already have subsystems evolving in time? For example, in the Mott case we justify treating the alpha particle as a plane wave because of the alpha particle's high velocities. Velocity is, of course, change with time. And just above I partly justified the semi-classical approximation by saying that changes in the quantum degrees of freedom happen *rapidly* with respect to changes in the gravitational degrees of freedom at many scales.

Anderson (2007) points to over 20 approximations necessary for the semiclassical time program to get off the ground. Let's be clear. *Not one* of those assumptions explicitly invokes time. The formalism has no 't' in it. We're not going to find a smoking gun, a formal assumption invoking time to get time. Mott can certainly write down the formula for a plane wave without using time. The worry is more subtle: is time smuggled in when justifying these formal conditions? If the only reason for using a plane wave approximation is that the alpha particle travels relatively rapidly, the time is still smuggled in even if not present in the formal condition used in deriving semiclassical time. Temporal talk like this would need to be cleansed away if we're really to get time from no time.

Personally, I'm undecided about what to make of this objection. We seem to have something like a chicken and egg problem. To get objects and subsystems, it seems we need time; but to get time, we seem to require objects and subsystems. My objection is reminiscent of Baker's (2007) charge against the Everett interpretation of quantum mechanics that it illicitly uses probability in its crucial employment of decoherence. The Everett interpretation infamously posits "many worlds," i.e., many macroscopic structures that "emerge" from the wavefunction of the universe. To see these structures emerge, the Everettian points to a process known as decoherence, whereby a wavefunction loses coherence in phase angles amongst the components in a quantum superposition. Decoherence requires probability, however, yet the Everettian isn't yet entitled to non-trivial probabilities. The theory is deterministic, so it has only trivial probabilities. On most schemes of explaining where quantum probabilities originate, one requires observers and other macroscopic structures.

Here is the chicken and egg. The emergent structures justify the probability, but the probability justifies the emergent structures. In fact, given the tight connection between probability and time in quantum mechanics (explained in Chapter 4), on the one hand, and decoherence and the 20 approximations necessary for semiclassical time (these approximations are essentially a complicated version of decoherence), on the other, the two problems are deeply related. Baker is saying to Everettians: you assume probability to get your structures and then assume structures to get your probability. I am similarly pointing out to advocates of semiclassical time: you're assuming time to justify the decoherence that gives you time.

Perhaps the most natural reply to both worries is to assume that each theory offers one a "package deal." Don't think of it as *deriving* time or emergent structures from decoherence. Rather, "simultaneously" assume what is needed and the resulting package is a coherent theory. Put like this, I begin to feel the same queasiness that I did when I was an undergraduate and first read Wittgenstein's claim that one could throw the ladder away after using it. Here no ladder is being thrown away, but two need to be climbed at the same time. I'll leave the verdict on whether that makes sense to the reader.

Independently of its current fortunes, the semiclassical time program should take pride of place in the pantheon of attempts to describe a fundamentally timeless world. Zeno, McTaggart, and Gödel all proposed fundamentally timeless worlds, but in each case the argument relies on a contentious understanding of time. That is not the case here. Even if it turns out to be nothing more than a complicated toy model, we have in "semiclassical time" a very real conceptual advance. We have what is to my knowledge the first rigorous and physically motivated explanation of how time can arise from timelessness.[14]

5.3 Conclusion

This chapter has investigated two rigorous and yet divergent possible paths for time in quantum gravity, one where it seems to disappear altogether, and another promising to restore time to its pre-relativistic glory. Whether quantum gravity goes one way or another, or yet a third course, remains to be seen. However, we've learned that the minimal time of relativity is harder to kill and more difficult to improve upon than is commonly thought.

---

[14] The reader may be curious whether other quantum gravitational approaches produce similar results. Given that loop quantum gravity is a descendant of canonical quantum gravity, one suspects that a similar semiclassical time can be recovered in that theory. Although there are suggestive results (e.g., Bojowald et al., 2005), so far that is not the case. Presumably this is one reason why Rovelli (2009) seeks to recover time-dependence via his "thermal time" hypothesis instead. Superstring theory, by contrast, begins with a background spacetime and no comparable result as above exists.

6

The Differences Between Time and Space

The past few chapters have taught us that physical time doesn't make room for important properties of manifest time. Many philosophers and physicists have gone further and claimed that physics (mistakingly, in their view) "spatializes time," in the words of Henri Bergson. Here is the physicist Lee Smolin:

> Around the beginning of the seventeenth century, Descartes and Galileo both made a beautiful discovery: You could draw a graph, with one axis being space and the other being time. A motion through space then becomes a curve on the graph. In this way, time is represented as if it were another dimension of space. Motion is frozen, and a whole history of constant motion and change is presented to us as something static and unchanging. If I had to guess (and guessing is what I do for a living), this is the scene of the crime.
>
> (Smolin, 2006b, pp. 256–7)

Like Bergson, Smolin traces the "crime" of spatializing time to pre-relativistic physics and the idea of representing time spatially with graphs and so on. For obvious reasons, the spatialization charge became especially popular after the advent of relativity. Einstein, Emile Meyerson, Milič Čapek, Mary Cleough, Bergson and others debated whether relativity made time another dimension of space (or whether relativity instead "temporalized" space).

Smolin and others may be right that representing time spatially makes it psychologically tempting to treat time like space. Certainly there is evidence that we recruit spatial concepts and metaphors when discussing time (e.g., Boroditsky, 2000). But that choice of representation doesn't strictly imply anything much about time. We graph the rise and fall of happiness too, but that doesn't spatialize happiness in any interesting sense nor has it led research astray in any obvious way. Until it's shown to lead to a clear mistake in physics, this point about spatial representation is something to bear in mind but otherwise not a worry.

Besides, because physics doesn't support manifest time doesn't mean that physics treats space and time alike. It doesn't. Physics posits sharp and important distinctions between the timelike and spacelike directions of spacetime. In many respects time is special in contemporary physics, even if its temporal structures aren't congenial to would-be defenders of manifest time. Thinking that physical time is just like space not only is mistaken, but it deprives one of potentially useful resources when explaining why we have a picture of manifest time so deeply entrenched in us. These differences may provide objective "pegs" onto which creatures like us "hook" when navigating the world. Missing the differences between time and space makes "closing the gap" between manifest time and physical time that much harder.

We need to investigate how time differs from space in physics. Such an investigation is crucial to the present project, but it also is independently of interest. At a time when researchers in quantum gravity regularly propose speculative theories containing no time at all, a better understanding of time in current physics is all the more important—even if only to see what is lost by its absence.

What is the difference between time and space? This question, once a central one in metaphysics, is now little studied. Changes in philosophical methodology and in physics together sapped some of the spirit out of this project. After reconfiguring the original project, this chapter and the next two make two arguments.

First, I survey time's role in physics and make the case that some features held to be inessential to time are in fact of crucial importance to time—and always have been. The metric difference between the timelike and spacelike in particular turns out to be of extraordinary significance. With these features in hand, I then identify a problem that I call the *fragmentation of time*. Is there a "glue" that links these otherwise disparate temporal features together, or is it just an accident that they hang together in our world?

Second, in the next chapter I propose a novel difference between time and space: with suitable caveats, the temporal direction is that direction on the manifold of events in which our best theories can tell the strongest, most informative "stories." Put another way, time is that direction in which our theories can obtain as much determinism as possible. Time is not only the "great simplifier" (Misner et al., 1973), but it is also the great informer. To support my proposal, I try to show that the direction of informative strength helps answer the problem of fragmented time. Strength is linked to time because it is deeply connected to the other temporal features of our universe. If I am right, "strength" is the glue that binds together many otherwise detachable features central to time. The third chapter (Chapter 8) then develops surprising senses in which this is—and isn't—rigorously true in physics.

6.1 The Project Reconceived

The question "What makes time different from space?" has not been treated kindly by recent history. What was once an active subject of research has dwindled to almost nothing over the last thirty or so years.[1] I submit that this is no accident. Though never

---

[1] Examples of the work I have in mind include: Taylor (1955, 1959, 1983); Thomson (1965); Webb (1977); Romney (1977); Schlesinger (1975, 1991); Huggett (1960); Mayo (1961, 1976); Dretske (1962); Meiland (1966); Shorter (1981). Skow (2007) is a notable recent exception.

articulated, the reasons for this question's demise are readily discerned. Two changes in the late twentieth century seem to have given it a one-two punch that knocked it from the ring of contention. The first hit was methodological. The question was pursued with what is now considered an old-fashioned and possibly discredited methodology. A gross caricature is as follows. One searched for statements whose truth-value or meaning was not invariant under substitution of temporal and spatial terms. When such statements were found, a difference between space and time was discovered. One then consulted one's intuitions to discern which of these differences constituted *the* difference between the two.

This methodology is odd. It assumes, first of all, that ordinary language and intuition hold the key to the difference between the two; and secondly, it assumes that there is some essence of the difference between space and time, some feature or features necessary and sufficient for the difference. Suppose one asked the comparable question: what is *the* difference between protons and neutrons? The best place to look for an answer is a physics textbook (and the knowledge it draws upon), not one's intuitions about what features are essential and accidental to each particle. And it's not clear that there would be much profit in trying to decide (before their internal quark makeup was known) whether it's the charge or mass difference that constitutes *the* difference between protons and neutrons. Philosophers of time looked in the wrong place for an answer, and they expected to find what may not exist.

If the would-be researcher persevered through this change and turned from ordinary language analysis to analysis of science, he or she would then encounter the next blow. Science appears unfriendly to the question. When Minkowski joined space and time together into spacetime, space and time were doomed to become mere "shadows"; i.e., non-fundamental entities. Metaphysics may deal with non-fundamental entities, of course, but the purportedly fundamental has all the glamour. Just as bad, relativity states that matter fields and spacetime affect one another. Our question, however, seems to presuppose that there are some features intrinsic to time, that we can separate features of time from features of the evolution of objects in time. Relativity is a threat to these intrinsic features because it makes many properties once considered intrinsic to time (e.g., being closed or open) dependent on the contingent distribution of matter and energy—and so extrinsic.

Neither of these threats imply that there isn't an interesting project in the neighborhood of the original. The threat to the question from modern physics is in fact no threat at all. That some features which were intrinsic become extrinsic (or vanish entirely) doesn't mean that they all do. The four-dimensional block envisioned by relativity still draws sharp and fundamental distinctions between the timelike and spacelike directions of spacetime, a distinction that doesn't vary with the mass–energy distribution. Moreover, the matter fields on this block vary with time differently than they do space. In fact, looking at physics as a whole, few equations are invariant under a transformation of spatial and temporal variables—and no important ones. As for the methodological obstacle, we need to modernize the question in at least two respects.

The data needs to change. It should be scientific, not linguistic. For our questions we are better off simply investigating time and space in our best scientific theories. Second, we should abandon the assumption that there is a clear simply expressible essence distinguishing time from space. Maybe there is such a difference, but how would we find it? The concept *time* includes many important features, as we're about to see, most of them logically detachable. We are not likely to find necessary and sufficient conditions for the concept. Nor is science likely to help us by highlighting some temporal features as essential and others not. We ought to acknowledge that the goal of previous researchers on this question may not exist: there may not be *the* difference between space and time.

Once freed from that assumption, a new question emerges. Contemporary physics distinguishes time from space in a variety of ways. Are there connections among these features? Is it just *accidental* that they coincide in our world? By identifying features special to time and finding connections amongst them, we learn something deep about the nature of time in physical theories. This chapter takes up this rehabilitated project and the next two propose a novel and I hope important distinction between space and time.

6.2 Time in Physics

Bergson claimed that physics was "spatializing" time. If so—whatever exactly that means—we'll see that physics is certainly taking its sweet time doing so. Precious little physics is invariant under a change of what are commonly interpreted the spatial and temporal directions. Simply listing the equations non-invariant under a transformation of timelike and spacelike directions is hardly likely to be illuminating. Instead, let's begin our investigation by focusing on properties commonly attributed to time but not space, and also, those features that do not vary with initial conditions. Thus we'll ignore features like being Hausdorff, connected, and so on, since they are also features associated with being spatial. And if thanks to the boundary conditions time is closed and space is open, we won't chalk that up to time and space being intrinsically different. Reflection on physics then reveals at least five features of time *possibly* surviving the above winnowing process: the metrical difference between time and space, the one-dimensionality of time, the "direction" of time, the mobility asymmetry, and the natural kind asymmetry. Let's briefly discuss each, pausing to notice theories wherein said differences might not hold.

6.2.1 The metric

The metrical structure of a spacetime is arguably the most fundamental and central feature of spacetime. It determines the distances between all pairs of events and the angles between any two spacetime vectors, and hence, its causal structure and much more.[2] Given its significance, it is critical to note that the metric of spacetime, *whether relativistic or classical*, distinguishes time from space at a very basic level. If I believed in *the* difference between space and time, the metrical distinction would have my vote.

Philosophers divide over how significant it is. Reichenbach judged it to be crucial to the difference between space and time:

> We shall find even that the Minkowskian world is incorrectly interpreted if one looks to it for support of the parallelism of space and time; on the contrary, the world of Minkowski expresses the peculiarity of the time dimension mathematically by prefixing a minus sign to the time expression in the basic metrical formulae.   (1957, p. 112)

By contrast, others have been less impressed with this "minus sign." Skow (2007) finds it "too formal and abstract" to be a central difference between space and time, and moreover, finds fault with it because it allegedly doesn't share enough with classical time to count as *the* essential feature of time. How deep this metrical difference is will hang on the rest of our story. As we'll see, I think this little "minus sign's" importance for understanding time has been badly underestimated.

To begin, forget talk of a "minus sign." With a change of coordinates we could eliminate it.[3] Reichenbach's minus sign—as he knew well—is a symptom of a much deeper geometrical fact, the *signature*. The signature is determined as follows. Given a smooth semi-Riemannian metric g, we can find an orthonormal basis $v_1, \ldots, v_n$ of the tangent space at each point p of M. This is a basis such that $g(v_\mu, v_\nu) = 0$ if $\mu \neq \nu$ and $g(v_\mu, v_\nu) = \pm 1$ if $\mu = \nu$. Let the number of basis vectors with $g(v_\mu, v_\nu) = 1$ be p and the number of basis vectors with $g(v_\mu, v_\nu) = -1$ be q. Then the metric has signature (p, q) at that point. Assuming that g is non-degenerate and continuous—as we do in relativity—it's a theorem that the signature of spacetime is then constant on M so long as M is connected. We can therefore speak of *the* signature of a spacetime $\langle M, g \rangle$.

One may get the impression that the signature is not a geometric invariant because it's possible to choose a convention (sometimes called "West Coast" convention) wherein the signature is $(1, 3)$ rather than ("East Coast") $(3, 1)$. This element of choice makes the distinction look shallow rather than deep. However, it is possible

---

[2] As a sign of this significance, witness the crucial role played by the metric signature in the many theorems proved throughout Malament (2012). Also, note the connections between the signature and the matter fields discussed in Chapter 8.

[3] For instance, if we employ "lightcone coordinates," a coordinization popular among string theorists, we express the usual Minkowski metric in terms of two null and two spatial coordinates: $-ds^2 = -2dx^+ dx^- + (dx_2)^2 + (dx_3)^2$. Here no coordinate is picked out as special. "Time" and "space" are thoroughly mixed in lightcone coordinates thanks to the transformation of the usual (x_0, x_1, x_2, x_3), where $x_0 = t$:

$$x^+ \equiv \frac{1}{\sqrt{2}}(x_0 + x_1')$$

$$x^- \equiv \frac{1}{\sqrt{2}}(x_0 - x_1')$$

The specialness of time is "spread out" over two coordinates. Nonetheless, this disguise doesn't change the signature.

to describe spacetime intrinsically in a way that doesn't require a choice between the two signatures. Stein (1968), for example, writes "for any four linearly independent and mutually orthogonal vectors, the inner products with themselves of three of these vectors have one sign (positive or negative), and the inner product of the fourth with itself has the opposite sign" (p. 7). It is then possible to characterize a vector as timelike just in case its inner product with itself is opposite in sign to that of any non-zero vector orthogonal to it.[4] Alternatively, stipulating that p are the positive eigenvalues and q the negative eigenvalues of metric g (a coordinate independent fact, as we know from linear algebra) one can define the signature as the absolute value of the number of positive eigenvalues minus the number of negative ones. Lorentzian four-dimensional spacetimes therefore have a signature equal to 2.

The signature describes an extremely deep difference between space and time. Recall that since Lorentzian metrics are indefinite metrics, they divide spacetime vectors exhaustively into three classes. For any semi-Riemannian metric g, a vector v is spacelike, timelike, or null depending on whether $g(v, v)$ is positive, negative, or zero, respectively. This division of vectors into these three classes is of course one of the most significant features of relativity. One "sees" this distinction between time and space built into the very foundations of relativistic geometry. The difference between space and time emerges when composing a Lorentzian metric. As is well-known, any paracompact (metrizable) manifold admits a positive definite metric. One can ask under what conditions such a manifold admits a Lorentz metric. The answer is that it's a theorem that a paracompact manifold admits a Lorentzian metric iff it admits an unordered, non-vanishing line element or *direction field*, that is, an assignment of equal and opposite vectors $(n, -n)$ for some direction n to each point of M. As a result, one can regard a Lorentzian metric as a positive definite metric with a direction field added to it.[5] How is this relevant to the difference between time and space? The answer is that one can show that the vectors in the direction field, n and $-n$, are the eigenvectors for the unique negative eigenvalues q above. *Hence the direction field runs in the timelike directions.* What distinguishes Lorentzian metrics from Riemannian metrics is a direction field aligned with the timelike directions! Although relativity banishes "time" and "space," there is nevertheless a sense in which it committed at its very core to a difference between the timelike and spacelike directions.

That the same goes classically may also be surprising. In Chapter 2 we built up Galilean spacetime from the Cartan formulation, one that bequeathes separate temporal and spatial metrics to spacetime. But I also mentioned that it was possible to represent Newtonian spacetime—Galilean spacetime with a preferred rest frame, as one closest to what Newton actually posited—as a manifold endowed with a *single*

---

[4] Thanks to David Malament for bringing this definition to my attention.

[5] Let $(n, -n)$ be the direction field and h_{ab} an arbitrary Riemannian metric, normalized such that $h_{ab}n^a n^b = 1$. Then $g_{ab} = h_{ab} - 2(h_{ac}n^c)(h_{bd}n^d)$, where g_{ab} is a Lorentzian metric.

spacetime metric on it as in relativity. What I didn't mention is that if one does one can posit a metric of *Lorentzian* signature too (Penrose, 1968)!

Here is how. Recall that whereas Minkowski spacetime possesses one spacetime metric η_{ab}, Galilean spacetime possesses two, a "temporal" metric t_{ab} and a "spatial" metric h_{ab}. Add to Galilean spacetime the structure that Newton demanded, namely, a distinguished preferred rest frame. This addition can be achieved by imposing a smooth timelike vector field λ^a on the manifold such that it satisfies $\nabla_a \lambda^b = 0$. The timelike vector field distinguishes Newton's preferred rest frame. Now it turns out that a combination of this rest frame and spatial metric is fully equivalent to the Minkowski metric:

$$\lambda^a \lambda^b - h^{ab} = \eta^{ab}$$

In a precise sense, one can regard full Newtonian spacetime as already containing the structure of Minkowski spacetime in it (Barrett, 2015).[6]

Hence Newton's own theory, written in modern covariant form, can pick out via a Lorentzian metric a direction field $(n, -n)$ in the timelike directions just as modern relativity theory does. True, the Cartan formulation of Galilean spacetime doesn't do so, as Skow mentions. Yet even here one might make a case that this spacetime contains an echo of this direction field, for one can explicitly show how a single Lorentzian metric can, in an appropriate classical limit, split up into two degenerate metrics one "spatial", one "temporal," as in Cartan spacetime (Malament, 1986, p. 407). Classical *and* relativistic physics have *always* singled out time from space metrically, and arguably, have done so in roughly the same way.

The metric difference between time and space has many deep ramifications. For example, one can't continuously rotate (or boost) a timelike vector pointing in the direction t to a timelike vector pointing in the direction $-t$, whereas one can do the spatial counterpart in the spacelike directions (and boost in the mixed spacelike–timelike directions).

Physically, these differences really matter. For instance, consider two events p and q on M that can be connected by a timelike curve. If that can happen, then they can also be connected by a spacelike curve. The reverse statement, however, is not true. If one restricts to so-called "reasonable" spacetimes, then there are more consequences. For instance, in any "chronological" spacetime (see Box 2 on page 72), there are no closed *timelike* curves; but there may be closed *spacelike* curves. The Friedman spacetimes discussed in Chapter 2 represent such a case. Additionally, it is a major theorem that in past–future distinguishing spacetimes (see Box 2 again) the class of continuous timelike curves determines the topology of spacetime (Malament, 1977b). Geometrically, this may be puzzling: why should topology "care" about the difference

---

[6] Thanks to Jim Weatherall for discussion of this point, and to Thomas Barrett for sharing his work while unpublished. I stress the word "can" in the above, as the inference may not be forced upon one; however, if we regard classical spacetime as emerging from relativistic spacetime, then it is the natural choice to make.

between timelike and spacelike curves? No comparable "timelike" theorem exists because there is no spacelike version of past–future distinguishing—and that naturally is due to the lightcone structure bifurcating the timelike but not spacelike set of events. Intuitively, timelike curves are forced by the asymmetry in the lightcone structure in reasonable spacetimes to "see" all of M whereas spacelike curves can wind around and become trapped in local regions, blinded to the rest of M. In any case, these sorts of powerful facts—the source of much of what's right about the causal theory of time, causal set theory (e.g., Rideout and Sorkin, 1999) and Maudlin's recent work on directed timelike lines (Maudlin 2010)—all have their origin in the above metrical difference.

The metrical difference is not merely "formal and abstract" but deep, concrete, and physically significant. It will play a large role in what follows.

That said, let's pause to observe that it is *conceivable* that the signature be demoted to a less than fundamental aspect of spacetime. Some theorists have entertained the possibility that the signature changes. This result is achieved by relaxing certain assumptions on the metric, namely, allowing the metric to be either degenerate or discontinuous (see, e.g. Dray et al., 1997). Consider the discontinuous metric $ds^2 = t\, dt^2 + a(t)^2\, dx^2$. Here the spacelike hypersurface at $t = 0$ is the point of signature change, the time at which the world evolves from a Lorentzian metric to a positive definite one. In even more speculative theories, it may be possible to understand spacetime's signature as the result of some dynamical process, such as spontaneous symmetry breaking (Greensite, 1993). On this view, the signature may not be such a basic aspect of spacetime. Nevertheless, in less speculative physics the signature is as central to spacetime as it gets.

6.2.2 Dimensionality

Every successful physical theory purporting to be fundamental has judged time to be one-dimensional. What does it mean for time to be one-dimensional?

Classically, we can understand this claim in a very flat-footed fashion. There the instants of time form a continuum under the *earlier than or simultaneous with* relation. This order relation can determine the open sets that form a basis for a topological structure. We can then ask of this structure, the structure representing the set of instants, how many topological dimensions it has. There are several definitions of topological dimension, all of which understand dimensionality as a topological invariant. Using one of these definitions, say, small inductive dimension, we can then point out that the set of instants is topologically one-dimensional, i.e., $\mathbb{R}^1$.

Relativistically, there isn't *the* set of instants. One can't grab a set of events corresponding to the "temporal" ones and check its dimensionality because there is no such set. Yet there are various senses in which time remains one-dimensional in relativistic physics. Here is one. Consider a point p on a timelike curve and a particular four-velocity field. Take a vector w^a at p. Then w^a can be decomposed into components parallel to and orthogonal to that four-velocity field. The set of orthogonal vectors form a three-dimensional subspace in the tangent space M_p at p. The set of parallel vectors form a one-dimensional subspace in M_p. Interpreting the first set as spatial and the second as temporal, we then have the sense in which the set of timelike directions at a point is one-dimensional, using the definition of dimension found in linear algebra, where the dimension is given by the number of basis vectors in the vector space. In effect, we have sliced up spacetime into space and time on an infinitesimal simultaneity slice; on this slice time is one-dimensional. This method will work equally well across relativity and classical Newton–Cartan models. Here is another sense. As we've seen in Chapter 3, in many spacetimes one can prove a "splitting theorem" that decomposes $\langle M, g \rangle$ into the pseudo-Riemannian product of $\langle \mathbb{R}, -dt^2 \rangle$ and $\langle \Sigma, h \rangle$. The first piece of this product is straightforwardly topologically one-dimensional. Although not unique, whenever this splitting is possible time will be one-dimensional.

Observe that neither manner of understanding the one-dimensionality of time is independent of the signature difference. In the local version, for instance, what vectors are orthogonal/parallel and (obviously) what vectors are timelike/spacelike depends on the signature of spacetime. The sense in which time is one-dimensional, therefore, hangs on the metric distinction. This fact bolsters the previous section's conclusion that the "minus sign" does indeed reflect a deep fact about the difference between time and space.

Before moving on, let's pause to consider models without this feature. Plenty of speculative physics and philosophy consider the possibility of more than one time dimension. This work is a bit on the "fringe" of physics, and if my later theory is correct, I think it helps explain why.

Philosophers Thomson (1965) and MacBeath (1993) argue that there are *conceivable* situations in which it would be rational to come to believe in an extra temporal dimension. MacBeath imagines a world populated by two groups of people, Zoe and her friends, and Adrian and his friends. Zoe lives until 80 years old and Adrian lives until 40 years old. But Adrian and his group live life fast: they pack into a year what it takes two years for Zoe and her friends to do. Adrian uses an alternative time measure, yonks. By yonks, Adrian lives 80 yonks and Zoe lives 40 yonks, but by this standard, Zoe packs in twice as much "life" as Adrian and his friends per yonk. In such a world, MacBeath claims, it is natural to posit two times.

This type of phenomenon actually already happens in real life. Interestingly, science sticks with one time. The discovery that mesons entering the atmosphere "shouldn't" make it to Earth, given their half-life, is one of the most renowned confirmations of relativistic time dilation. By our terrestrial time reckoning, mesons survive (don't decay) more than ten times as long as they should. Like Adrian and his friends, mesons pack a lot into their lives thanks to time dilation. But time dilation is symmetrical. If mesons carried little scientists, they would see all sorts of terrestrial processes lasting longer than they "should," by their time reckoning. They would view us as Adrian and his friends view Zoe and her cohort. Yet confronted by more or less the very facts MacBeath imagines, scientists did not posit two times. Instead scientists posited indefinitely many reference frames all related by symmetry transformations. (See King (2004) for essentially the same point.) I agree with MacBeath and Thomson that two or more time dimensions is conceivable, but in actual theorizing it seems that we don't quickly jump to this possibility.

We get a glimpse of why not when turning to physics. Physicists assume that temporality supervenes upon q in the spacetime signature (p, q), so when talking about multi-dimensional time one is assuming that $q \geq 2$ in the signature. Plenty of speculative physical models posit this, as one can find $(4, 2)$ brane worlds, $(3, 3)$ Kaluza–Klein theories, $(11, 2)$ F-theories, and more in the literature. So far none of these theories have caught on. The reason is that they are beset with a variety of challenges :

- Stability. Dorling (1970) shows with an elegant geometrical argument that the principal mechanism enforcing the prohibition of certain types of decay is removed if more than one time dimension exists. The extra time dimension wreaks havoc with existing theory.
- Causality. Models with multidimensional time are hard pressed to avoid closed timelike curves. Suppose the spacetime has signature $(2, 2)$. Then unless very contrived, curves running up the "timelike" direction along x_2 wander back and forth and hence time travel along the "timelike" direction x_3, and vice versa. This time travel is different than that found in "ordinary" $(3, 1)$ solutions such as Gödel's and Taub-NUT, as it is possible even in flat topologically Euclidean spacetimes too. In flat Euclidean spacetimes with one time dimension, closed timelike loops don't appear because such a loop would have to be at some point spacelike—which is not so if we have two time dimensions.
- Unitarity. It is desirable in a probabilistic physical theory—like quantum mechanics—that probabilities be conserved with time, i.e., unitary (see Chapter 4). Ynduráin (1991) shows that unitarity is violated in quantum mechanics if two of the spacetime dimensions are temporal. Even a tiny extra dimension would lead to violations with current experiment, e.g., more matter should disappear than is detected in experiments of proton decay. His paper suggests that observations rule out a compact timelike extra dimension if it is rolled up to even $1/10$th the Planck radius.
- Tachyons. If we go to quantum field theory, then the presence of an extra timelike direction causes trouble even with a scalar field in a $(3, 2)$ "Minkowski" spacetime. The field has tachyonic modes, i.e., modes that travel faster than light.

There are plenty of other "technical" problems with extra timelike dimensions, e.g., the presence of "ghost modes" in the Feynman propagator for massless vector particles in a second timelike direction (Foster and Müller, 2010). None of these problems eliminate the *possibility* of extra timelike dimensions. Craig and Weinstein (2009) show that the simple free wave equation is well-behaved, even well-posed (see below), in systems with extra timelike directions. But their exception proves the rule, for the addition of interactions to this system spells trouble. In general, when the physics gets more realistic extra time dimensions seem to add severe obstacles to any reasonable physics. Theories are postulated, but physics in this vein always seems to have to resort to some ad hoc restriction along the "new" timelike direction. The extra timelike dimension's radius needs to be shrunk so small as to be experimentally innocuous, gravitons are allowed to explore one timelike direction but not the other, or the dynamics is "thermalized," i.e., ergodic, along one timelike dimension and not the other. See Gogberashvili (2000), Foster and Müller (2010) and references therein for a sense of the options.

Why do extra timelike directions cause such trouble? Technical problems pop up all over the place by switching the metric signature. Given the centrality of the signature, this fact is hardly surprising. The problem is that the physical dynamics is tuned to evolving in the "minus" direction. From this dynamics all sorts of nice physical properties hold, especially conservation principles. In classical and non-relativistic theories, particle number is conserved. In quantum mechanics we saw that probability current is conserved through time too. These principles and many others are derivable straight from the dynamics. If you now insist that these properties hold in two directions on M, directions unrelated by a symmetry transformation, you are asking for trouble.

Take a toy example. Imagine building a theory with two classical times in it, with particles having the ability to sneak into either time's past. Allow that but also give consistent dynamical stories from the perspective of each timelike dimension. How do you now impose conservation of particle number on your dynamics? In Fig. 6.1, we see the potential for mischief: at $Time_1 = 1$ there are four particles, but at $Time_1 = 2$ there are two; similar problems arise in the $Time_2$ dimension.

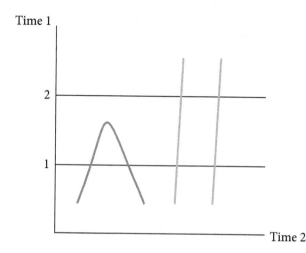

Figure 6.1 Two times

Suppose we want conservation of probability to hold in both timelike dimensions. We'll need the weights on outcomes to normalize to unity. How do we guarantee this if particles can sneak away? We face the same problem. Either we're going to have to prohibit particles sneaking into the past or we'll have to carefully balance the comings and goings of particles from each time's past. Both strategies mandate imposing intricate restrictions on the laws or initial states to make this work. Matters are even worse if we allow interesting interactions.

With enough sophistication, one can steer around the above difficulties. Yet the core problem remains. It's very hard to tell a physical narrative in two metric directions when all the worldlines, dynamics, and conservation principles are tuned to just one.[7]

6.2.3 Mobility asymmetry

One of the most obvious differences between time and space is that we have relatively free mobility in the spatial but not temporal directions. How best to put this is a little tricky. Suppose I never make it to Machu Picchu. That would be a pity, as it's on my "bucket list." Yet the reasons for my not going probably would be relatively mundane. I wouldn't have the free time, the money, the will, and so on. The reason my current future lightcone doesn't contain the event of me hiking Machu Picchu has nothing to do with the deep structure of spacetime. Contrast that with going back to the Jurassic Age to watch dinosaurs. I don't have that on my "bucket list" because I know that even if I summoned the time, money, and will, still none of the airlines fly there. The reason that they don't is that I suspect that there aren't any future timelike paths in our spacetime connecting me now with events in my past lightcone.

Confined to future timelike paths, we are sensitive to and constrained by the orientation of the lightcones defined at each point. The actual world is such that the lightcones aren't "tipped over" with respect to one another, thereby allowing for the "back-and-forth" behavior that we enjoy along the spacelike directions. As a result, we cannot time travel. If you were a five-dimensional creature like one of Kurt Vonnegut's Tralfamadorians and could look at our four-dimensional spacetime, you would immediately see the mobility asymmetry: massive worldlines go this way and that in three directions but never the fourth, as in Fig. 6.2.[8]

What I just said might not be true. Einstein's field equations allow plenty of solutions that permit time travel, e.g., Taub-NUT, Gödel (see Earman, 1995, for discussion). In these worlds there is no mobility asymmetry. In Gödel spacetime, for example, one

---

[7] I haven't explored this point, but it seems to me not a coincidence that many of the technical problems we've run into here with multitemporal physics also plague attempts to devise a consistent physics in which tachyonic and non-tachyonic matter non-trivially interact. I suspect that the connection is due to the fact that particles sneaking backward in time are going faster than the speed of light according to one "timelike" dimension.

[8] Beware! The picture just painted is nothing more than impressionistic. Since the lightcone structure is imposed on the tangent space and can anyway vary point by point, there is no need for the timelike curves on M to line up so neatly as in the picture. Nevertheless, there is a real asymmetry, assuming that our world doesn't have time travel in it.

Figure 6.2 Mobility asymmetry

can in principle travel an always future timelike path from any event to any other on the spacetime manifold. The lightcones tip over such as to make these journeys possible and the spacetime is chronologically vicious in the language of Box 2. If the field equations delimit the physically possible, then I *am* free to go back to the Jurassic in the same sense I am to Machu Picchu. If I don't go to either, it's simply because the contingent matter–energy distribution didn't fix the actual world up so that I do. The mobility asymmetry becomes *factlike* rather than *lawlike*. The contingent distribution of matter–energy is to blame for my actual inability to leave my office in the north direction (there is a wall in the way), to get to Machu Picchu (too many kids, too little time, etc.), and to catch a flight to the past (too much fuel required[9]). The inabilities are chalked up to contingent boundary conditions in all cases.

That said, our collective inability to go into the past is far more pervasive than it is to go to the north. The walls, rocks, and so on that prevent my northward travels only stop *some* northernly pursuits, not all of them. By contrast, the matter–energy distribution plus laws manage to prevent all known movements to the past. This fact makes some physicists and philosophers suspicious of relativity's time travel solutions. Some think that time travel ought to be ruled out with new nomic constraints, just as we discussed in Chapter 3 in the context of cosmic censorship. Right now the best we can say is that whether the mobility asymmetry is lawlike or factlike is up in the air.

One intermediate possibility deserves recognition. Even if there is no law ruling out time travel, still it might be that our inability to travel freely in temporal directions is a result of theory *stacking the deck* against time travel. Not all actions permitted by the laws are equally permitted, given the type of physical object I am. We have free mobility in the "up" direction, but given the laws and our mass and locations, movement in this

9 Cf. Malament (2012, p. 138).

direction is severely curtailed compared to the east or north directions. If the Earth were more massive, upward travel would be even harder and thus more infrequent. Similarly, it's possible for the air in my office to spontaneously concentrate in the corner, but this doesn't happen. The laws can allow behaviors but still stack the deck against them.

The more careful way to say this is to state that "most" solutions of the Einstein field equations don't allow time travel, just as "most" solutions of Hamilton's classical equations don't have the air concentrating in the corner of my office in the next ten minutes. The problem with putting it this way is that unlike in the classical statistical mechanical case, we don't have a well-defined natural measure over the space of all solutions to Einstein's field equations (see McCoy, 2016). Thus we simply don't know whether *most* solutions of Einstein's field equations inhibit time travel.

What we do know is that there is some kind of difference between time and space with respect to mobility. Time travel's absence might be due to contingent boundary conditions, the laws stacking the deck against it, or the laws ruling it out altogether. At our current stage of knowledge this is an open question. In any case, once again we find in the mobility asymmetry a major physical difference between space and time that at least conceivably might not exist.

6.2.4 Direction of time

The temporal directions possess an asymmetry the spatial directions do not. The temporal directions have "arrows of time" and the spatial directions don't have similar "arrows of space" as counterparts. Thus, for example, we find thermodynamic entropy increasing in time, radiation using the retarded but not advanced solutions to Maxwell's equations, time-asymmetric K^0-meson decay, black holes but not white holes, and more in time (see, e.g., Albert, 2000; Callender, 2011b; Price, 1996). In the spatial directions there are asymmetries. In the Earth's northern (southern) hemisphere, moss grows preferentially on the north (south) side of trees. But apart from the existence of parity-violating particles, we don't find patterns of spatial "arrows" worth dubbing laws. Certainly there is nothing spatially comparable in scope and influence to the thermodynamic or radiation arrows. To have a concrete example, consider the classical heat equation

$$\frac{\partial T}{\partial t} = k \left(\frac{\partial^2 T}{\partial x^2} + \frac{\partial^2 T}{\partial y^2} + \frac{\partial^2 T}{\partial z^2} \right) \tag{6.1}$$

where k is a structural constant, independent of x, y, z, and t, and the temperature T is a function of where and when the temperature of a solid body is taken. The equation's intended interpretation distinguishes time from space in several respects; e.g., there are more spatial variables than temporal ones, and the former have a different sign than the latter. The difference we care about here is that the equation is invariant under spatial but not temporal reflections. Phrased in admittedly coordinate-dependent language, the equation remains the same under substitution of $-x$ for x, say, but not $-t$

for t. As with the free mobility example, to what extent this difference between space and time is the result of laws or boundary conditions distinguishing time from space is a matter of controversy.

The conventional understanding of the "arrows of time" is that they are ultimately due to asymmetric boundary conditions. K^0-meson decay is the only one of the arrows widely believed to be a result of a purely nomic preference. The asymmetry is inferred from the combination of the CPT theorem and the violation of CP. However, there are nonstandard interpretations of the CPT theorem that make even this standard claim controversial (see Arntzenius and Greaves, 2009). The thermodynamic and radiation arrows are typically regarded as not being due to intrinsic features of time. Or are they? Focus on the thermodynamic arrow (for electromagnetic, see Earman, 2011). Theorists such as Ilya Prigogine have wanted to see the thermodynamic arrow result from deeper time asymmetric laws. Others have hoped that it would arise from an asymmetry in time itself (the so-called "anisotropy" of time). I'm skeptical of both approaches. But even the conventional explanation arguably makes the arrow lawlike. The "Boltzmannian" explanation is that the fundamental laws may be time symmetric, but the constraint of low initial entropy and lack of a comparable future constraint is supposed to be responsible for the thermodynamic arrow. (The heat equation's temporal asymmetry is therefore thought to be a result ultimately of this boundary condition asymmetry.) There are no similar constraints in the spatial directions. Are these constraints nomic or factlike? Since one normally regards the thermodynamic arrow as projectable (e.g., we expect heat to flow from hot to cold next week), that is reason to suspect that these constraints are lawlike. The laws stack the deck in one direction of time rather than another.

As with the mobility asymmetry, it could be that the direction of time asymmetries are the consequence of a built-in asymmetry or a boundary condition asymmetry or both. Also, it's easily conceivable that these asymmetries between the past and future be local asymmetries that are not global. Price (1996), for instance, points out that the world could be a so-called Gold universe (in honor of the cosmologist Thomas Gold). A Gold universe has low entropy constraints at both the beginning and end of time, making entropy increase away from both times. Again, it's conceivable that a deep feature of physical time does not exist fundamentally or globally.

6.2.5 Natural kind asymmetry

Although the distinction is so deeply entrenched it's easy to ignore, physics makes a basic time–space split in its choice of natural kinds. In theories with particles, people, and other sorts of objects, we "thread" together some events as being instantiated by the same object and others as not. These are the threads of identity, what Hans Reichenbach called *genidentity* (Reichenbach, 1956). Two events are genidentical if they are part of the history of the same object. The event of you reading this sentence and the event of you being born are tied together by these threads. At a more basic level, a photon emitted by a particle may be identified with a later one caught in a

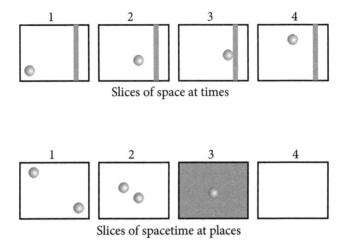

Figure 6.3 One or two balls?

trap; but a neutrino event now and an electron event later are not part of the same fundamental object's history. Some objects are seen as persisting, surviving changes, others not. In theories without individual objects, such as field theories, we sacrifice some of these threads of identity, but not all of them; e.g., we devise laws for massive but not imaginary mass fields (see below).

To vividly see how our choice of kinds interacts with our time/space split, consider a ball bouncing off a wall (Fig. 6.3, based on Callender, 2010). The usual slicing of the world takes slices of space at successive moments of time, creating a "movie" of a ball's motion as in the top set of pictures. Reading left to right, we see one ball starting in frame 1 head toward a gray wall, impact the wall in frame 3, and rebound back in frame 4. If we drew these slices on a pad, one per page, and flipped, then we would see a rather boring movie of a ball bouncing off a wall. The laws of physics determine what will happen in each frame, depending upon what happened just prior. But we can also imagine an alternative slicing of the world, not from past to future but instead along one spatial direction. Each slice is partly spatial, partly temporal. Going from left to right, we here see the incoming ball and its later post-bounce version together in frame 1. Frame 2 shows an "earlier" version, where the incoming and outgoing temporal parts of the ball are close to one another. Frame 3 displays the impact with the wall and only one ball. Frame 4 shows the region without a ball on the other side of the wall. Notice that in describing the second version of events, I describe the two balls depicted as different versions of one and the same ball. I have made a choice about the threads of identity. I did not describe it as two balls becoming one and then none. I can of course *illustrate* spacetime any way I like, but the laws of nature make choices about identity and kinds and distinguish the top frame over the bottom one.

Consider our concept of a worldline. Our discussion has taken worldlines for granted, but what is a worldline? Geometrically, there are of course all kinds of curves on a spacetime manifold. Of these, only those curves such that at each point their velocity is $v = \frac{ds}{dt} < c$ count as worldlines of massive particles, and only those curves such that at each point their velocity is $v = \frac{ds}{dt} = c$ count as massless particles. "Tachyonic" curves such that at each point $v = \frac{ds}{dt} > c$ are not worldlines, nor are curves whose velocities change classification. By calling strings of events such that each point $v = \frac{ds}{dt} \leq c$ "worldlines" and not calling strings of events such that $v = \frac{ds}{dt} > c$ "worldlines" there is a presupposed time versus space split.

"Worldline" is only a word, you might reply. Not so! We haven't merely dubbed one type of curve a "worldline," for the designations are tied to the basic properties: *massive* systems are always timelike, *massless* systems are always null, and spacelike systems, if such there be, would be systems with *imaginary mass*. The laws of nature then inherit this difference. They take as input some properties (and not others) and yield as output values of those same properties (and not others). The properties invoked by the laws are properties attached to objects that persist through time. Strangeness, charm, charge, mass, and so on, are the basic predicates in which our laws are formulated, not imaginary mass or any other "gruesome" property. You can call the events that compose the edge of your desk—right now in your approximate rest frame—a worldline if you like. But we have no laws "evolving" the events of one part of the edge of the desk into another. That "worldline" is not an object of physics.

6.3 The Fragmentation of Time

If any features are distinctive of physical time (as opposed to space), it's the five described above. A priori most of these temporal features are logically detachable from one another. We can imagine possible worlds having or lacking many combinations of the features mentioned. Call these *temporally fragmented* worlds. For instance, consider:

World 1:   a manifold with metric of $(+++-)$ signature, no mobility asymmetry at all, and a thermodynamic arrow running along the metric's "minus" direction.

World 2:   a manifold with metric of $(++--)$ signature, where both the mobility asymmetry and thermodynamic arrow run along x^3.

World 3:   a manifold with metric of $(++++)$ signature, no mobility asymmetry, but a thermodynamic arrow running along x^3.

World 4:   a manifold with metric of $(++-+)$ signature, where the thermo-dynamic arrow runs along x^0, the mobility asymmetry runs along x^1, and the Lorentzian metric and worldlines distinguish x^3.

World 1 is simply our world except with time travel. Whether inhabitants of such a world would believe in manifest time is not clear, but they would still have reason to think the "minus" direction special. World 2 breaks the connection between the metric signature and two of the other temporal features, mobility and thermodynamics. At first blush, this also appears conceivable. Here is a world with two "times" if you like, one of which allows for time travel and athermodynamic behavior and one of which doesn't. World 3 is a Riemannian world, but where something worth calling thermodynamics operates in one direction rather than the others (for no metrical reason). The science fiction author Greg Egan vividly imagines and derives physics in a Riemannian signature world in his *Orthogonal* trilogy. Of course it has drastic effects on physics, but it still seems like something thermodynamical makes sense. Like our world, the mobility asymmetry and thermodynamic arrow run in concert, except here distances amongst events can go negative in the (say) locally "east" direction. World 4, we might say, is maximally fragmented. I'm not certain we can conceive of this world; but insofar as the properties are logically detachable, we can appreciate it as an abstract possibility.

The conceivability of temporally fragmented worlds suggests that the project of summoning intuitions to discern *the* difference between time and space is bankrupt. Even if one can really imagine what it's like to live in these worlds, still one has little reason to dub one direction as genuinely temporal and the others not. Philosophers with more fine-grained intuitions than I have are free to pursue this project. Much better, I think, is to reflect on these temporally fragmented worlds and ask whether there exist non-trivial connections amongst these different temporal features. If we pay attention not only to what's logically but what's physically possible, or even better, to what's physically possible given broadly similar boundary conditions, then is it still possible for these features to vary freely? Our world is not temporally fragmented, as best we can see. Is this simply a coincidence? Or is there some glue sticking these features to one another? What unifies time? Call this the *Fragmentation Problem*.

There may not be an answer to the Fragmentation Problem. The temporal features are each deep and in some cases probably basic in our physics; and as we witnessed, some might be accidents due to boundary conditions. As such, we can't *expect* that we'll find much (if anything) connecting these deep features. Nevertheless, I think it is clear that we would better understand time if we saw connections amongst the various temporal features. Compare the situation here with the pre-relativistic equality of inertial and gravitational mass. Because the two bear the same name in English, students of classical physics typically don't puzzle over why the two types of mass are equal—yet classically there is no particular reason why they should be the same, as the two types of mass could vary independently of one another. Relativity yoked these two features to one another. Now we feel that this mystery has been explained. The Fragmentation Problem likewise shows that another basic physical concept is associated with a loose collection of independent fragments. If we could find some glue that would connect these fragments, we would make a genuine advance in our knowledge of time.

I said that the Fragmentation Problem may not have an answer, but equally, it may enjoy more than one correct answer. Below I will propose a new distinction between time and space and use it to try to glue together the fragments. That doesn't mean that there aren't additional constraints that stick together different fragments using a different glue. For instance, one may point out that the mobility asymmetry and thermodynamic asymmetry aren't entirely independent. The thermodynamic asymmetry implies the existence of a nearly monotonic increase in entropy for long periods of time. It is difficult but not impossible—thanks to fluctuations and recurrence—to square this near monotonicity with travel along a closed timelike path.[10] Hence if the thermodynamic arrow is at all lawlike, it's not entirely surprising that macroscopic creatures like us find it difficult to time travel.

6.4 Conclusion

There may not be *the* difference between space and time. We've identified many deep and significant differences—from the natural kind asymmetry to the direction of time—and there is little reason to think that one's intuitions when these features come apart are terribly reliable. That said, the metric signature is perhaps the deepest division between time and space. One can think of Lorentz metrics as singling out time for special treatment, treatment that (we'll see) has ramifications throughout the rest of physics.

What I'm going to do in the next two chapters is point out a new difference between space and time and use it to forge some connections amongst the fragmented features discussed in this chapter. The new difference strives not only to be interestingly novel, but also to play the role of the glue mentioned above. I will show with a mathematical argument in Chapter 8 how this informal argument can be improved.

---

[10] Consider a "cylinder" spacetime wrapped up along a spacelike axis. Closed timelike curves could go around the cylinder. By delicately tailoring the cylinder's radius with the recurrence time for such systems on these curves, we could get the entropy to go up and down consistent with thermodynamics along such a trajectory. This silly example seems very unstable to perturbation, but it does show that thermodynamics' Second Law doesn't *strictly* imply that there aren't closed timelike curves.

7

Laws, Systems, and Time

Time, I'll tell you about time. Time is the simplest thing.

Clifford Simak, *The Fisherman*

The previous chapter surveyed various important physical differences between space and time and left us with a problem, namely, the fragmentation of time. We looked for differences and we found some. But the connections amongst the features of time are thin. Maybe there is nothing more to be said. Still, it would be wonderful if we could find something deeper or at least more general to say about these differences. That is what I want to try in this chapter and the next.

Philosophers and others have posited many important metaphysical differences between time and space. The biggest, no doubt, is the idea that time is irreducibly tensed, or flowing, or passing, or becoming, or branching. Since space is not tensed, flowing, becoming, or branching, these sorts of positions do indeed distinguish time from space. If properly motivated, they might offer compelling answers to both the question of what distinguishes time from space and the question of the nature of time. For the purposes of this book, however, we've assumed that time isn't distinguished from space in any of these ways.

Even without adopting the view that the difference is metaphysically primitive, there remain other positions compatible with a "block" universe that distinguish time from space. An old and deep one connects *physical modality* with time. Perhaps the first version of this (although it may go back to Aristotle) originates in Gottfried Leibniz's causal theory of time. According to the causal theory, temporal relations (e.g., *earlier than*) but not spatial relations are reduced to causal relations. In the wake of special relativity, many empiricists advocated such a theory, e.g., Hans Reichenbach, Adolf Grünbaum, and Bas van Fraassen. They analyzed causation in empiricist-friendly terms, usually through variants of Reichenbach's "mark method" theory of causation. If correct, there is a deep metaphysical difference between time and space: time but not space is reduced to causation. A tree blossoming is different than the spatial variation of colors on a shirt because the first process is causal in a way the second process is not. Unfortunately, this grand project founders on many serious objections, both of detail (it doesn't seem to work for relativity as a whole) and of motivation (the concepts the empiricists used—e.g., genidentity—don't seem clearer than the concept explained). See Earman (1972) for details and references.

More recently, non-empiricist metaphysicians such as Mellor (1998) and Tooley (1997) have advocated a causal theory of time. The first adopts a tenseless theory of time and the second a tensed theory, but both treat causation as a primitive and do not analyze it in terms of empiricist-friendly concepts like Reichenbach did. As a result the causal theory perhaps buys itself immunity from some of the objections of Earman and others. Yet to my mind it does so at the cost of clarity. Reducing time to causation has always seemed to be a case of explaining the obscure in terms of the obscurer. To the extent Reichenbach et al. escaped this charge, it was by explicating the causal relation in what seemed an illuminating way. Offering no explication of causation, the current project seems guilty of this long-standing charge. And for the question at hand the answer is ultimately unsatisfying: time is different than space because we posit a primitive that makes them different. Maybe that is the right move—perhaps we need a primitive notion of causality anyway to make sense of things, and we may as well put it to use. However, if possible, it would be preferable to say something more informative in the present context.

A related idea appeals to the laws of nature. Arguably the laws of nature distinguish time in some way.[1] Skow's idea is that the direction in which the laws govern *defines* the temporal direction (2007, p. 238). To me, this tie is too strong. Skow is at pains to deny that all imagined or real laws governing in a spacelike direction are really laws. What support is there for this?

We need not resort to imaginary cases, for arguably there are plenty of actual laws that govern in a spatial direction in contemporary physics. The constraint equations of relativity seem to be an example: these operate across spatial directions and cannot be derived from the dynamical equations of relativity. For example, the ten vacuum Einstein field equations separate into six "evolution" equations $G_{ij} = 0$ and four "constraint equations," $G_{00} = 0$ and $G_{0i} = 0$, with $i = 1, 2, 3$. The latter impose nomic conditions across a spacelike slice. To decree that four of the ten equations that constitute Einstein's field equations are not nomic without good reason is unacceptable. Skow even goes so far as to deny theories of lawhood that might classify such spacelike generalizations as laws.

Here we see an echo of my worry about non-empiricist causation being the difference maker. Skow seems almost to be defining laws via the time/space split as much as finding the time/space split in the laws. One can make this brute metaphysical assertion, of course. But one won't find much support from it in physics.

Still, I find the idea of linking laws of nature with time very attractive. Can we make sense of the idea that the laws of nature "prefer" time in some way? Better, can we find

---

[1] Sider (2001) and Loewer (2012) mention this possibility in passing, and Skow (2007) and Maudlin (2010) develop it. Maudlin comes at the idea through his monumental attempt to find new underpinnings for all of physics. We won't tackle that ambitious project here.

connections between laws and the different fragments of time discussed previously? Using the "systems" account of laws described below, I think we can.

7.1 System Laws and Time

The topic of laws of nature is fraught with controversy. Some philosophers believe that there are no laws, others believe that laws "govern" (in a rich sense) the events of the universe, and still others feel that laws are simply the best summary of the facts. The last view of laws is typically associated with Humeanism. Humean theories seek to explain the laws *given* the distribution of actual facts, rather than going the other way round and explaining why the facts are what they are in virtue of the laws. The most attractive such theory is arguably the *best system* theory of laws (see, e.g., Lewis, 1973; Cohen and Callender, 2009; Loewer, 2004). Whatever its other costs and benefits, it is interesting that best system accounts of laws suggest two very deep features of time. The first feature is that time is the so-called "great simplifier." The second and more novel feature is that time is also the great informer.

A rough sketch of the best system theory is as follows. Consider various deductive systems, each of which makes only true claims about what exists. Consider the theoretical virtues had by various systems. The laws literature focuses on two very coarse ones, simplicity and strength. Simplicity is measured with respect to a language that contains a primitive predicate for each fundamental property. Strength is informativeness about matters of particular fact. The theory could be generalized to include more theoretical virtues, e.g., unification, probabilistic fit, or more precise understandings of particular virtues, e.g., Akaike simplicity. Many of these virtues will trade off against one another. A simple theory might not be that informative. An informative one— such as a huge list—might not be so simple. The important insight behind the systems theory is that laws arise from trying to balance many theoretical virtues. Imagine scoring each system on each virtue. The laws of nature are claimed to be the axioms that all the "best systems" have in common. The motivation for the theory is the idea that physical laws seek to describe accurately as much of the world as possible in a compact way.

Now turn to time. Time, it is often said, is the "great simplifier." What this means is that the temporal metric is chosen to make motion look simple. According to Poincaré (1913) and others, we pick a measure of duration that yields the most powerful and simple physics. If we used a pendulum on a boat to define time, then we would have to devise a physics that explained why the sun sped up or slowed down as the boat went closer or further from the Earth's poles (because gravity, humidity, etc. will affect the pendulum's period). A very complicated physics would result. Instead, one defines time intervals, i.e., the temporal metric, in such a way that the laws of physics come out strong and simple. In evaluating time metrics, one looks not merely at the proposed metric and how it connects to the phenomena, but one also weighs its effect on the whole theoretical system.

This attractive idea about time is more or less implied by a "systems" approach to laws. There is nothing special about time here. The same goes for any other aspect of the laws of nature. Space, for instance, is also the great simplifier. What goes for clocks and temporal duration goes for meter sticks and spatial duration. If space expands or shrinks at the Earth's poles (alternatively, the meter stick shrinks or grows) then one will have to rewrite physics so as to be more complicated. Time is defined in such a way as to make the equations of mechanics as simple as possible, but so is space. (And in spacetime theories, spacetime intervals are so defined.) For this reason simplicity alone will not distinguish the timelike from spacelike.

Nonetheless, I believe that the other "half" of the systems approach, informativeness, may well play a role in distinguishing time from space. In balancing simplicity with strength, a best system will not include a random catalogue of everything that happens. It will instead contain a way to generate some pieces of the domain of events given other pieces. In other words, it will favor algorithms, and short ones at that. The more of what happens that is generated by small input the better. Further, it might be, as Barry Loewer once suggested (p.c.), that the distribution of basic properties on the manifold picks out one set of directions as special in this regard. This idea, like the best system theory, is a little murky, and it's not clear how to translate it into the setting of contemporary physical theories; however, I think we can dimly perceive the outlines of the core connection.

To help fix ideas, take a not-so-toy example. Modern cosmology begins with the remarkable fact that the universe is approximately isotropic and homogeneous in three of its four directions (relative to co-moving coordinates). The universe looks the same no matter where you are and what direction you look. However, it isn't isotropic and homogeneous in all *four* directions. The cosmological principle holds three-dimensionally, not four-dimensionally. The "perfect" cosmological principle is false. Thanks to this asymmetry, one can write simple but strong laws, i.e., the Friedman equations. As we saw in Chapter 3, these equations result from exploiting cosmological symmetries and finding a convenient time parameter to march the universe forward.

In the above example, the system "finds" time by discovering a powerful algorithm. Familiar from computer science, algorithms are ways of getting back lots of information about the world from minimal input:

No human being can write fast enough, or long enough, or small enough† (†…"smaller and smaller without limit . . . you'd be trying to write on molecules, on atoms, on electrons") to list all members of an denumerably infinite set by writing out their names, one after another, in some notation. But humans can do something equally useful, in the case of certain denumerably infinite sets: They can give explicit instructions for determining the nth member of the set, for arbitrary finite n. Such instructions are to be given quite explicitly, in a form in which they could be followed by a computing machine, or by a human who is capable of carrying out only very elementary operations on symbols.

(Boolos and Jeffrey, 1989, p. 19)

With the small initial input of a few values—initial spatial scale, initial matter density—the Friedman equations are a way of getting back the universe's properties for any other time step on the manifold of events M. If these algorithms are adopted, they add a lot of inferential power to a system at a small cost to simplicity. Cosmologists happily accept this as a great buy. Algorithms and systems have nothing essentially to do with time. What is picking out time as special is the distribution of physical properties. The algorithms and system do the rest, exploiting this asymmetry by including an algorithm "marching" information forward across M in one set of directions rather than another. The system picks the algorithm that picks the time.

Simplicity and strength together pick out time as special. If we didn't care about simplicity, then we might find algorithms that "evolved" the matter density in the "sideways" direction too. Or we might find a bunch of different algorithms generating the data in different directions in different patches of M. Neither of these prescriptions pack as much inferential punch in as few lines of code, however. Although simplicity is important, we'll relegate it to a background role in what follows. Simplicity has already had its day; it is time for strength to shine.

7.2 Time is the Great Informer

Earlier I said that if we could find a way of saying that the laws "prefer" a time direction, then that would be better than simply insisting that laws—to be laws—govern only in the temporal directions. The systems theory offers a solution. Above we saw that the system might seize on an asymmetry in the distribution of events and utilize that in forming an algorithm or algorithms that form the best balance of strength and simplicity. In some cases one or more of these algorithms may play especially general and powerful roles in the system. Think of Newton's Second Law, for instance. In Newton's scheme, the second law is not the only law, of course, and we wouldn't want to insist on that. But there is still a sense in which it stands out as the most general and powerful of the famous three laws. The other two laws are constraints that give us information about the events across a hypersurface of M whereas Newton's Second Law is an algorithm that will recover, in principle, all of the events on M. Same goes with the example of Friedman equations above. To generate strength, the laws seize on a "time" parameter.

These considerations motivate the following proposal as another important difference between time and space. Time is that direction on the manifold of events in which we can tell the strongest or most informative stories. To phrase it slightly more carefully, suppose we have a manifold M^d of events endowed with metric g. The points of M^d are interpreted as elementary events. In none of this do we presuppose a prior picking out of the temporal directions. Then the very general proposal is:

> A *temporal direction* at a point p on $\langle M^d, g \rangle$ is that direction $(n, -n)$, where $(n, -n)$ is an unordered pair of nowhere vanishing vectors, in which our best theory tells the strongest, i.e., most informative, "story."

At this point, we need the metric g to make sense of there being "directions" on M. Apart from that, we don't assume that the "timelike" directions of g, if any, are themselves temporal directions. That's something we hope emerges from the analysis, not something put in. Nor do we insist on a particular direction over its opposite. Like the line element field mentioned in the previous chapter, it is a local "arrowless" time. Since local, we could in principle have arrows of strength pointing willy-nilly. If the world permits, we may also have a global temporal parameter too.

Here is a useful way to understand the proposal. If an algorithm, given some input, could get back everything that happens, that would be best. A deterministic theory is maximally strong in that respect. Another way of thinking about the proposal, therefore, is that time is that direction of M in which the best system can get the most determinism. Call a history H a map from $\mathbb{R}$ to tuples of the fundamental properties, where for any t in $\mathbb{R}$, $H(t)$ gives the state of the fundamental properties at t. Then a theory is deterministic iff for any pair of histories, H_1, H_2, that satisfy the laws of physics, if $H_1(t) = H_2(t)$ at one time t, then $H_1(t) = H_2(t)$ for all t (Earman, 2006). Note that this definition presupposes a time versus space split; in fact, it presupposes that we have a global time function. Assuming the world behaves, we can easily turn this around and define time in terms of determinism: a smooth map $t \colon M \to \mathbb{R}$ is a global temporal function if it's true that for histories that satisfy the laws of physics, if any pair agree at one value of t then they agree for all values of t.[2]

Let me emphasize that my proposal in *no way* commits me to asserting *determinism true* of our world. The above is simply an illustration. Indeterministic theories can be strong too, of course. Most stochastic processes in physics are Markovian. A process $q(t)$ is Markovian if $q(t_1)$ alone predicts $q(t_1 + dt)$, where no previous values $q(t_n)$, $t_n < t_1$, are needed. Learning about the future based on knowing the present and not also the past is certainly a kind of informative strength. Hence, a natural probabilistic analogue of my claim is that time is that dimension in which the laws are Markovian. Being deterministic or Markovian are merely *marks* of strength. Confining attention to the marks of strength and not strength itself would be a mistake. The degree to which a theory is informative is determined by how much of the actual world it manages to imply, not (in the general theory, at least) by formal characteristics.

That said, one can't help but notice that of the theories we've considered fundamental all seem *nearly deterministic* in what we call the temporal direction and not the spatial ones. Newtonian mechanics is indeterministic unless one suitably restricts the force functions or "enables" the theory with special boundary conditions (see Earman, 1986, 2006). Doing so, one can prove finite in time existence and uniqueness of solutions, i.e., determinism. Alternatively, Newtonian particle mechanics may be

---

[2] Because "bad" choices of foliation can ruin determinism, some care is needed; see Earman (1986) for the necessary modifications. Also, it's possible to distinguish a direction as temporal even where no global time function is definable. For example, there exist spacetimes with closed timelike curves (and hence no global time function) that nonetheless are perfectly deterministic (Friedman, 2004). And there exist other spacetimes that cannot be foliated via everywhere spacelike hypersurfaces, e.g., Gödel spacetime, yet where one may locally wish to distinguish directions of strength.

"almost deterministic" in the sense that for almost all initial conditions there exist unique solutions (Saari (1977) showed this for global existence for $n = 4$; it's an open question for $n > 4$). In any case, deterministic, almost deterministic or indeterministic, the theory is remarkably powerful and it's not comparably powerful in the directions we call spatial. The same goes for quantum mechanics' Schrödinger equation. Here determinism is on safer ground, so long as the quantum Hamiltonian is essentially self-adjoint (see Earman, 2006).[3] In relativity general results are hard to come by.[4] What one has is a smattering of results related to determinism. We'll return to this topic in Chapter 8.

Dramatic alterations to the proposal are possible. We'll consider one in Section 7.4. Right now, however, let's see whether the directions picked out by strength deserve the denomination "time."

7.3 Binding Time

Why believe that the direction of strength picks out the temporal directions? One can dub any set of directions one wants "temporal." To be meaningful this denomination must have something to do with time as we ordinarily understand it in physical theories. Earlier we saw that time was fragmented, that is, that there were many different detachable features associated with physical time. One way for a set of directions to warrant the title "time" is to show that these features are connected with this new feature. In the next chapter we'll show this formally to be the case in a certain regime. Here we'll make do with an informal and therefore looser set of connections.

7.3.1 One-dimensionality

We saw that there are many pressures for time to be one-dimensional in a physical theory. Systematization helps explain this tendency. The best system craves the optimal trade-off between informativeness and simplicity. Suppose the fundamental theory is deterministic, that the system is maximally informative. Then what incentive is there for the system to achieve determinism again in another set of directions? None. The theory is already maximally strong, i.e., all events follow. It's hard to imagine that one could even find two such directions, especially in a complex world like ours. And it is even harder to imagine this happening without a compensating loss of simplicity. Informativeness in one set of directions is enough. Weakening the example from deterministic to indeterministic evolution doesn't alter this logic. If

---

[3] If so, one can use its unique self-adjoint extension as the generator of a one-parameter group of translation operators. Interpreting the translation operators as what we call time translation operators yields complete determinism for the quantum state for all time. (Interpreting the translation operators as what we call spatial would yield complete determinism for the quantum state for all of space, but at the price of making the theory false.)

[4] Einstein's equations are nonlinear partial differential equations, which implies that there will be differences between the local and global questions of determinism. Furthermore, there exist deep and complicated issues arising from "gauge freedom" of Einstein's equations.

strength is achieved in one direction, more algorithms in other directions will only add complexity with no compensating gain in strength.

Interestingly, in relativity there is a sense in which we have multiple time dimensions and directions. Many of the global time functions we defined are non-unique, as we saw in Chapter 3. Imagine one is the time function Adrian above uses and the other is the time function appropriate for Zoe. Because the physically genuine properties of a relativistic world look the same from either perspective, the two time functions are related by a symmetry. The symmetry increases simplicity and eliminates the need for more than one time. The laws can be written so that "evolution", i.e., our algorithm, uses a set of directions, not a particular one. But this set, understood via the metric, is one-dimensional.

There is no guarantee that the best theory will contain algorithms that permit such a nice time. Dozens of examples spring to mind where this would not be so. If the universe is compact, the best theory might give data on the surface of the four-dimensional manifold and have laws that give solutions for the interior. Or maybe some patches of the universe require algorithms that can't be sewn together to provide a unified time, or whose times are at right angles to one another. Nature must be kind. So far, it has been.

7.3.2 Closed timelike curves

> I see a lot of men end up as their own uncles. Super-easy to avoid, totally dumb move. See it all the time. No need to go into details, but it obviously involves a time machine and you know what with you know who.
>
> Charles Yu, *How to Live Safely in a Science Fictional Universe*, p. 46

Time travel can be very costly. The above Charles Yu quote hints at its possible personal and social costs. Yet closed timelike curves also can harm theories by inhibiting strength and simplicity. Consider a simple example. Identify two regions that are timelike related to one another, R_1 and R_2. If you like, imagine these regions are the "mouths" of a wormhole extending "behind" the spacetime. Suppose a particle enters region R_2, as in Fig. 7.1. Then because R_2 *is* R_1, it will emerge from R_1 traveling at the same velocity it had as it entered R_2.

First consider the blow to strength. Made famous by Thorne (1994), the trouble with such spacetimes is that the local initial state S before the first wormhole mouth isn't sufficient to determine what will happen in the time travel region between mouths. Suppose a particle at S has a position and velocity that will make it enter the time travel region from the left, "miss" region R_1, but enter region R_2. Depending on the size of the R's and the velocities and positions, the particle will then go back and forth in time in the time travel region until eventually emerging from R_1 heading rightward. For example, in the above picture the particle enters R_2, goes back in time three times, and then emerges from R_1 to the right. That's one solution. There exist an

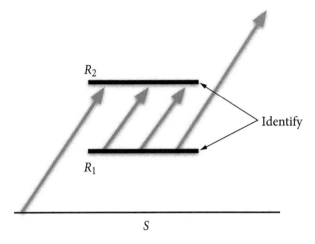

Figure 7.1 Wormhole: no self-interaction

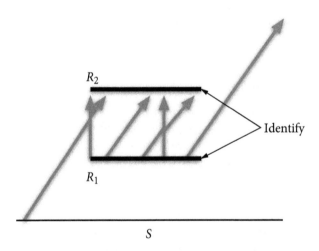

Figure 7.2 Wormhole: with self-interaction

infinite number of others too, however, each fully consistent with the laws. Consider the solution diagrammed in Fig. 7.2. Here, when the particle is midway to R_2 it gets hit by a later version of itself. The impact brings the particle to rest. It enters R_2 and emerges from R_1; then it gets knocked to the right by an earlier version of itself. It goes through R_2/R_1 again and exits from R_1 to the right, just as in the first picture. The inputs and outputs are the same as in the first scenario, but what happens in the time travel region is dramatically different. Clearly, given suitable initial conditions, an infinite number of such solutions are possible.

The problem with regard to strength is that the data on the local surface S isn't sufficient to determine what will happen in the time travel region. The local data is consistent with many different global solutions. Arntzenius and Maudlin (2002, 2009) conjecture that this problem is likely to be generic in spacetimes with closed timelike curves in it. As they note, "What happens in the time travel region is constrained but not determined by what happens on S, and the dynamics does not even supply any probabilities for the various possibilities." The lack of probabilities for events in the time travel region is especially startling. In such a case, even knowing *everything* on S, we get virtually *no information* about what is happening at a later region of spacetime. If one thinks of strength loosely as the number of possibilities eliminated, then these regions are maximal blows to strength as an infinity of solutions remain compatible with the initial data. When searching for strength, looking for the directions in which worldlines don't close is a good start.

A comparable claim for simplicity can perhaps also be defended. At an intuitive level, the problem with time travel is that it seems that there must be a kind of pre-established harmony in the initial conditions to ensure that time travelers execute consistent behaviors. If I go into a time machine and try to kill myself as an infant, then logic dictates that it can't be done. I can't exist and not-exist at the same event on M. Suppose I try. We know it won't work, for if an older version of me is around to kill, then a younger version wasn't killed. Something must happen to prevent me from executing my plan: a last minute reconfiguration of neurons, police interference, or a few well-placed banana peels. The time travel literature thus talks about the existence of closed timelike curves as imposing "consistency constraints" on the physics. These constraints are restrictions on the boundary conditions that prohibit impossible scenarios like killing my earlier self. Sometimes consistency constraints can be fairly simple, e.g., Maxwell's source equations. However, "the consistency constraints imposed by the existence of closed timelike curves (i) are not local, (ii) are dependent on the global structure of space-time, (iii) depend on the location of the spacelike surface in question in a given space-time, and (iv) appear not to be simply statable" (Arntzenius and Maudlin, 2002, p. 191). Consequences (i), (ii), and (iii) are problems for any theory prizing strength in application, and (iv) is obviously a threat to simplicity.

Given the dual threats to simplicity and strength posed by closed timelike curves, it is no surprise that any systemization looking to maximize these virtues avoids closed timelike curves. How this avoidance plays out depends on how radically one interprets the present theory (see Section 7.4). Beginning with a manifold of events suffering a mobility asymmetry, the avoidance may simply reflect exploiting this asymmetry when writing the laws. Or less conservatively, the avoidance may come into play when drawing the very threads of worldline identity. Choosing worldlines that do not "bend backward" might be a boon to a system.

7.3.3 The direction of time

Heat, radiation, and much more are asymmetrically behaved along one set of directions of spacetime. The cause of these temporal arrows is much debated, but most agree that many or all of these temporally asymmetric phenomena do not follow directly from what we normally consider the fundamental laws of nature. Instead they are thought to arise from a combination of statistical features and temporally asymmetric boundary conditions.

However, arguably a best system would want to recover, for the sake of strength, thermodynamic (etc.) generalizations. Thermodynamics, for example, conveys a staggeringly large amount of significant information. Getting this information to fall out of the best system will mean adding time asymmetric laws to the best package. So the laws, properly conceived, will be time asymmetric after all. On a Boltzmannian understanding of thermodynamics, what we need to add to the fundamental dynamical laws are a low entropy initial state and a uniform probability distribution over this state. The first is a restriction on boundary conditions such that the "initial" entropy of the universe is extremely low compared to now. The second is a probability measure over the space of possible microstates at a time, guaranteeing that most microstates execute thermodynamic behavior. Much in this story is controversial (see Albert, 2000; Callender, 2011c and references therein), but I suspect that all ways of recovering thermodynamics will distinguish time, and moreover, that because we deem thermodynamic generalizations projectable, there will be great pressure to understand this distinction as *nomic* in nature.

If this is right, the idea represents another way in which the laws distinguish time from space. Both special posits, the probability measure and the low entropy constraint, are applied across spacelike hypersurfaces, not timelike hypersurfaces. With regard to the probabilistic posit, we certainly don't assume that outcomes in the spatial directions are normalizable (sum to unity). Nor do we posit a comparable (say) "East Hypothesis" (one across a ++− surface). These putative laws are usually discussed in a context where one stresses the asymmetry *within time* that they produce, but of course these laws also distinguish time from space too. Once again, this aspect of time is motivated by strength. It is for the sake of strength in the macrophysics that the best system makes adjustments.[5]

7.3.4 Natural kind asymmetry

On Lewis (1994)'s version of the systems approach, what the laws systematize are the fundamental properties of physics. We're not told what they are, but Lewis insists that these properties be intrinsic, non-modal, monadic, and perfectly natural. On other ways of understanding the theory, what the system does is find the best package of

---

[5] Interestingly, in this case we have an asymmetry of strength even within our asymmetry. The "new" laws get extra strength in the timelike versus spacelike directions, and within the timelike, they get more retrodictive strength than predictive strength.

laws and natural kinds together in some way (Cohen and Callender, 2009; Callender and Cohen, 2010; Loewer, 2007b). The natural properties are not a given on this second view. Either way, the natural kinds with their time/space split are essentially connected with what the laws are according to a systems perspective. The laws—and the measures of strength, simplicity, and balance between the two—are written in a special vocabulary (whether chosen by us or by the world) and the predicates of that vocabulary are the natural kinds of that world. This metaphysical connection in systems theory enshrines as metaphysical the epistemic point that we appear to choose the kinds we do in part because they are the ones with which creatures like us can make good explanations and predictions.

Arguably the laws are also connected to the persistence conditions of objects. Which temporal parts can we string together as persisting objects and which can we not? Sider (2001) expresses a common answer that traces its origins to Reichenbach and Carnap, namely, that "[A] sequence of temporal parts counts as a continuant only if that sequence falls under a causal law" (p. 227). The laws are relative to different candidate worldline "stitchings," but equally, the stitchings are relative to the laws. So if the laws make an essential space/time split, the threads of identity that compose the worldlines also will.

7.4 Metaphysical Variations

The core idea that the best system picks out time is ambiguous between more and less radical theses. To see what's at stake, ask: what does the system *systematize*? Does it, for instance, systematize the events e on a given Lorentzian manifold $\langle M, g \rangle$ and arrive at laws L that privilege some time (Fig. 7.3)? Or does it, more radically, systematize just

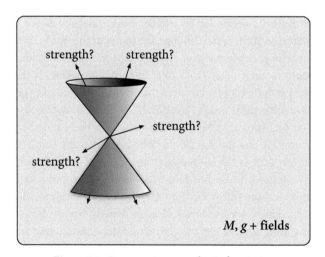

Figure 7.3 Conservative metaphysical variation

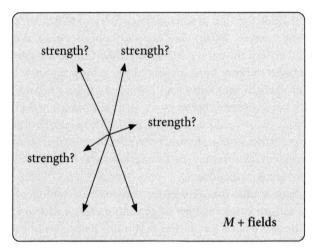

Figure 7.4 Radical metaphysical variation

the events e on M and arrive at the spacetime metric g and laws L together as the best system (Fig. 7.4)? One can dig further too: what are the events e? Are they characterized in some distinguished vocabulary (Lewis' "perfectly natural" fundamental properties), one already equipped with a time–space split, or does the natural kind structure come with the package too? As I mentioned, some philosophers have argued that the last option, that the natural kinds come with the laws as part of a joint package, is the most attractive version of the best system. The first question about the spacetime geometry $\langle M, g \rangle$, however, more directly connects up with our concerns here, so we'll focus on that.

If the first more conservative version of the thesis is right, then "time" is a kind of honorific applied to the directions on which the best system settles to make its principal algorithms. Its origin lay in whatever asymmetries in the distribution of events makes that algorithm best. Although the tamer of the two versions, this picture doesn't lack surprising consequences. For instance, if the best system doesn't have any such algorithms, then there is *no time* in such a universe. If the best system's algorithms distinguish directions that aren't relativistically timelike (i.e., $g(v, v) > 0$), then the timelike directions aren't "temporal." If the best system uses a multitude of algorithms all pointing in different directions, then there may be a *multitude* of times. Furthermore, if one follows Lewis—one needn't but I think one should—in thinking that the system is the source of all the physical modality in a theory, then this conservative approach also has ramifications for other philosophical topics, in particular, chance and causation. For Lewis, chance and causation, like laws, depend on the best system: chances are theoretical terms added to the system to gain strength and the system's laws ground the counterfactual dependence that underlies causal dependence. Both distinguish time. For instance, the chances are normalized at times and not places and the causal asymmetry is typically understood as running along the timelike direction. Although the label "time" might be viewed as a kind of honorific, still an awful lot of metaphysics can hang on it.

If the more radical version of the thesis is right, then the choice of metric geometry hangs on systematizing too. That would mean that the *metrical difference* between the timelike and spacelike—the centerpiece of all our physical theories—also depends on the system. To be clear, this difference will remain objective because the laws are objective on a best system theory (see Cohen and Callender, 2009, p. 30 for discussion of this subtle but important point). Nonetheless, the difference is not "out there" prior to systematizing, and this consequence may be surprising.

In many ways this radical perspective is the most natural development of the theory. In relativity spacetime is dynamic, dependent on the matter–energy distribution. In this context the idea of systematizing the material events of the world while they sit on an independent given background seems wildly inappropriate. It's hard to see how the best system could even come up with Einstein's field equation if so restricted. Even in theories with non-dynamical spacetimes we seem to have a problem. We move to Galilean spacetime from Newtonian because the dynamics doesn't require absolute rest; similarly we move from Galilean to Minkowskian because the dynamics doesn't require absolute simultaneity. We want the dynamical symmetries to match the spatiotemporal symmetries in a theory. Clearly they come together as a package, so it's welcome if the best system respects this practice. If so, then the best system finds the best package of geometry and laws together. Similar reasoning suggests that the Cohen and Callender and Loewer modifications are welcome too and that consequently the natural kind division comes with the system too. "Center of the universe" is a natural kind of Aristotelian physics and "absolute rest" of Newtonian physics. It's odd to force one of these to be systematized without guidance from the rest of the system, in particular, the laws and geometry. If these moves are correct, then the best system systematizes "at once" the kinds, laws, and geometry.[6]

If so, then the metrical distinction between time and space and everything that hangs on it—for instance, the lightcone structure in the tangent space of any point— all fall on the "system" rather than "systematized" side of things. In particular, the lightcone structure that plays a crucial role in our understanding of time *emerges* from the laws that best systematize the events on M.

The physicist Bob Geroch, when discussing the speed of light (Geroch, 2011) paints a picture that nicely expresses the idea. My discussion requires a few technical concepts, but readers without mathematical background should be able to see the point. Suppose

---

[6] Incidentally, this way of looking at things perhaps aids Huggett (2006) against Belot (2011). Huggett offers a best system development of spacetime relationism. Buttressed by some historical examples of disputes over simplicity in geometry, Belot objects that these diverging attitudes suggest that there isn't a simplicity metric for geometry that will serve the systems theory. (He also doesn't like the idea that the geometry of the world could be species relative, but this is more a question of taste than an official objection.) That may be right, but if the view considered above is correct, he is looking in the wrong place for a simplicity metric. An enlightened systems relationism shouldn't look solely at geometrical dimensions of evaluation, but rather at the package of kinds, laws, and geometry that is simplest.

that all there is to the world is a four-dimensional manifold of events, M, and one or more matter–energy fields on M. No metric geometry is placed on M. We're going to get the metric, lightcone structure, etc. from the fields. These fields obey laws. In fact, we'll assume that they obey at every point a first-order quasi-linear system of partial differential equations:

$$k^{Aa}_{\ \ \alpha}(\nabla_a \phi^\alpha) = j^A \qquad (7.1)$$

where ϕ^α represents the fields and the index α runs through an abstract field space and k and j are smooth functions of the field and point of the underlying M. The limitation to such equations is not overly restrictive. In fact, as Geroch notes, the vast majority of physical systems in our world can be described by equations of this form, including Maxwellian fields, Klein–Gordon fields, and fields described by Einstein's equations. The restriction is neither idiosyncratic nor irrelevant. So we have a simple world, one with fields evolving according to (7.1) on a manifold M.

An important feature of systems of equations describing such fields is that they admit a *hyperbolization*. That means that a special type of tensor $h_{A\beta}$ can be defined.[7] Readers with the expertise will wish to consult Geroch (2011) for more details about this tensor; readers who are not, and I will cater to you, can suffice with its implications. One implication is that one can use this tensor $h_{A\beta}$ in concert with Eq. (7.1)'s $k^{Aa}_{\ \ \alpha}$ to define the notion of the *causal cone* C of a system of partial differential equations. In particular, for a point $p \in M$ and field value φ^α at that point, the causal cone is the set of all tangent vectors ξ^a at p, such that $\xi^a n_a > 0$ for every n_a for which $n_a h_{A\beta} k^{Aa}_{\ \ \alpha}$ is positive-definite. C is a nonempty, open, and convex cone of tangent vectors at the point p, so the appellation "cone" is well-deserved. What about "causal"? That also is earned, as Geroch shows that these hyperbolized systems admit well-posed initial value problems. We'll discuss such problems extensively in the next chapter, but for the moment observe that a special type of determinism follows. The field values at points on M connected to p via a curve (that always has its tangent vector in the causal cone of each of its points) *depend* on the field value ϕ^α at p. Hence, given a field evolving according to Eq. (7.1) on M, there will be cones of dependency amongst the field values on M.

For Geroch, the significance of this formulation is that the causal cones can in principle be wider than the null cones provided by relativity, and also, that different fields could have causal cones of different widths. For us, however, the significance is that Eq. (7.1) defines its own set of causal cones, and moreover, that in our world all the causal cones coincide with one another, whether defined from Einstein's field equations, Maxwell's equations, Klein–Gordon equation, or others.

Imagine what life is like in such a world (easy, because we live in one). Creatures will notice the causal cones. Their meter sticks, flashlights, and clocks are all aspects of

---

[7] A *hyperbolization* of a system of equations is a special type of tensor $h_{A\beta}$ that is symmetric in the indices α and β when combined with k and is such that for some covector n_a in M, the symmetric tensor $n_a h_{A\beta} k^{Aa}_{\ \ \alpha}$ is positive-definite.

the fields—as they are themselves—and all subject to the laws described above. Signals from flashlights can in principle reach some events but not others. Meter sticks will look a certain way to some aspects of the field and differently to other aspects. And so on. Due to the cones' agreement and robustness, I claim, these creatures will have reason to attribute these cones to the geometry of spacetime itself, not merely the fields. Fields behaving this way provide all the reason one has to posit relativity, etc. So these creatures will attribute to the world *a lightcone geometry that isn't fundamentally there* (or if you prefer, it's "there" in the fields, but the spacetime doesn't come equipped with it). All that exists are the fields and manifold. The laws governing the fields define causal cones (specifically, through $k^{Aa}{}_{\alpha}$) that trick inhabitants into believing that nature comes equipped with a lightcone geometry.[8]

Notice that this picture is a fleshed out version of the more radical metaphysical alternative picture described above. Here informative strength doesn't merely help justify a particular set of independently existing metrically defined directions as temporal. Rather, it plus the laws help define the metrical geometry that picks out a set of directions as temporal. Suppose we have a world composed of only a manifold M and a set of fields ϕ^{α} evolving according to Eq. (7.1). We look for informative strength. The hyperbolization gives us extreme strength—determinism—in one set of directions and not others and it also provides the basis of the metrical difference between time and space. Here in $k^{Aa}{}_{\alpha}$ we find the principal difference between time and space. And that difference arises thanks to the choice of kinds and form of the laws of nature.

7.5 Questions and Connections

Let me briefly draw some connections between the present idea and more familiar ones.

Objective—really? Is the difference between time and space objective according to this theory? Doesn't the reliance on our standards for strength make the theory anthropocentric?

Aristotle asked a similar question about his own theory of time as a measure of motion:

> Whether if soul did not exist time would exist or not, is a question that may fairly be asked; for if there cannot be someone to count there cannot be anything counted, so that evidently there cannot be number.   (*Physics* IV, part 14)

If there is no one to measure motion, it was asked, is there still time? Thomist commentators on Aristotle answered: even if no one is around to count the motion in the world, still a world with motion in it is *countable*. Anthropocentric factors enter

---

[8] Obviously this picture intersects in interesting ways with many topics in philosophy of spacetime. It has affinities with a relationist, as opposed to substantivalist, interpretation of spacetime, for the null cone structure is derived from features of the matter fields. Yet the theory is not entirely relationist, it seems, since the fields are defined on a background manifold.

into the conditions defining the temporal, but whether anything meets those criteria is perfectly objective. In baseball, given certain (admittedly, shifting) standards for what counts as a strike, a given pitch by Mark "The Bird" Fidrich in the 1970s objectively is or isn't a strike. Same here: given certain standards of strength, a given direction on M objectively is or isn't temporal.

Fair enough, the reader may think, but isn't the proposal up to its neck in relativism? Not only does it make time relative to a standard but also relative to a choice of natural kinds. Answer: the proposal is *knee deep* in relativism, but this is a virtue not a vice. If Martian scientists cared about an alternative carving of nature into kinds, then presumably they might devise a different system than we do. Let's suppose the direction of strength in that system is orthogonal to ours, so the two temporal directions don't line up. The Martians think that a direction we consider spatial is temporal. Is one of us wrong? This question is a reflection of the more basic one about whether there are objective natural kinds. Does the world "prefer" some properties? If so, then we and the Martians aren't both right. One or both of us are using illegitimate kinds; therefore, at least one of us is wrong about time. If an abundant conception of properties is right, and it's only *we* who pick out the natural kinds, then the Martians can pick out a distinct temporal direction without error. We can't decide these issues here (see Cohen and Callender, 2009; Kitcher, 2003, etc.), but for what it's worth I prefer the abundant properties option. The Martian "time" should count as time for them if that set of directions plays the time role in their vocabulary, just as ours should count as time for us if it plays that role in our vocabulary. Radical? Yes, but not unmotivated. In any case, this is an aspect of a broader philosophical problem.

Ties? Suppose no single set of directions emerges as strongest. Are there more than one time directions, or is there then no such thing as a temporal direction?

This question is essentially about what to do if we have a "tie" in the best system competition between two (or more) equally strong and simple systems—each distinguishing a time direction orthogonal to the other's. I don't believe that this is the case in a world like ours. Here the choice of time direction seems lopsided in favor of the one we pick. That said, I do believe that there is a conceivable state of affairs where this happens, and in fact, something like it once was considered as a live scientific possibility. The cosmologist Arthur Milne posited a cosmology with two times in it (Milne, 1948). In it, the ideal clock of atomic phenomena doesn't line up with the ideal clock for gravitational phenomena. One is logarithmically related to the other. Put in our terms, Milne found no good system wherein the one type of phenomena could be seen as marching to the same time as the other. In such a world we might consider there being a time for particle physics and a time for gravitational physics.

Causal theory of time. For centuries philosophers have sought to reduce temporal relations to causal relations. There seem to be affinities between the present position and the causal theory of time. How are the two related?

The current proposal is not a causal theory of time, although it offers resources for those interested in defending the theory. Let me explain after I remind readers of what the causal theory maintains. A causal theory of time cashes out temporal order in terms of physically possible causation. One event is deemed later than another just in case a physically possible chain of causal dependence links the first to the second, e.g., a possible signal by a light ray. Remarkably, beginning with Robb (1914), philosophers and physicists were able to derive the Lorentz transformations and more from such modest foundations. Today the core challenges remain the same as always, namely, explicating the core concepts of physically possible, genidentity, and causation in a satisfactory manner. Causal theorists have a hard time explicating the modality invoked in physically possible causation consistent with their empiricist scruples. (Actual signals do not suffice for generating a spacetime; possible ones are also required.) Extending the theory to full blown relativity is also a problem, for—as we've witnessed—relativity has solutions with closed timelike curves, no global spacelike slices, and other features that wreak havoc upon the causal theory. Due to these problems, plus the background trend toward spacetime substantivalism, enthusiasm for the causal theory has waned since the 1970s.

Nonetheless, headway can be made on both the relativistic and philosophical fronts. On relativity, proofs by Malament (1977b) and Hawking et al. (1976) show that for the set of spacetimes that are past–future distinguishing, the causal structure determines the topology of spacetime and the metric up to a conformal factor. Intuitively, that means that the causal structure of spacetime contains all the information we care about except for the scale or "size" of spacetime. If we let $\ll$ be the causal connectibility relation, the same one previously discussed in causal set theory (Chapter 5), then we nearly can get $\langle M, g \rangle$ from $\langle M, \ll \rangle$ in these spacetimes. If the restriction to past–future distinguishing spacetimes is justified (and maybe it is on "systems" grounds), then perhaps in this limited set of solutions relativity doesn't cause as much trouble as originally thought. And if we can get $\ll$ from the matter–energy fields, as suggested as a possibility above, then so much the better for the causal theory. On the philosophical challenges, we can see the best system as possibly coming to the rescue. The best system is supposed to be the source of all physical modality. What is a possible signal is decided as the result of trading off theoretical virtues in developing the best theory of the actual world. An event could be defined as later than another just in case the best system states that a physically possible chain of causal dependence links the first to the second. The idea, in other words, would then be to obtain the modality in the $\ll$ relation from the best system, and get $\langle M, g \rangle$ from $\langle M, \ll \rangle$ and the time/space split. I'll leave it to interested readers to further explore ways the best system and the present stress on strength might aid the causal project.

Governing. Is the present theory a kind of "systematized" version of the Tooley–Skow–Maudlin idea that the laws (or causes) govern in the temporal direction.[9]

To some extent, yes. The core intuition in that view is right. However, I feel that the "systematized" version has a number of additional virtues over that view; moreover, I

---

[9] Tooley (1997); Skow (2007); Maudlin (2007).

also think I've done a lot more to make the general view (systemized or not) plausible. One, since it is in some sense a Humean theory, we have an understanding of why the laws pick out time. The difference between time and space is not found simply in a metaphysical primitive one has and the other doesn't. The difference ultimately lay in the distribution of physical properties. Two, in contrast to the Tooley–Skow–Maudlin line, we are not committed to laws of nature not operating across spacelike hypersurfaces. The system may "prefer" the time direction, but that is compatible with particular axioms of the best theory being written on spatial hypersurfaces. A direction can be privileged even if a few axioms impose constraints across spacelike hypersurfaces. Three, and most importantly, given the way systems work, we have connections between system laws and all of our other temporal features. System laws help bind together many otherwise disparate temporal features. In so doing the directions singled out by systematizing warrant being dubbed "temporal."

The last benefit stands apart from the others. Even if one rejects the rest of the picture, the connections forged in answering the problem of fragmented time may still be of use. The Humean thinks explanations and predictions and good theory are constitutive of modality. The non-Humean, by contrast, doesn't ignore good explanations, predictions, and theory. Rather than *constitutive* of modality, they are viewed as *symptomatic* of it. Here I've proposed a Humean extension of the original idea to time: the best system, in manufacturing modality, also picks out time; moreover, this process binds together otherwise fragmented aspects of time. The advocate of the governing picture, however, may also have wondered about the connections among the laws, the metric of spacetime, the mobility asymmetry, and more. Why do the laws govern in the "minus" direction, and so forth? Little in the present analysis is preventing them from using the above explanation and the fruit of the next chapter to answer these questions too.

7.6 Conclusion

Even in the "block" universe of relativity, there remain sharp and important differences drawn between the spacelike and timelike directions. These differences are important features of the physical world, especially the metric signature. What "makes" these features line up and march together? Could these aspects of time instead be fragmented? My reply is a bit indirect. First, I make a case for the idea that the laws of nature single out time as a distinguished set of directions. Time, loosely put, is the direction in which physics tells its best stories. Second, using this novel picture, I reveal informal connections amongst all our other temporal features. It begins to "glue" together the different fragments of time. In so doing, it justifies the thought that the asymmetry the system seizes upon to write its most powerful algorithms is rightly regarded as temporal. We consequently gain insight into what is special about time.

8

Looking at the World Sideways

When physics tells its story of the world, it writes on spatial pages and we flip pages in the temporal directions. The present moment contains the seeds of what happens next. Time is productive. The philosopher Tim Maudlin writes that

> the passage of time underwrites claims about one state "coming out of" or "being produced by" another, while a generic spatial . . . asymmetry would not underwrite such locutions.
>
> (Maudlin, 2007, p. 7)

This idea is found not only in common sense but also in typical presentations of our physical theories. For example, in Minkowski spacetime one can narrate what happens as a sequence of three-dimensional spatial slices successively happening (within an inertial frame); and in many other spacetimes one can tell this story via one of the cosmic time functions previously discussed. Indeed, to the extent that one can speak of the physical state in a region of spacetime "determining" another, one can often show that the state of the universe at one of these "times" produces the next. Relativity challenges many of our pre-theoretical thoughts about time, yet even this would-be destroyer of time adheres to the idea that production or determination runs along the set of temporal directions. We might think of this fact as one of the last remnants of manifest time in physics.

Is even this residue of manifest time safe from physics? The famous RPP argument (see the discussion near Fig. 2.7) against manifest time exploited the lack of a preferred foliation in Minkowski spacetime to show that the "present" is dependent upon inertial frame. Distinct inertial observers have presents "tilted" at an angle from yours. These hypersurfaces are good sources of production. Recall that one could only tilt alternative presents so far. Lorentz-boosted simultaneity planes always stay spatial. They never contain events lightlike or spacelike related to one another. Yet one can ask what would happen if we considered such three-dimensional hypersurfaces, that is, "mixed" hypersurfaces that are only partly timelike? Could one take the data on such hypersurfaces and "evolve" it into non-timelike directions? Looking at the world sideways, can we march "initial" data from "east" to "west" as well as from earlier to later? Or put yet again even more loosely: can physics tell its stories if we write on non-spatial pages and read in non-temporal directions?

Either answer—yes or no—is deeply interesting. If we can do basic physics sideways, then it seems physics stamps out the last vestige of manifest time. Or put more carefully, those seeking an arrow of production vindicated by physics will be left unsatisfied. By contrast, if we cannot do basic physics sideways, then that shows one important way in which time is special in physics. The laws would be tuned to the temporal directions in ways they aren't to the other directions. That would be relevant to the claims made in the previous chapter. That chapter made a "philosophical" argument that connected time with laws via informative strength. Time was said to be the direction in which physics tells best stories. In addition, I made a somewhat contentious case for the idea that informative strength binds together many of the otherwise detachable properties we associate with physical time. Now it is time to look under the hood and see what physics says.

The answer to our question is a qualified No. One cannot tell the stories of physics in the non-temporal directions; or rather, one can, but the stories aren't as good. What I wish to do in this chapter is make these claims precise. Remarkably, when the proposal of Chapter 7 climbs down from the lofty heights of philosophy it looks even more attractive. That is what I'll show in this chapter, namely that informative strength picks out the temporal directions and binds together many otherwise detachable temporal features. The price of rigor is applicability. We won't be able to bind as many temporal features as we did previously, but those we do will hold that much more tightly. I'll close with two "loopholes" that are perhaps just as surprising as the argument. In one case the pages upon which we write the story of the world are lightlike and in the other the pages are partly timelike. There is a sense in which one *can* narrate the story of the world "sideways," but this sense does not threaten the overall claim that time is special.[1]

8.1 Strength and Well-posed Cauchy Problems

To make the above claims, I need a precise notion of informative strength. Speaking of "stories" won't do. Many choices are available, and a lot hangs on this choice. Not every notion of strength will distinguish a time/space split. The most I can do is motivate a natural definition.

To get a sense of the possibilities, consider *transcendental determinism*. The transcendental poet Ralph Waldo Emerson wrote:

> Each particle is a microcosm, and faithfully renders the likeness of the world... Every natural form to the smallest, a leaf, a sunbeam, a moment of time, a drop, is related to the whole, partakes of the beauty of the whole.   (1836)

The idea is that each particle, droplet, etc., carries within it the entire universe. By studying a small dew drop one can learn about the whole cosmos. Take this romantic

---

[1] Readers with little technical background may find the foregoing difficult. I ask them to bear with me, as the overall logic should be plain if they trust the technical assertions. In any case, I'll try to provide impressionistic pictures of what I'm doing at various points.

idea literally. Suppose you were given the physical data describing all the fine features of a dew drop and also provided the laws of physics? Ignore computational limitations. Would knowledge of the dew drop's state tell you anything? Where your lost socks are? When the next observable supernova will happen? Could you deduce the rest of the universe?

Make this question more precise. Suppose that you live in a classical gravitational universe and know your trajectory. Can you determine the trajectory of another system? In general the answer is no. Yet unexpected riches await you if you make a few assumptions. The gravitational force extends infinitely, so you feel the other particles' pull on you. And you can of course use the laws to figure out where they may be, especially if you knew some extra information about your universe. That is the key here. If this information includes only a few facts, such as the total particle number, plus a few constraints, such as that particle speeds and accelerations are bounded, that there is no sourceless gravitational radiation, that the trajectories are smooth, and that there are no collisions, one can prove some remarkable results. In particular, one can show that in a classical universe, an accurate enough specification of any *piece of a worldline*—even a drop!—is enough to predict the *entire* worldline. Moreover, with this information and a few more assumptions one can predict *all the other worldlines* too. By looking at a small piece of the universe accurately enough, the whole universe follows. Schmidt (1997, 1998) contains these astonishing theorems and conjectures for various classical theories and even extends some to the relativistic domain. A water droplet in Concord, MA in 1731 contains (with some help) enough information to deduce your whole life!

A special case of Schmidt's theorems is that one can take a timelike trajectory as input and use his methods to deduce the rest of the worldlines, as pictured in Fig. 8.1. Here "strength" is aimed in the spatial directions, contrary to the idea that strength picks out the temporal directions. On Schmidt's understanding of prediction, strength can't possibly distinguish any directions because each part determines the whole in many classical and relativistic theories.

The sense of determinism underlying Schmidt's in principle predictions is fascinating, but it won't be the sense used here. Here I want to focus on strength for creatures like us. Determinism and predictability are different, of course. The canonical example of a chaotic classical system teaches us that the two can come apart. Nonetheless, some types of determinism are better exploited by creatures like us than others. I want to adopt an understanding of strength closer to strength *in application* à la Earman (1984) than Lewis' strength in itself.[2] As amazing as Schmidt's determinism is, it's very far from anything we can use. As Bishop (2003) notes, one of Schmidt's

---

[2] Earman writes, "Lewis suggests that strength be measured by information. But the practice of science speaks not in favor of strength per se but strength in intended applications . . . . [S]trength as measured by the amount of occurrent fact and regularly explained or systematized relative to appropriate initial/boundary conditions seems closer to actual scientific practice than strength as measured by information content per se" (1984, p. 198).

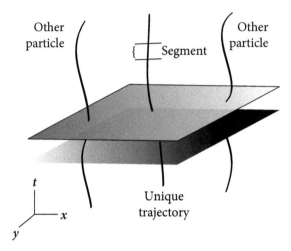

Figure 8.1 Transcendental determinism

principal theorems assumes that one's inaccuracy in the limit goes to zero. It's hardly clear that is nomologically possible for coarse creatures like ourselves. Even if it is, there is then the question whether creatures with all that information stuffed in our heads (or computers) could exist. One might also wonder whether we're entitled to all the assumptions on bounds on particle number, speeds, and so on—especially the false ones, e.g., no collisions. A Laplacian demon may need no more than a water droplet and a few assumptions to deduce the universe, but there is a reason why actual scientists ponder more than droplets.

If Schmidt is too generous about what we might predict, others may be too stingy. Manchak (2008), for instance, produces a theorem that prediction is *impossible* in relativity (see also Geroch, 1977). The notion of prediction used in the theorem is, loosely, that one predicts an event p to the future of event q if all the past causal curves intersecting p also intersect the past lightcone of q. The idea is that you, at q, have access to your past lightcone, hence you can "see" all the possible influences on p. For this to be the case, part of what you see in your past must be a Cauchy surface whose future domain of dependence includes p. Geroch and others then point out that for a large class of spacetimes, including Minkowski spacetime, prediction in this sense is simply impossible. Manchak goes further and asks, if you *really* know a future event, shouldn't you *also* know that that past surface is a Cauchy surface? If so, then his theorem proves that you can't predict anything in any relativistic spacetime.

We have a Goldilocks problem. Allow ourselves "too big" a set of what we can know, like Schmidt, and we can predict the whole from an arbitrarily tiny part; allow ourselves "too little," like Manchak, and you can't predict anything. What is the "right" amount, if such there be? In practice, we do make predictions in relativity. Numerical relativity does all the time. Scientists assume the universe has Cauchy surfaces, add reasonable data to these surfaces, and then run this forward in accord with the laws. When we have confidence that the initial data approximates the actual universe, we have a prediction. Practicing physicists help themselves to information that the epistemologically scrupulous don't allow (knowing Cauchy hypersurfaces, data across those surfaces) but not nearly as much as the epistemologically generous grant (say, the 10,104th derivative of the solution at a point on the surface). Where do we draw the line?

I have no idea. Here I can only state that I am going to *stipulate* a certain sense of strength. It is far from the only sense of strength, nor is it necessarily even the best. Readers opting for other notions are free to explore the ramifications for time. I insist only that my choice is not arbitrary. It is the most widely used sense of determinism in physics, a sense true of many putative fundamental laws, a goal of much work in physics, and furthermore, it is a nice kind of compromise between the two approaches above. With it, many surprising connections to time can be proven. How well these claims hold up with other notions of strength I dare not guess.

To begin, consider the traditional notion of determinism. In mathematical parlance, a deterministic theory is one for which a Cauchy problem has a unique solution, one where there is a one-to-one correspondence between solutions and "initial" or Cauchy data. A Cauchy problem is that of finding a solution u of some partial differential equation (see below) given certain prescribed Cauchy data. The Cauchy data is placed on an $(d-1)$-dimensional hypersurface $S \subset M^d$ that satisfies various continuity conditions. The data for a second-order equation consists of u and first derivatives of u. (For higher-order equations we'll need more than first derivatives as part of the initial conditions.) If we label as "t" the dimension into which S "evolves," then the initial data is:

$$u(x, 0)$$
$$u_t = \partial_t u(x, 0)$$

where $x = (x_1, \ldots, x_{d-1})$. See Fig. 8.2.

If given arbitrary "initial" data the solution *exists* and is *unique*, then we have a deterministic theory. When the Cauchy data is restricted to an initial time $t = 0$ surface, then we say the theory has an *initial value formulation*. If we find a solution only in some neighborhood of the initial data at $t = 0$ then we say we have a local Cauchy problem; in a global Cauchy problem we seek a solution for all x and t. (For linear equations, usually if one has local solutions for a Cauchy problem, then one also has global solutions too; but this is not true for nonlinear equations.)

If we focus on theories that are maximally strong *for us*, we can hope for something better. Hadamard (1902) introduced *un problème bien posé*. A problem involving a partial differential equation is *well-posed* if a solution of the problem exists, the solution is unique, and the solution depends continuously on the data of the problem. More precisely, a partial differential equation defined over a certain domain, possibly supplemented by boundary conditions, is well-posed if

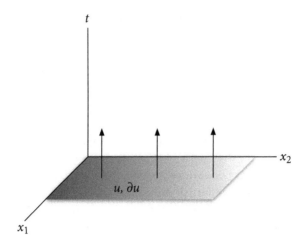

Figure 8.2 Cauchy problem

1. There is a solution u for any choice of the data, D, where D belongs to an admissible set X.
2. The solution u is uniquely determined within some set Y by the data D.
3. The solution u depends "continuously" on the data D, according to some suitable topology.

To make sense, condition 3 needs a precise characterization of the topology involved. If induced by a norm $\|\cdot\|$, condition 3 implies that there is a nondecreasing nonnegative function $F(x)$ such that $\|u(x)\| \leq F(x)\|u(0)\|$, $x > 0$, for any solution $u(x)$.

A *well-posed Cauchy problem* differs from a Cauchy problem by the addition of the third condition. The reasons why we want existence and uniqueness are clear: we want our theory to tell us something (condition 1), and even better, precisely what will happen (condition 2). If we understand strength in terms of strength *for us*, it is equally clear why we would want continuous dependence of solution on data. If u depends continuously on the data, that means the solution $u(x, t)$ is equal to some continuous function G of the initial data, i.e., $u(t) = G(u(0), u_t(0))$. Suppose one needs to find the position of some asteroid at a particular time t within accuracy $\varepsilon > 0$. In practice, we measure the initial data with a certain error, which we can often make indefinitely smaller by investing enough energy and time. Both are in demand, however, so we narrow down u and u_t only so far. Thanks to the continuity of G, we're still guaranteed that there are positive numbers p_1 and p_2 such that if $|u(0)\quad u'(0)| < p_1$ and $|u(0)\quad u'(0)| < p_2$, then $|u(t)\quad u'(t)| < \varepsilon$. Hence the hope: if we can just make our initial errors small enough, we'll be able to get a solution good enough for our goals.[3]

---

[3] Actually, the "strength scale" (if such there be) includes many categories between Hadamard well-posed Cauchy problems and mere unique existence. There are many different notions of stability available,

If the solution depends on the data in a discontinuous way, then that will mean that small errors in data can create large deviations in solution. Rounding off numbers, noise from perturbations, and so on, may imply very different solutions. The partial differential equation's predictive value will plummet if it is not well-posed. If we have to specify the initial data with infinite precision to get anything out of the theory, then it is worthless to us. That Laplace's super-intelligence might not care about the difference between uniquely solvable problems and uniquely solvable problems whose solutions depend continuously on the data doesn't mean that we don't care. We plainly do.[4]

Hadamard claimed that ill-posed problems are *dépourvu de signification physique*. I don't want to endorse this claim. Probably more problems in physics textbooks are ill-posed than strictly well-posed, and a lot of these have physical significance. However, it is clear that a well-posed problem is an excellent ideal. This is the sort of problem physicists routinely try to solve, computer scientists try to simulate, and the kind of setup we indeed have for many of the putatively fundamental theories of our world.

Cauchy data is one type of boundary condition, but there are others too. With two variables, x and y, the three kinds of boundary condition are:

- Dirichlet: $u(x, y) = f(x, y)$
- Neumann: $\frac{du}{dy}(x, y_0) = g(x, y)$
- Cauchy/Mixed: $u(x, y_0) = \phi(x)$ and $du/dy(x, y_0) = \psi(x)$.

Dirichlet problems begin with the solution u on the boundary of the domain. Neumann problems instead give one the derivative of u (in the normal direction) on the boundary. Cauchy or Mixed problems start with a conjunction or linear combination of both. The third case is usually called "Mixed" if the data is on what is intended to be a timelike hypersurface and "Cauchy" if the surface is intended to be spacelike. But since we're trying to determine the difference between space and time it would be question begging to distinguish the two. We'll stick with "Cauchy." There do exist "spatial" Cauchy problems, such as the "sideways heat equation," which is the two-dimensional heat equation (Eq. (6.1)) with x and t exchanged, wherein one determines the temperature inside a rod given the temperature and heat flux at one end. What is interesting (as we will see) is that such problems are ill-posed. In any case, we can individuate problems by the types of information on the boundary, and

---

and well-posedness hangs on the sense used. For example, the initial value problem for the backward heat equation is well-known to have solutions that do not continuously depend on the initial data; but that is based on instability in the sense of Lyapunov, whereas in the sense of Hölder continuous dependence holds. Also, there is the phenomenon of "almost uniqueness" where some important equations are known to have exactly two solutions, such as the Monge–Ampere equation, and these systems of course seem very strong even if they are not officially well-posed. See Ames and Straughan (1997) for more on these notions and others.

[4] We might also care about very sensitive but continuous dependence of solution on data, as in chaotic systems. How much of what follows could be repeated with this in mind is something I won't pursue.

this information by itself does not smuggle in any particular time/space split. It's perfectly coherent to have the Cauchy data march "sideways" in an "eastern" direction.[5]

We have a mathematically precise notion of "strength" that is independent of any particular direction on the manifold.

8.2 The Worlds

A notion of strength in hand, we now must delimit the possible worlds that we wish to consider. To do anything precise, we need to undertsand our laws as equations. By selecting some general form, we in effect pick out some possible worlds. The possible worlds are the solutions of these equations. Clearly, the more general we are here the better. Fortunately a tremendous amount of physics is described via second-order linear partial differential equations.[6] These look like:

$$\left[\sum\sum a_{ij}\frac{\partial}{\partial x_i}\frac{\partial}{\partial x_j} + \sum b_i\frac{\partial}{\partial x_i} + c\right]u = 0 \tag{8.1}$$

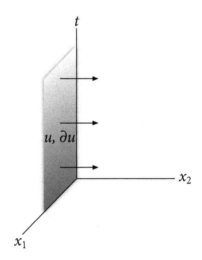

Figure 8.3 Sideways Cauchy problem

[5] McCabe (2005) complains that picking Cauchy and not Dirichlet or Neuman problems stacks the deck in favor of Lorentzian signature. McCabe is right that we cannot run the same argument with those types of problems. Lacking a grand theory of strength, I also don't wish to claim that Cauchy problems are in some sense overall stronger than Dirichilet or Neuman problems. I can only reply that the choice of Cauchy problems is hardly idiosyncratic, especially since most of our fundamental equations are hyperbolic (see below).

[6] The only philosopher who has discussed the structure of partial differential equations and the nature of time, so far as I am aware, is Lautman (1946). The idea that the equations themselves help distinguish a time variable is in the background of this work—as I read it in my almost non-existent French—but Lautman's main concern is that partial differential equations determine a priori the two roles time plays, that of coordinate and that of parameter.

Here a_{ij} is a matrix, b_i a vector, and c a scalar; all three are differentiable functions of the coordinates. a_{ij} is sometimes called a "coefficient matrix" and it can be assumed to be symmetric without loss of generality.[7] Scores of the most important equations in physics fit the form of Eq. (8.1), including the wave equation, the heat equation, the Schrödinger equation, the Klein–Gordon equation, the Euler equation, the Poisson equation, and parts of the Dirac equation.

A few comments about generality. First, many equations not in the form Eq. (8.1) can be put in that form. For instance, it is well-known that one can take higher-order equations and reduce them to lower-order equations through various means. One can take a third-order equation and put it in second-order form by introducing new auxiliary fields. Thus third-order or higher equations can be reduced, without loss, to equations of form Eq. (8.1).[8]

Second, nonlinear equations can be approximated by linear ones if the neighborhood chosen is small enough. What holds for Eq. (8.1) will therefore have some ramifications for nonlinear equations.[9]

Third, one can regard many of the equations that are not of the form (8.1) as effective equations that are "truncations" of the more fundamental equations that are of form (8.1). Because not fundamental, we wouldn't want these equations to dictate the basic time/space split. Geroch (1996) writes, "A case could be made that, at least on a fundamental level, all the 'partial differential equations of physics' are hyperbolic—that, e.g., elliptic and parabolic systems arise in all cases as mere approximations of hyperbolic systems." While I'm unsure that all fundamental equations are hyperbolic, the second point is sometimes demonstrably correct: many equations not of form (8.1) can be derived from ones that are by taking limits or making simplifying assumptions. The elliptic Poisson equation, for instance, is a truncation of the linear hyperbolic Maxwell equations. The Schrödinger equation is the non-relativistic limit of relativistic field theories. And Newtonian mechanics is the classical limit of the Schrödinger equation. One can continue along these lines indefinitely, e.g., the parabolic Navier–Stokes equation holds only in the macroscopic limit, the elliptic Euler equation holds only in the incompressible limit, and so on. If we assume these

---

[7] We can assume the matrix is symmetric, i.e., $a_{ij} = a_{ji}$, because $\frac{\partial^2 u}{\partial x_i \partial x_j} = \frac{\partial^2 u}{\partial x_j \partial x_i}$.

[8] The problem runs the other way. In general, a partial differential equation of second order is equivalent to a system of first-order equations (of special form), but the converse is not true. And in the special cases where it is true, the equivalence usually obtains by restricting the set of solutions by selecting special initial conditions. Aspeirsson's theorem, needed below, works on second-order equations, so for those first-order equations not transformable into second-order equations, we are left without a crucial premise.

[9] If we know the linear equation for that neighborhood is ill-posed, then adding nonlinear terms to (8.1) will not make the equation well-posed over a larger neighborhood. At best, additional nonlinear terms will make well-posed problems ill-posed, not vice versa. (Note also that there do exist extensions of Aspeirsson's theorem (see below) to some types of nonlinear systems.) The only question, then, is whether nonlinear terms added to (8.1) will make the well-posed problems ill-posed. No doubt sometimes this will be the case; however, it is by no means the rule. For quasi-linear second-order hyperbolic equations, for instance, one has the extension of the theorem proving well-posedness for the linear case. Leray (1953) proves that these equations have a well-posed Cauchy problem when the Cauchy data is placed on a spacelike hypersurface.

theories inherit the space/time split from their more fundamental parents, they aren't problematic.

Any choice of worlds will be restrictive. My hope is that the current choice doesn't strike the reader as too restrictive to be interesting.

8.3 Proposal

We have our worlds and our technical notion of informative strength. Now we can state the more rigorous counterpart of last chapter's proposal:

> For worlds described by Eq. (8.1), a *temporal direction* at a point p on $\langle M^d, g \rangle$ is that direction $(n, -n)$ in which our laws allow a well-posed Cauchy problem.

Here, $(n, -n)$ is an unordered pair of nowhere vanishing vectors on manifold M, and by "allow" I mean that a temporal direction at point p is that direction normal to the $(d - 1)$-dimensional hypersurface intersecting p upon which Cauchy data can be prescribed to obtain a well-posed Cauchy problem. For this claim to be plausible the temporal directions had better mesh well with the directions physics normally picks out as temporal, and ideally, we could connect these directions to other features we normally attribute to time. The idea behind the claim is that there is a sense in which well-posed Cauchy problems pick out temporal directions. Put very generally, a partial differential equation determines various invariant structures, e.g., a field of Monge cones, and these structures determine which hypersurfaces one can put data on to obtain well-posed Cauchy problems. What I suggest is turning this fact around and letting it implicitly define the timelike direction.

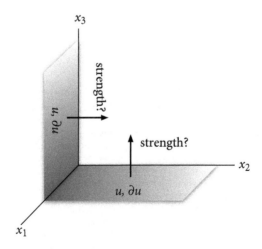

Figure 8.4 Which direction?

Imagine we have a manifold scattered with events. The events are compactly describable according to an equation of form (8.1). But we're blind as to which axis is the temporal axis, i.e., we're not told how to carve this into spacelike hypersurfaces evolving in the timelike directions. Instead we're told to seek informativeness, and in particular, a well-posed Cauchy problem. Turn $\langle M, g \rangle$ this way and that, searching for a direction in which (8.1) is well-posed (Fig. 8.4). If this procedure singles out a set of directions, the proposal asserts, those directions are "temporal." Will this procedure single out a direction or set of directions? If so, do they correspond to what we ordinarily deem temporal, i.e., connect to different facets of time? These are the questions to which we now turn.

8.4 The Argument

I claim that well-posed Cauchy problems for systems described by (8.1) pick out the temporal direction. The idea behind the argument is simple: only some types of equation of form (8.1) possess well-posed Cauchy problems, and those that do are *finicky* about where the initial data is placed. In particular, they demand that the data is placed on what we would associate with spacelike hypersurfaces, which means the Cauchy development is "temporal."

To begin, it will help to note that partial differential equations can be classified into various types, depending upon how many positive and zero eigenvalues the matrix a_{ij} in (8.1) has. Let P be the number of positive eigenvalues and Z be the number of zero eigenvalues a_{ij} has. Then we say that (8.1) is

- hyperbolic if $Z = 0$ and either $P = 1$ or $P = d - 1$
- parabolic if $Z > 0$
- elliptic if $Z = 0$ and either $P = 0$ or $P = d$
- ultrahyperbolic if $Z = 0$ and $1 < P < d - 1$.

The classification of a partial differential equation is a coordinate-independent fact. Since a_{ij} is symmetric, its eigenvalues are real. It is then a theorem of linear algebra that, at a particular point, the number of positive and zero (and negative) eigenvalues of the coefficient matrix is invariant under smooth nonsingular transformations of coordinates. Since these are the only transformations we allow, changing the coordinates in which the equation is written will not alter its classification.

With this classification in hand, we can ask which types of equation (8.1) admit well-posed Cauchy problems? The answer is that *only* hyperbolic partial differential equations do.[10] That the hyperbolic has well-posed Cauchy problems is well-known. In fact, it is a theorem that *all* linear hyperbolic second-order systems have well-posed

---

[10] So long as we place our Cauchy data on non-closed hypersurfaces, which seems reasonable.

Cauchy problems, given certain mild assumptions.[11] That *no other type* has well-posed Cauchy problems is also well-known, although it is too messy a fact to warrant composing as a theorem. Yet it is true that *none* of the other types of equations of form (8.1) defined over a non-closed domain are well-posed for Cauchy data. There is no single reason for this fact. Elliptic equations—such as the Laplace equation—suffer a variety of fates: non-unique solutions, lack of existence, and lack of continuity. Parabolic equations—such as the heat equation—have too many solutions given Cauchy data and so are non-unique. And ultrahyperbolic equations are ruled out as an indirect consequence of Asgeirsson's "mean-value" theorem (1937), as we'll see below.

For a concrete example, consider the hyperbolic Klein–Gordon equation

$$\frac{\partial^2 \phi}{\partial t^2} = \frac{\partial^2 \phi}{\partial x^2} + \frac{\partial^2 \phi}{\partial y^2} + \frac{\partial^2 \phi}{\partial z^2} - m^2 \phi \tag{8.2}$$

This equation has a well-posed formulation. However, what the above facts mean is if we change the sign in the Klein–Gordon equation and let it go "Laplacian," the equation goes from being well-posed to ill-posed. Same goes if we allow the Klein–Gordon equation to turn into a parabolic diffusion equation or an ultrahyperbolic equation.

We can safely restrict attention to the hyperbolic case. Although all hyperbolic versions of Eq. (8.1) possess a well-posed Cauchy problem, that doesn't imply that one can put the initial Cauchy data anywhere and get a well-posed problem. As mentioned, hyperbolic examples of Eq. (8.1) are picky when it comes to choice of "initial" Cauchy surface. They require that the initial data be placed on a "spatial" surface. Before seeing the argument, let me characterize a "spatial" surface in a non-prejudicial manner. For those new to partial differential equations, this will be slightly technical, but it will very quickly become familiar from relativity. When we have all the pieces, I will put it all together in a simple illustration.

We will use the *characteristics* of Eq. (8.1) to define the spatial. Characteristic curves (surfaces, etc.) are curves along which a partial differential equation becomes an ordinary differential equation. We're interested in surfaces. A precise definition is as follows: $\phi(x_0, x_1, \ldots, x_d) = c_0$ describes a characteristic surface for an equation of form Eq. (8.1) if

$$\sum a_{ij} \frac{\partial y}{\partial x_i} \frac{\partial y}{\partial x_j} = 0$$

for all $\phi(x_0, x_1, \ldots, x_d)$ on the surface $\phi = c_0$. Characteristics are of singular importance in the study of partial differential equations. They are important theoretically for the study of the structure of solutions and practically for their use in solving partial

---

[11] Theorem. Consider a linear, diagonal second-order hyperbolic system. Let $\langle M, g \rangle$ be a globally hyperbolic region of an arbitrary spacetime. Let Σ be a smooth spacelike Cauchy surface. Then one has a well-posed Cauchy problem. See Wald (1984, pp. 250–1) for details. Note that for the theorem to hold, it is crucial that Σ be spacelike in the sense that $g(v, v) > 0$.

differential equations, e.g., the "method of characteristics." Like the classification of equation, they are "geometrical" structures and hence coordinate-independent (Courant and Hilbert, 1962, p. 423). For hyperbolic equations, characteristics are those level surfaces with constant $\phi(x_0, x_1, \ldots, x_d)$ whose normals to some hyperplane C, $\phi_{x_0}, \phi_{x_1}, \ldots, \phi_{x_d}$, satisfy the first-order equation

$$\phi_{x_0}^2 = \sum a_{jk}\phi_{x_j}\phi_{x_k} \qquad (8.3)$$

for some x_0. If for variable x_0 and hyperplane C the normals to C satisfy the inequality

$$\phi_{x_0}^2 > \sum a_{jk}\phi_{x_j}\phi_{x_k} \qquad (8.4)$$

then C is spacelike. In this way one sees the characteristic as a kind of limiting case of spacelike surfaces, where the inequality of Eq. (8.4) goes to the equality of Eq. (8.3).

To get a familiar picture of the spacelike versus timelike, consider the characteristic conoids for hyperbolic versions of Eq. (8.1). The characteristic conoid is a characteristic surface along which the distance between any two points on the surface is zero. For hyperbolic versions of Eq. (8.1), the characteristic conoids consist of two sheets ("cones") emanating from each point of d-dimensional space. These sheets divide the space into three disjoint regions. Call surface elements at the vertex of these three regions *spacelike* if they lie in the region bounded by both sheets, where Eq. (8.3) is satisfied. Directions pointing into one of the two regions bounded by a single sheet are timelike (Garabedian, 1964, p. 178). If this looks suspiciously familiar to the lightcones of relativity, it should. The paradigmatic hyperbolic equation is the wave equation, and the Lorentz transformations at the heart of relativistic physics are precisely the group of linear transformations that do not alter its form. The sense of timelike given in relativity coincides with the sense defined in partial differential equation textbooks.[12] Dispel any suspicions, however, of smuggling in this connection. The desire for a well-posed Cauchy problem has and will do all the work.

It remains to show that hyperbolic equations have well-posed Cauchy problems *only if* the initial data is placed on a spacelike surface, in the above sense of spacelike. We also have unfinished business with ultrahyperbolic equations, for we claimed that this variation of Eq. (8.1) possesses no well-posed Cauchy problem. We can secure both claims through Asgeirsson's "mean value theorem" (1937). The details of Asgeirsson's theorem and how it manages both tasks can be found in Courant and Hilbert (1962, pp. 754–60) (see also Garipov and Kardakov, 1973; Kaistrenko, 1975). The surprising argument requires more mathematical sophistication than we have developed here, but it is possible to give a quick sketch.

In a well-posed Cauchy problem one can freely specify data on the "initial" surface. The question for us is whether one can put arbitrary data on a not-everywhere spacelike plane. Because ultrahyperbolic equations require this to be the case, they are

---

[12] To be clear, the "cones" defined here are more general than those found in relativity. They can be wider or narrower than those defined by relativistic physics. See the discussion surrounding 7.4.

implicated here too.[13] Asgeirsson's theorem rules out such choices of initial condition. In the case where the initial surface is mixed, timelike and spacelike, one can use the theorem to show that the initial values over an arbitrarily small region uniquely determines everywhere the initial values on a much larger region. Thus the initial conditions are not arbitrary, since what values you place somewhere on your initial surface determine the values elsewhere on the surface. And that means that unless one picks the *right* initial values across the surface one doesn't get a consistent solution. In other words, assuming arbitrary initial conditions, one often will lack existence of solution, contrary to the definition of well-posedness.

Intuitively, the problem, as Hadamard himself noticed, is that by grabbing a mixed timelike–spacelike hypersurface, one has chosen a hypersurface wherein some points on the surface are in the domain of dependence of other points on the surface. As Hadamard puts it, there are an "infinite series of relations" among the points on the hypersurface. Often one thinks physically of the characteristics as the lines along which causal propagation goes. To the extent that this is correct, what one has done is choose an initial surface that intersects the characteristic cones. A "pulse" from one part of this initial surface will be carried along by the characteristics to another part of that same surface, thereby determining its values. Hence the lack of freedom. This unwelcome result happens for every surface except for those carefully bisecting the two lobes of the characteristic cones associated with hyperbolic equations. It is not possible, therefore, to place arbitrary initial Cauchy data along a non-spacelike surface and get a well-posed problem.[14]

8.5 Illustration

Without getting mired in the details, let's understand what's happened in a toy example (Fig. 8.5).[15] Consider the Cauchy problem for the one-dimensional nonhomogeneous wave equation,

---

[13] Craig and Weinstein (2009) show that the ultrahyperbolic free wavefunction has a well-posed formulation. That isn't a counterexample to Courant and Hilbert's argument because Craig and Weinstein prove this for initial data satisfying a certain "non-local" constraint. See below for some discussion of constraints and well-posedness.

[14] Is this surprising result merely a mathematical curiosity? I don't think so. Suppose we weakened what we mean by strength. The most obvious weakening would be to look at equations for which unique existence holds, but not stability. Initially, this weakening seems unpromising. The fundamental theorem in the field regarding existence and uniqueness, the Cauchy–Kowalewskaya theorem, does not require a spacelike versus timelike split. However, Geroch (1996) shows that for quasi-linear symmetric first-order equations, which he says are capable of representing virtually every system of physical interest, one can make plausible an existence and uniqueness claim. And the important point for us is that one of Geroch's conditions for unique existence is what he calls a *hyperbolization*, the condition met in the previous chapter, and this demands a spacelike versus timelike distinction once again. The space versus time split enters as part of a sufficient condition for getting unique existence. This result is not as strong as that here, where we establish that putting data on a "spacelike" slice is also necessary; but with first-order equations the scope is enlarged and yet again we find a non-trivial connection between the time/space split and strength.

[15] I came across this example in a partial differential equations text while browsing in the library. Try as I might, I have not been able to locate it. I apologize to the author and welcome identification from readers.

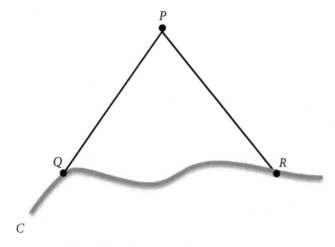

Figure 8.5 Defining spacelike in a toy example

$$u_{tt} - u_{xx} = f(x, t)$$

Normally we might specify the Cauchy data on a $t = 0$ slice of the x-axis. However, we are interested in how the Cauchy problem distinguishes one direction as temporal; so rather than assuming the "initial" data fall on the x-axis, let us prescribe data on a arbitrary curve C. Take a point P and a curve C not containing P. If we have a solution of our equation at P, $u(P)$, then there will be *characteristic curves* (defined below) intersecting P and C, at different points on C, Q, and R. The solution $u(P)$, if it exists, will be determined by the values of u and u_x (where x is the locally normal direction) on the arc QR of C and the values within the "triangle" created by QP, PR, RQ.

Let us now define what it is to be "spacelike." A curve C is spacelike if as P tends to a point on C, the points Q and R also tend to that same point on C. Notice that this definition *doesn't presume a prior time/space split*. Yet it turns out that $u(P)$ is consistent with the Cauchy data on C only if C is spacelike in this sense. Assume the temporal direction is locally orthogonal to the spatial directions and we have found time. In this way we see how strength can "find" the temporal direction. It would then remain to be shown that this distinguished timelike direction coalesces with what we know about time—which would be silly in this toy two-dimensional example. Generalizations of the above ideas hold for all second-order hyperbolic equations.

8.6 Is It Time?

Does this implicit definition of the spacelike/timelike correspond with that used in physics? It's no surprise that the answer is "yes" given the connection mentioned between the timelike in the above sense and the timelike in relativity. Better than that, this sense of timelike welds together otherwise disparate features of time too. To see how this is accomplished, we need three Facts:

1. For any partial differential equation, the submanifold upon which ones places initial data must be $d - 1$ dimensional (where M is d-dimensional) if one wants a well-posed Cauchy problem.

2. The signature of $\langle M, g \rangle$ is connected to the signature of the fundamental equations of free particles and fields. In many cases—Klein–Gordon, Dirac, Weyl— these equations are the coordinate space representations of what are known as the irreducible unitary representations of the local spacetime symmetry group. Change the metric signature and you change the spacetime symmetry group and the corresponding configuration space expression of that symmetry. See McCabe (2005) for an explicit derivation via the energy–momentum Fourier transform of the Klein–Gordon equation of how flipping the metric sign switches the signature of the Klein–Gordon equation. For covariant field equations the matrix a_{ij} in Eq. (8.1) will have the same eigenvalues as the metric tensor. Not all equations used in physics will reflect the signature, of course, but the fundamental particle or field equations will.[16]

3. In spacetimes with closed timelike curves, strictly speaking, there are no well-posed Cauchy problems because there are no Cauchy surfaces. A subset of a manifold is a Cauchy surface if every inextendible worldline of the spacetime intersects the surface exactly once.

Fact 1 tells us that if we define time via well-posed Cauchy problems, time is bound to be one-dimensional. One cannot get a Cauchy problem if one places the "initial" data on a $(d - 2)$-dimensional hypersurface or less. This fact is not surprising, but it does imply that if one wants maximum strength and $d = 4$, then we're going to have to place data on 3-dimensional hypersurfaces.

Fact 2 is even more interesting. Fact 1 tells us that the Cauchy data must be placed on a three-dimensional submanifold (assuming $d = 4$) if we are to have a hope of finding a well-posed Cauchy problem. Which submanifold? Fact 1 leaves this entirely open. However, Fact 2 coupled with our earlier argument tells us that it must be on the $(+++)$ submanifold and not the "mixed" submanifolds (e.g. $(++-)$). Time *must* march in the "$-$" direction. Hyperbolic versions of (8.1), Lorentzian spacetimes, and well-posed Cauchy problems are linked. As we saw, if we want strength in the form of a well-posed Cauchy problem, this means that the equation will be hyperbolic and

---

[16] Tegmark (1997) takes Fact 8.6 to provide an anthropic justification for the signature of spacetime, or as he puts it, the "one-dimensionality of time." Creatures able to predict successfully will find themselves in Lorentzian spacetimes, just like fish caught with nets with one foot holes in them tend to be over a foot long. Obviously I think that there are some deep insights in Tegmark's argument, for mine depends on many of his central observations. However, I don't wish to go as far as he does. I'm content to discover the existence of constraints among various temporal features, thereby helping justify applying the denomination "time" to a set of directions. The additional claim that prediction in the precise sense required here is crucial to the existence of observers is not something I want to defend.

moreover that the "initial" data must be imposed on a "spacelike" hypersurface. Fact 8.6 then implies that the signature of spacetime is Lorentzian.[17]

Fact 2 also explains why the "productive" aspect of time marches in step with the "minus" direction and the direction of strength. The basic laws are the source of production—however this is understood—and we now have a link between this production and both the strength and the signature of spacetime. Coupled with Fact 1, these links help us understand why our physical theories aren't productive in more than one set of directions. In the last chapter I said the systems view would understand seizing more than one set of directions as the dimension of strength as a needless waste, for it would add complexity to a system without any corresponding gain in strength. Here we see an echo of that, for production in two senses must wreak havoc with our basic physical theories. The connection between the signature of the basic laws and the signature of the metric would have to be undermined for this to occur.

Fact 3 warns us not to expect closed timelike curves in spacetimes with well-posed Cauchy problems. If time is the direction of maximum strength à la Proposal 2, then we can't get maximum strength in directions in which worldliness "bend over" onto themselves.

In sum, in the arena of worlds described by Eq. (8.1) and limited to desire for information via well-posed Cauchy problems, one can rigorously demonstrate that strength distinguishes time from space; in addition, one can strictly connect three features we initially associated with time: the ways in which it is special topologically and metrically and with respect to production and free mobility. In this way we have a rigorous example of many of the informal connections suggested by Chapter 6's arguments.

8.7 Turning Pages in Non-temporal Directions

As surprising as the above results are, departures from it may be even less expected. Relativity shocks new students when they learn that multiple simultaneity planes can be used as one's now, each sufficient to tell the story of the world. One can "tilt" the now over, compatible with Lorentz invariance, so long as it remains spacelike. If we loosely regard the surface upon which we place "initial" data as our now, can one tilt even more, such that the "now" is null or lightlike, or even more than that, such that the "now" is partly timelike? Can we, in other words, march this data "sideways" into non-temporal directions and still recover the full four-dimensional world? As it turns out,

---

[17] One can imagine other arguments that might supplement this one. There exists in physics a tiny program of finding out why Nature uses a (3, 1) signature. In that literature one will find proofs that spinor structure and even electromagnetism require a Lorentzian signature. Connected to strength via the laws used in the proofs, we would then have another rigorous (but less direct) connection between strength and the signature. See van Dam and Ng (2001), Borstnik and Nielsen (2002), and Itin and Hehl (2004).

if we're allowed to "cheat" it may be possible to turn the world's pages in non-temporal directions.

8.7.1 Pages of light

Characteristic problems are "initial" value type problems in which one places the initial data on null surfaces. Philosophers, to my knowledge, haven't studied null initial value problems in relativity, but they are potentially of deep significance. For reasons to be discussed, they are commonly used, and yet, prima facie, pose a challenge to the idea that we must always "march" data on spatial slices in the temporal direction.

The characteristic initial value problem began life in relativity with Bondi et al. (1962)'s use of null coordinates to deal with gravitational waves.[18] Used extensively in numerical relativity, it also arises in discussions of the holographic principle and problems devoted to radiation and quantum gravity. Theoretical work on the characteristic problem is very much ongoing, as counterparts of many basic questions answered in the Cauchy approach are to date unknown in the characteristic approach. Nevertheless, because of certain perceived advantages, the characteristic approach remains popular.

I speak of *the* characteristic approach, but in fact there are many distinct setups. We won't dive into the details, but let me sketch the core idea. First we must pick the initial surface. To be a null problem obviously this surface must be partly null. That leaves many options, for the rest of the surface could be null, timelike, or spacelike in various configurations. The most common surfaces are either the lightcone or two intersecting null surfaces. In the first case, the lightcone can be either the past or future lightcone. The second case divides into many other subspecies depending on the type of surfaces and intersection. One option is choosing past and future lightcones and their intersection across a spacelike two-sphere. Another are two null "wedges" that intersect on a two-dimensional spacelike submanifold. Another is the future null surface emerging from the diameter of a spatial two-sphere, plus information on the timelike worldtube developing from those points. See Fig. 8.6 for two possible choices.

Having chosen a surface, naturally the second step is to place the "initial" data on it. Doing so is a challenge. In the Cauchy setup, the four-dimensional metric induces a natural spatial $(+++)$ metric on the initial Cauchy slices. In the null case, by contrast, the four-metric induces a geometric structure on a degenerate hypersurface. Much of the work on the characteristic problem has been coordinate-dependent, and getting natural geometric objects as initial data is not nearly as easy as in the Cauchy case. In general, however, no matter the initial surface, the characteristic problem uses just the solution of the relevant systems of equations on the surface, plus initial data on the

---

[18] That was part of a flurry of activity on the problem: see Sachs (1962); Penrose (1963); Dautcourt (1963), and many others. The work continues today in papers by Rendall (1990), Luk (2011), and Choquet-Bruhat et al. (2011).

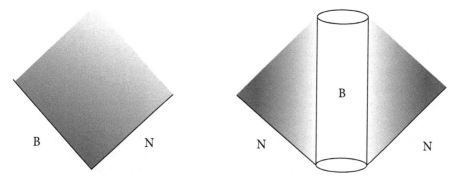

Figure 8.6 Varieties of initial null surfaces and boundaries

intersection, and not, as in the Cauchy case, the normal derivative. So if you choose to work with the future lightcone as surface, you would get the solution on that lightcone plus the vertex, and push this into the interior of the cone. If you choose the wedge surfaces, you get the solution on the two wedges plus data on the two-dimensional spacelike intersection, and push this in the neighborhood of the intersection or along the whole wedges. In a sense, one is replacing the information missing from the Cauchy derivatives with extra boundary information.

The characteristic problem is seen as having some advantages over the usual setup. It is sometimes advertised as being more in line with the observational data. But more importantly, the initial data is often unconstrained or constrained only by ordinary differential equations. In the Cauchy case I mentioned that the Einstein equations imposed constraints along the spacelike surface. What I didn't mention is that these are nasty elliptical constraints that are hard to solve. Getting rid of those is a huge relief and explains why the characteristic problem is popular in the numerical relativity community. Though these setups may seem odd to the lover of manifest time, the absence or abatement of constraints shows that at least in one precise sense the setup is actually quite natural.

Now to the punchline: there are some cases where the characteristic problem is provably well-posed. Rendall (1990) and developments from this work do this in the intersecting null surfaces approach. And while shy of well-posedness, Choquet-Bruhat et al. (2011) recently demonstrates existence in the lightcone approach.

These results are very interesting. Earlier I argued that the temporal arrow and the arrow of lawful production are aligned. The result of the previous section demonstrates a sense in which this is correct. However, if we can get well-posed null problems, then arguably the two arrows aren't aligned. The laws that underwrite production march forward in directions that aren't intuitively temporal.

Consider turning on a flashlight. The photons leave the event and travel up the future lightcone. Even though the metric interval between any two events on the lightcone has zero distance between them, still there is an intuitive sense in which it still takes "time" for the one photon to get from one event to another. Typically we choose coordinates according to which the distance in coordinate time is non-zero. And when we measure the speed of light, we adopt a reference frame with respect to which we can say that the light took time and therefore had a certain speed. What I'm saying is that usually our laws "march" initial data in this very direction, partially vindicating the intuitive sense in which it *takes time* for photons to get from one place to another. However, if we could take as "initial" data the entire lifetime of photons, and then push that data "forward" to determine the "next" slice, that would be quite astonishing. How odd from the perspective of manifest time to place data on a "moment" that itself intuitively takes time. Yet that is precisely what we can do in relativity in these nulllike cases. These results mean that there are precise senses in which the "arrow of production" and "arrow of strength" conflict with the intuitive time/space split.

How is this possible? The null case is a kind of loophole to the argument of (8.4). Tegmark (1997) dismisses this case, claiming that to obtain the Cauchy data on a null surface, one would need to "live on the lightcone." This claim is too quick, for there is a sense in which we *do* live on the lightcone, or at least, our data does. After all, much of the evidence we obtain about the cosmos comes in the form of light. What we're seeing when we look back in time is essentially the surface of the past lightcone. Since it's natural to use what one sees as data, perhaps one could insist that we do "live" on the lightcone (see Komar, 1965). So I don't wish to dismiss the null case. Yet the existence of well-posed characteristic initial value problems threatens, with one stroke, both my idea that the best system "seizes" on one set of time directions in its formulation and the idea that the laws govern only in the timelike direction (if governing is tied to the physics and not a metaphysical dangler).

Or does it? Actually the verdict is still open. To my knowledge, the null problem has never been solved in characteristic coordinates. Rendall and others typically reduce the characteristic problem to a Cauchy problem in their proofs. The existence theorems of Choquet-Bruhat et al. use normal spacetime coordinates, not characteristic coordinates. Moreover, as Jeff Wincour (p.c.) reports, the characteristic problems used in physics typically have enough hyperbolicity to ensure that a well-posed Cauchy problem can be given. All of this raises the suspicion that the success of the characteristic problem in relativity is *parasitic* upon the success of the Cauchy problem. What would alleviate this worry would be the demonstration, for some realistic system of equations, that the characteristic problem is well-posed but the corresponding Cauchy problem ill-posed. That is not something we have, again, to my knowledge. Instead we possess innumerable cases where we know the Cauchy problem is well-posed and don't know whether the characteristic problem is too, and no cases where the characteristic problem is well-posed but Cauchy ill-posed; and the characteristic problem seems well-posed only when reduced to the Cauchy problem. With the evidence we currently possess, the characteristic problem seems to be in some sense parasitic upon the Cauchy approach in relativity. The "specialness" of time isn't yet threatened.

8.7.2 Pages of time

We just tried to tell the story of the universe while "tilting" it over and writing on lightlike pages. We learned that this is indeed possible, but not likely too much of a threat to the idea that physics still "prefers" the timelike direction. Let's try something even more radical, namely, tilting the initial surface *all the way over* so that the pages are *partly timelike* and the story is read in a spacelike direction. To be slightly more precise, can we place initial Cauchy data (solution, derivatives) on a mixed surface, say $(++-)$, and "evolve" it in the remaining $+$ direction (Fig. 8.7)? If we could, then we could tell the story of the world "sideways."

Amazingly, there are some indications that we can look at the world sideways and still tell its story. And we can do this with Cauchy data on a slice, so the extremes of information needed in the "transcendental determinism" described in Section 8.1 aren't needed. How can this be, given the mathematical argument of Section 8.4? The argument of Courant and Hilbert is perfectly valid. However, as is well-known, it is possible to take a system of equations that is ill-posed and make it well-posed by insisting on some further constraint being satisfied among the initial data. That is precisely what is done in classical electromagnetism. Classical electromagnetism has no well-posed Cauchy problem in the sense used in Section 8.4. That is, if the equations are Maxwell's and we consider arbitrary electric and magnetic fields at a time, then we don't have a well-posed problem. Nonetheless, well-posedness can obtain if we first impose Gauss' laws as constraints on the initial fields. Hence electromagnetism

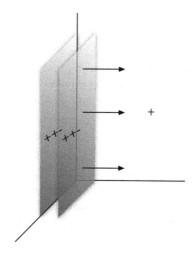

Figure 8.7 Telling the world's story sideways

is well-posed, so long as these constraints are imposed. Perhaps the same thing can happen here: add a constraint on the initial data and we may well have a well-posed problem, Courant and Hilbert's result notwithstanding.

Robert Geroch (p.c.) has suggested that the same may be true with the timelike Cauchy problem. Consider the Klein–Gordon equation (8.2), which is an equation of the form (8.1), and a partly timelike hypersurface of one dimension less than the space. Assume the usual initial data plus the following:

> *Constraint*: for any point t on the surface, there exists a neighborhood of that point and a Klein–Gordon solution in that neighborhood.

It is well-known that local uniqueness holds for the Klein–Gordon equation. But local uniqueness plus Constraint implies global uniqueness in the sense of a neighborhood of the whole hypersurface. The intuitive reason is that one gets existence of solutions in the neighborhoods of each point, by Constraint, but by local uniqueness these solutions must be compatible when they overlap. One can then "patch" together these solutions to cover the entire surface. Given the similarities between the Cauchy problem for the Klein–Gordon equation and that for Einstein's field equations, this argument also makes plausible the claim that timelike well-posed Cauchy problems exist in general relativity as well.[19]

As mentioned, this result is compatible with the argument of Section 8.4 because it rejects the assumption that arbitrary initial data can be placed on the hypersurface of interest. Here the data must be highly constrained. From our earlier perspective, such constraints "cheat" by referring to off surface neighborhoods.

How bad is this result for the overall argument? Strictly speaking, the existence of Constraint will mean the timelike initial value problem is not as powerful as the spacelike one without invoking Constraint. That said, Constraint is local, so it's not so clear that prediction would be seriously inhibited, nor is it clear that the difference in the two scenarios is something that would be noticeable to beings like us. Better, I think, to remind the reader that the general argument prizes simplicity too, even if it has been in the background until now. The thought is that Occam's razor may well prune such examples due to the ugly constraints necessary to realize them.

Certainly that seems the correct reaction (from the current perspective) to Craig and Weinstein's nice example of a well-posed "sideways" Cauchy problem. There they show rigorously that one can get a well-posed Cauchy problem with data on a "mixed" (e.g. $(++-)$) hypersurface. This result is very interesting. The price to be paid for such strength, however, is a serious loss of simplicity. Instead of constraints that the solution exist off surface, as in Geroch's case, here the constraint must be highly non-local. This cost to strength is especially bad if one interprets strength in terms of usable strength for creatures like us.

---

[19] See Garipov and Kardakov (1973); Klibanov and Rakesh (1992); Kaistrenko (1975) for more work on the Cauchy problem with nonspatial initial data.

8.8 Conclusion

The reconceived project of Chapter 6 has borne fruit. Instead of pursuing a possibly fictitious goal—finding what constitutes *the* difference between space and time—we have allowed that there may be many differences, no one central enough to count as the difference. Instead we have sought connections among the various features associated with physical time. Not only have we found novel ones, but we have seen that some old ones are deeper than expected and some act as a kind of "glue" holding together these logically detachable pieces of our concept of time.

In particular, we have discussed seven aspects of time in physics:

a. one-dimensionality
b. metrical distinction
c. mobility asymmetry
d. direction of time
e. asymmetry of strength
f. natural kind asymmetry
g. production

With an informal argument based on the systematization account of laws, we have discerned connections of various degree among all seven features in Chapter 7. This argument was then bolstered in this chapter by a formal one that rigorously linked features a, b, and c to feature e (and perhaps also f).

We also learned that it's possible for the threads of physical modality to tie the world together in non-temporal directions. This possibility doesn't threaten the idea that time is special: these stories told in these directions aren't quite as good as those told in the temporal directions.

To me, the lesson of the foregoing considerations is that metaphysically speaking, as it were, one can get robust counterfactual dependencies in all directions, null, timelike, and spacelike. The world doesn't "unfold" along the time direction as the tenser hopes, at least not in any way suggested by the physics. Nevertheless the tenser's intuition holds a central insight into the world. There really is a difference between space and time embodied crucially in its connection to the laws. If one forgets about simplicity and strength, this difference tends to vanish. If focused on finding the simplest and strongest system of the world, however, this core difference between space and time stands out. It is an objective difference between space and time that survives the onslaught against time from modern physics.

9

Do We Experience the Present?

The non-perspectival structure which, as realists, we conceive to underlie and support perspectival temporal discourse is, as yet, a partially covered promissory note the cash for which is to be provided not by metaphysics . . . but by the advance of science.

(Sellars, 1962b)

We began the book articulating the problem of time, the fact that physical time doesn't recognize many of the properties we attribute to manifest time. Subsequent chapters devoted to physics confirmed this view. Einstein was right to be worried. Yet he was resigned to the idea that science would always be unable to deal with the Now. Reflecting on this resignation, the physicist Brian Greene writes

> This resignation leaves open a pivotal question: Is science unable to grasp a fundamental quality of time that the human mind embraces as readily as lungs take in air, or does the human mind impose on time a quality of its own making, one that is artificial and that hence does not show up in the laws of physics? If you were to ask me this question during the working day, I'd side with the latter perspective, but by nightfall, when critical thought eases into the ordinary routines of life, it's hard to maintain full resistance to the former viewpoint. (2004, p. 141)

Greene's nightfall softening to the idea that the Now is fundamental is natural. Manifest time is lived time. Even if we regard the world as a four-dimensional "block," it is virtually impossible not to think of ourselves and a Now as crawling up this block in a timelike direction. Some events, we think, have *happened* and others *haven't*. The Now is the pivotal moment marking this crucial difference. The ensuing differences among past, present, and future events matter to our choices, actions, feelings, language, and more. Granting that resistance to manifest time is almost futile, psychologically speaking, I nevertheless want to develop the Carnapian explanation suggested in Chapter 1. Although I agree that the idea of a fundamentally tensed world is embraced as readily as lungs take in air, I think we *can* explain why that is the case for creatures like us.

Before I can turn to this constructive task, however, we must dispense with the idea that our experience directly confirms the existence of a Now—a real physical or metaphysical feature of the world—that is missed by science. Many philosophers and scientists have felt that experience itself shows that physical time is either inaccurate or incomplete because it doesn't capture "a fundamental quality of time" corresponding to the flowing now. Here I want to question the claim that we sense a (tensed) now or present. I'll argue that there is no good reason to think we do. And that question naturally suggests another one rarely asked: can we even say that we directly experience (tenseless) simultaneity? We obviously don't experience two events as happening at the same time when they are spacelike related. But do we even experience two events as being synchronous at all, or is it merely something that we infer later? I'll suggest a way of framing this question and will argue that, surprisingly, the answer is probably negative.

This chapter and the next have a Humpty Dumpty-like character. The present chapter aims to tear apart the idea that we experience the now in almost any of its forms. After breaking the now apart, the next chapter aims to put what's left back together to explain why we nonetheless believe the now is an objective feature of the world.

9.1 Metaphysics of Time

Philosophy of time is divided, very broadly, into tensed and tenseless theories of time. Tensed theories come in many forms (Broad, 1923; Dummett, 1960; Geach, 1979; Lucas, 1973; McCall, 1969; Prior, 1967), but what I want to stress is that most versions take the now as metaphysically distinguished. It is a genuine fundamental feature of the world, something that so far science has missed. Presentists believe that only the present exists. Advocates of the "growing block" hold that the past and present exist, yet the present remains special because it is the cusp of what exists. The "moving spotlight" picture posits an eternal "block," but one containing an ontologically irreducible now that crawls up worldlines. Theorists who advocate branching think the future admits divergent paths yet the past is a settled trunk, hence that there is a special time, the now, when alternative future histories drop off. What unites all these positions is that they all treat the now as having an objective metaphysical counterpart. This now is ontologically special—be it the only thing that exists, the point where branching occurs, the point where the non-existing future turns real, or what have you. Advocates of the tenseless theory (e.g., Grünbaum, 1973; Smart, 1966; Williams, 1951) instead hold that time itself only has the properties that we pick out when using the allocentric language of science. For detensers, the now does not carve out objective joints in nature.

I have reservations about the usefulness of conceiving philosophy of time as a grand debate between tensers and detensers (see Chapter 13). Putting these aside for a moment, the situation in the field seems to be as follows. Tensed theories of time have been the target of nearly a century's worth of attacks. Conceptual arguments, like McTaggart's famous paradox, and scientific arguments, like that from special relativity, seek to demonstrate that the above metaphysical theories are logically and physically impossible, respectively. Focusing on impossibility proofs, however, has had unfortunate consequences for the detenser. First, detensers have not succeeded in obtaining outright victory. Second, excessive concentration on outright victory leaves the tenseless theory hollow and undeveloped when it comes to explaining our experience and intuitions. This result was natural; after all, if the tensed theory could be shown impossible, detensers win. They could then explain manifest time at their leisure, confident in victory that it must be possible. However, if this "nuclear" strategy of obliterating the tensed theory fails and the debate instead becomes "conventional," that is, a dispute over which theory *best explains* the phenomena, detensers are left with little to say. Tensed metaphysical systems are commonly said to be more in tune with our experience of time than tenseless theories. Even many detensers admit this (e.g., Balashov, 2005; Falk, 2003).

Although that is the conventional wisdom, I believe it is mistaken. By positing objective counterparts to some egocentric temporal properties, the tenser obviously begins with a model of time that looks, on its face, more like manifest time. No surprise there. But whether this model is better confirmed by experience is another matter. Contemporary analytic philosophers of time typically point in a perfunctory way to various stock mental experiences as confirmation of their sometimes byzantine metaphysical systems. One is told that we feel time pass or that the present is sensed as special—as if we know exactly what that means—and that we can only explain this through a tensed metaphysics. Meanwhile, those who have studied temporal experience in detail, e.g., phenomenologists and contemporary cognitive scientists, typically remain silent about the metaphysics. Are there really experiences that better confirm the tensed hypothesis over the tenseless hypothesis? I don't think so. Indeed, I think it would be a minor miracle if our experiences could discriminate between these two metaphysical hypotheses (see Chapter 13).

It's time detensers stand up for themselves and challenge the claim that experience favors tenses. Doing so will remove an impediment to our project. Once experience directly confirming tenses is off the table, we will be able to direct attention to the actual psychological "now" and eventually to recovering some of manifest time from scientific time.

9.2 The Problem of the Presence of Experience

Consider some experience, say, reading this book. While you read this—*right now*—you're experiencing the present. Reading the beginning of the previous sentence is now fading into the past and new events are being "lit up" by the feature of presentness. These new events—reading *this* sentence—are now (temporally) present. Is there something in this prosaic experience that is problematic for the detenser? The philosophy of time literature often speaks of "the problem of the presence of experience." However, that is an umbrella covering many distinct problems. As we'll see, few have anything to do with experience, and none are damaging to the idea that time is objectively tenseless.

Experience is trouble for the detenser only if one reads the (tensed) theory into the data. One frustrating feature of the metaphysics of time literature is that it habitually does this. For example, Craig (2000) claims that we are "appeared to presently" (p. 139), where he means via a tensed present, and Schlesinger (1991) claims that the present, unlike the past and future, is "palpably real" (p. 427), as if we stand outside the realm of what exists and can touch the contrast with the unreal. What they mean is that we experience presentness, the metaphysical property, in some sense, and not merely experience events as happening at the same time, or as subjectively simultaneous. We may experience subjective simultaneity, but I am suspicious of claims that we experience presentness.

To evaluate such assertions, we must carefully disentangle the experience itself from judgments and descriptions of it. Doing so is a messy business. What I hope to do is only make room for the idea that we don't experience a property of presentness in a metaphysically loaded sense—that experience does not confirm the tensed theory of time. I admit that it's quite natural to think it does. Our mental representations of the world seem tensed. I think that *now I'm typing*. So when speaking about this mental token, it's not surprising if I describe my experience as if I were interacting with some feature of the external world called nowness or presentness. However, it's hardly clear that being present is a *phenomenal* property, a feature of my actual experience such that there's something it's like for me to have an experience with that feature. Right now I sense my computer humming and pain in my lower back, but I don't perceive a stamp of presentness on any experience. Do I have an experience of presentness or am I simply at present having an experience? Moreover, and more importantly, if we do have such experiences, it's even less clear what they imply.

Let's begin with some simple observations. As Mellor (1998) notes, objects look the same to us even if they are past. The experienced point of light from Jupiter is of an object an hour in the past. Other points of light in my visual field may look the same even if the source is roughly simultaneous with my experience or something a million years old. Since events appear the same despite tremendous differences in age, the view that we're sensing the property of being present of objects/events seems problematic. I'm not sensing Jupiter's presentness.

The philosopher William Lane Craig protests:

> as a result of physics and neurology, we now realize that nothing we sense is instantaneously simultaneous with our experience of it as present. But in most cases, the things and events we observe are contained within a brief temporal interval which is present ... and our basic belief makes no reference to instants, so that such a basic belief remains properly basic even for scientifically educated persons like ourselves. The fact that under extraordinary circumstances our basic belief in the presentness of some event/thing should turn out to be false is no proof at all either that we have no basic beliefs concerning the presentness of events/things in the external world or that such beliefs are not properly basic. Mellor is

therefore simply wrong when he asserts that we do not observe (defeasibly) the tense of events.   (2000, p. 143)

This response is untenable. The original claim is not that we don't have *beliefs* concerning the presentness of events in the external world, but that we don't *observe* the tense of external objects. Inasmuch as we hold manifest time dear, it's agreed that we do have beliefs in objects occurring now. Yet one cannot appeal to the common belief to justify the claim that the present is actually observed. Additionally, contrary to Craig, there is nothing "extraordinary" about seeing stars or hearing thunder after seeing lightning. We must be suspect of any theory that takes as a basic, unchallengeable belief a claim that makes these common events exceptional.

Craig aside, most readers are probably content to let the presentness of external objects go and rephrase the idea of perceiving the present as follows: the objects/events we perceive aren't present, but the *sensory experiences* we have *when we perceive them* are. In other words, conscious experiences are confined to the present. The metaphor of a spotlight moving along a wall, where the light is supposed to be present conscious experience, is a popular way of articulating this idea. Despite its popularity, this claim clearly begs the question if taken to be a problem for tenseless views. On this conception all times are on a par. My experience of my first day at school is an existent conscious experience on this view, but alas, it is an experience in the distant past and not present. To avoid begging the question, the claim must be that *in the present all my conscious experiences are present*. Yet this tautologous statement is something the detenser happily can endorse. To cause trouble for the detenser, the tenser must read the theory into the data again.

Presentness may be a phenomenological property in some sense. If it is, it seems a bit different from the paradigmatic ones. Normally phenomenal properties distinguish some experiences (or aspects of experiences) from others (Hestevold, 1990). The property *being loud* distinguishes some experiences from quieter experiences, and *being red* distinguishes parts of my visual field from other (say) blue parts. But every experience is present, and so the property doesn't play this crucial role. I don't deny that there is something it is like to have *any* given experience. There may be indexical phenomenological properties, such as *the way this experience feels*. Presumably these experiences don't have a something in common the way all red experiences have something in common. It's hard to see how one could go from such indexical phenomenological properties to a global objective present (see also Chapter 1).

In fact, that point can be made more generally: it's not clear that experience could tell between a tensed or tenseless present. Recall that the tensed theory takes presentness to be a monadic feature of events whereas the tenseless theory takes presentness to be a relation amongst events. Suppose we do experience the present—phenomenologically or otherwise—and represent it to ourselves as monadic. What follows? If Shoemaker is right, the answer is potentially nothing:

> But the way properties are represented in our experience is not an infallible guide to what the status—as monadic, dyadic, etc.—of these properties is. Reflection shows that

the relation to the right of is, at least, triadic, but do we experience it as such? And consider being heavy. What feels heavy to a child does not feel heavy to me. Reflection shows that instead of there being a single property of being heavy there are a number of relational properties, and that one and the same thing may be heavy for a person of such and such build and strength, and not heavy for a person with a different build and strength. But when something feels heavy to me, no explicit reference to myself, or to my build and strength, enters into the content of my experience. Indeed, just because one is not oneself among the objects of one's perception, it is not surprising that where one is perceiving what is in fact the instantiation of a relational property involving a relation to oneself, one does not, pre-reflectively, represent the property as involving such a relation.

(Shoemaker, 1996, p. 254)

I agree with Shoemaker and find the situation similar with time. Prosser (2006) argues that the nature of temporal experience makes it especially natural to substitute the one-place for two-place in the case of temporal predicates. The argument I'm about to develop in the next chapter can be conceived as a detailed explanation of why this might happen. Note that in the above cases we have "disagreement data." The fork is on my right but on your left. The hundred pound weight is heavy to me but not to a bodybuilder. There is no reason to believe that one person is right and the other wrong in either case. Similarly, disagreement data about the present exists from what we learn in both physics and psychology (as we'll see), yet there is no reason to regard anyone as being wrong. For a host of fascinating reasons—we'll see—we don't notice the disagreement data and so we represent the present as monadic. In sum, experience may not be decisive in favor of the tensed view.

9.3 The Temporal Knowledge Argument

Given the above snags, it shouldn't be too surprising that the quandary the literature often dubs "the problem of the presence of experience" in fact has little-to-nothing to do with experience. Here, for example, is a recent statement of the most prominent problem:

The challenge I wish to consider is that manifested by conscious experience.... [T]he present is experientially privileged in that we are only ever capable of experiencing that which occurs in the present. To put this observation another way, though we may know all week that the movie and Friday, 1:00 p.m. are simultaneous, when we learn that Friday, 1:00 p.m. is present and, therefore, that the movie starts now we seem to learn a new fact. Accordingly, tenseless relations cannot be all there is to time and the best explanation of the presence of experience is that the present is ontologically privileged, more real than other times. (Mozersky, 2006, pp. 441–2).

Mozersky is faithfully summarizing (not endorsing) the objection to a tenseless metaphysics found in (e.g.) Ludlow (1999) and elsewhere. Notice that the connection with experience is so deeply ingrained in the literature that even a detenser like

Mozersky provides an experiential gloss in the first two sentences of the quote—in contrast to the actual problem.

The actual problem is what Perry (2001) calls the *Temporal Knowledge Argument*. Perry suggests that it lies beneath almost every argument in favor of the tensed theory. Truly this is not much of an exaggeration. Scratch almost any argument and quickly the knowledge argument reveals its face. This argument is the temporal version of Frank Jackson's notorious Knowledge Argument in philosophy of mind, wherein colorblind Mary has an operation to recover color vision and thereby allegedly learns a new non-physical fact (Jackson, 1986). The Knowledge Argument is deeply controversial; few endorse its conclusion, Jackson included.[1] Perry declares the temporal counterpart of this argument to be the "heart" of the tensed theory. He is right. The argument comes in many forms, but the core idea is that tensed propositions give knowledge or make certain behaviors rational that tenseless propositions do not. To use his famous example (Perry, 1979), I knew all along that the meeting was at 5 p.m., but only my thought that "*now* is 5 p.m." explains my behavior. Evaluation of such arguments hangs on delicate issues regarding indexicals and contexts of representation, *not* phenomenal experience. Making the debate about the Temporal Knowledge Argument means that all the claims about the present being experienced as special and privileged are mostly rhetorical bluff unrelated to the actual argument for tenses.

Is the Temporal Knowledge Argument, whether experience-based or not, a good one? Not at all. Indeed, I believe the continued reliance on it in philosophy of time is something of a scandal. This argument and related puzzles (see Mozersky, 2006) plainly rely on quite general features of indexicals. We've known for a long time that indexicals such as "I," "here," and "now" are "essential" (Perry, 1979): replacing non-indexicals into indexical beliefs doesn't preserve the cognitive significance and explanatory power of the original indexical beliefs. I may know that someone's pushcart contains a leaky bag of flour but not know that that person is *me*. It's only when I learn that the culprit is me that I act on matters and close the bag. Similarly, I always knew that 5 p.m. is 5 p.m.; that doesn't explain why I get up to go to the meeting. Only learning that *now* is 5 p.m. does that.

It is true that the correct semantics for terms like "now" and "present" is a com-plicated and open project in linguistics and philosophy of language. I have neither the space nor ability to delve into it in any detail (see Mozersky, 2011). Even without possessing the final, true theory of temporal indexicals, a fairly standard account is on offer. Based on the work of Kaplan (1979) and Perry (1979), it is now standard to make

---

[1] Most philosophers (including Jackson, 1986) reject the knowledge argument. It's worth pointing out that even among those who *do* believe it sound, some still would not consider the corresponding temporal version sound too. The two arguments need not stand or fall together. Indeed, it's central to Chalmers' (1996) famous defense of the Mary argument that what Mary learns is *not* merely indexical knowledge. Also, it may be thought that "space zombies" and "time zombies" are less plausible than "color zombies." Chalmers argues that knowledge gaps due to indexicals disappear when others have omniscience whereas phenomenally based Mary-type gaps do not.

a *content* versus *character* distinction when treating indexical terms. The content of "I," "now," and "here" are, respectively, just the person, time, and place referred to by the term. The linguistic character, by contrast, is the rule by which we link the utterance of the term with its content. The rule for "I," for instance, is that the speaker of the utterance is the content of the term. The rule for "now" is analogous, namely, that the time of the utterance is the content of the term.

It's generally accepted that more is required for semantic evaluation than what is overtly linguistically encoded. You ask where the coffee is, and I see it on the right, think of it as on the right, and say it's on the right. Left out in all of this is the spatial frame with respect to which it is on the right. If you're Perry, what's left out is an unarticulated constituent, a syntactic representation that has an initially unsaturated variable later filled by some contextual provision. If you're Bach (1994) or Recanati (1993), by contrast, there is no semantic hole that needs filling—context will suffice. Understood either way (for a full menu of options, see Cohen (2009), section 4.1), there are many reasons why one might use the one-place predicate "on the right" instead of the two-place relation "to the right of X," just as there are many reasons why one might say that someone is a foreigner without explicitly stating to what country that person is foreign.

However the details are worked out, the Temporal Knowledge Argument can't be compelling because the argument would overgeneralize. Since spatial and personal indexicals behave the same way, they likewise can generate isomorphic spatial and personal knowledge arguments. The first would rely on the essentiality of spatial indexicals like "here" ("I knew the meeting was in Room 215, but didn't know that Room 215 is *here*") and the second on the essentially of personal pronouns, as in Perry's case of the shopper dropping flour. Just as the shopper example doesn't prove that there is a *me* over and above the person with the pushcart and the spatial case doesn't prove that there is a *here* over and above Room 215, neither does the temporal indexical case prove that there is *now* over and above the network of temporal relations. I suppose that one could try to find a relevant disanalogy; however, I certainly can't find one that doesn't beg the question. Even if one did, then it's not phenomenology or the Temporal Knowledge Argument that supports tensed time, but instead this currently non-existent argument.[2]

Despite this objection, the argument lives on. Today it continues in the form of complaints against what I called the standard account of temporal indexicals. Application of the Kaplan–Perry treatment of indexicals results in a particular account of the truth conditions of tensed statements. The linguistic character rule suggests that the truth conditions for the judgment that some experience *e* is present or now is that *e* is (roughly) simultaneous with the judgment. Independently devised by J. C. C. Smart and others, this treatment of the now is sometimes known as the

---

[2] In philosophy no position is so odd that it isn't occupied. Hare (2007) and Craig (1996) bravely endorse the generalization to persons. Fine also discusses it ("first person realism") in Fine (2005).

token-reflexive account of the present, which is based on work by Castañeda (1966, 1967). (There also exists the rival *date theory*, Le Poidevin (2003), but I can't see that much hangs on the difference.) The source of the main complaint is the feeling that there is some phenomenological fact ignored by the token-reflexive analysis. The detenser Balashov (2005) argues that the token-reflexive account misses the fact that some experiences are "more radically" present than it allows; Falk (2003) holds that "presentness is an inextricable part of all sensory awareness . . . and tensers are right to insist on an account of the experience of presentness that does not appeal to any reflexivity" (p. 221).

Now we're back to square one. Detensers admit that the representations are tensed.[3] Is there anything more than that? Balashov (2005), for example, tries hard to identify this experience not captured by the token-reflexive theory. He says tenseless accounts to date miss the crucial feeling, the feeling of events "simply occurring" (p. 295). What is it to simply occur? Balashov struggles to say. At one point he explains that the "hard problem" of the presence of experience is that "some experiences are known to be *occurring*, or *present*, as opposed to *not* occurring, or *absent*" (p. 296). But if this problem is about knowledge (and not experience), and in particular, knowledge at one time that other experiences don't exist—as opposed to don't exist now— then the claim is clearly question-begging. Later Balashov settles on the "distinctive aspect" of occurrence as the alleged fact that "present experiences are known to be occurring *simpliciter*, in addition to occurring when they are" (p. 298). This again pushes our question back: what is occurring *simpliciter*? The answer is more Latin— the experience, he says, is "*sui generis*" (p. 298)—but we never get more *lumen*.

Critics of the token-reflexive account like Balashov have a point, but it is misdirected if thought to bear directly on the metaphysics of time. What I think has happened is as follows. The critics are undoubtedly right that the token-reflexive account doesn't "account for" the psychology of an experience itself. The token-reflexive account explains a correlation between a type of judgment, belief, or other propositional attitude about presentness and an experience; and the character describes a correlation between a linguistic item and a content. Neither accounts for or describes a psychological experience itself. But that was never their job! In accounting for indexicals psychologically, we need more than the content versus character distinction.

This point (again) is already appreciated for other types of indexicals. As Recanati writes, when using the term "I," "I think of myself as myself, not as the utterer of such and such a token" (1993, p. 71). Psychologically we need more than the character. What is needed are some *psychological* modes of presentation (Recanati, 1993, p. 72ff.), or more generally, an "internal mental symbol or vehicle" (Nichols, 2008, p. 522) that corresponds to I-, now-, and here-concepts (see also Rey, 1997, p. 290).

---

[3] On its face, this seems natural. Hoerl (2009) recently argues, however, that we actually don't have tensed contents.

This is a point appreciated by Frege (1956), Evans (1982), and others. In coordinating a subject's location, time, and self with action, there is great value in having such descriptively thin concepts. One can act even if one doesn't have the description associated with your self (e.g., "the shopper with the leaky bag"), the time (e.g., "it's five o'clock"), or the location (e.g., "in room 215"). It seems very plausible that we have such concepts. Ismael (2007) makes a strong case for their importance in locating us and guiding action. Complaints about the token-reflexive account of the now missing something crucial about experience would seem to point only to the existence of a mental token corresponding to the now. How one could see through this extra layer all the way to the metaphysics of time is beyond me.

In sum, we have not found any distinctive aspect of experience that deserves to be dubbed experience of the tensed property of *being present*. Direct appeals to phenomenology are unconvincing. And even if they were compelling, there would be many more layers between the phenomenology and the metaphysics that would need to be crossed before one gets anywhere near concluding that what's experienced is the tensed metaphysical categories of our world. After all, hallucinations exist, yet it's wrong to infer from these experiences that pink elephants do too. So even if we found a phenomenal present, that wouldn't automatically be a point in favor of the tensed theory. We haven't, in any case. The literature focuses on various puzzles involving temporal indexicals and then adds a thin coating of phenomenological language to justify the use of the word "experience." To argue from the existence of unarticulated constituents of thoughts or the need for contexts to a particular metaphysics of time is to recklessly ignore many other possibilities. These arguments at best point to the need—independently posited elsewhere anyway—for a nowness mental symbol or vehicle. Despite many claims to the contrary, we have not found support for a tensed metaphysics in our experience of the present.

9.4 From Metaphysics to Psychology: Perceived Synchrony

We began this chapter reacting to some philosophers who believe that we directly perceive an aspect of manifest time, namely, presentness. We found such claims unconvincing. That raises the question, where do we get the idea of the manifest present? Does it originate in experience? And if so, which ones? My overall picture, recall, is that the model of manifest time makes sense for creatures like us, despite it not being an accurate fundamental picture, because the model is the natural by-product of solving various challenges facing us. Before getting to that, however, we ought to acquaint ourselves with some of our experiences, judgments, and challenges. We need to assemble the relevant material before putting it all together in the next chapter.

With regard to experience, it is contentious whether we experience the present, but it is not controversial to claim that we weld together some sensory input into coherent forms; nor is it controversial that we make judgments of simultaneity. We'll add some nuance to the idea momentarily, but for now I'm simply referring to the familiar fact that, for instance, I can experience that the doorbell rang at the same time as I saw my dog Odie get up, or that I perceive Odie barking while (simultaneous with) seeing his mouth moving. The former is what psychologists sometimes call *subjective simultaneity* or *perceived synchrony* and the latter is a case of audiovisual integration into a simultaneous percept.[4] The neuroscientist Pöppell (1988) believes that the foundation of the present lay in subjective simultaneity and multimodal integration:

> our brain furnishes an integrative mechanism that shapes sequences of events to unitary forms . . . that which is integrated is the unique content of consciousness which seems to us present. The integration, which itself objectively extends over time, is thus the basis of our experiencing a thing as present. (1988, pp. 62–3)

I'm not sure if we do experience anything as present, as Pöppell suggests. But one might think that such an integrative mechanism is nonetheless the basis of our claims of subjective simultaneity, and ultimately, belief in a manifest now. For me, it won't be the whole story (see the next chapter) but it will be a crucial part. In the next few sections I'll be at pains to display the evidence for such integrative mechanisms, how they are responses to challenges, how they vary, and finally, whether an experience of synchrony "pops out."

The grand challenge facing us is easy to see. Our brains are under siege from a confusing and chaotic barrage of information about the internal and external worlds. All of this information originates in events located in the backward lightcone of the brain, so we are obviously not experiencing physical simultaneity; nonetheless, many of these signals originate in the same event or events that were simultaneous with respect to your local inertial frame, and sometimes we get things approximately right. Consider an explosion. It may emit light, heat, sound, pressure waves, and various chemicals. These effects will impinge on our different sensory modalities enabling us to see, hear, feel, smell, and even taste the explosion. Each of these signals travels at different speeds. Some travel very quickly—light at 300,000,000 m/s—and others slower—sound at roughly 330 m/s in normal atmospheric conditions. Many of these signals are then integrated together or judged to be simultaneous. How do we do that?

In fact, the problem is even harder than just indicated. Most signal speeds depend on the ambient conditions, e.g., smell on wind. Even something relatively stable like sound waves are quite sensitive to humidity, pressure, and so on, as talking under water makes abundantly clear (sound travels as much as five times faster underwater). The

---

[4] Note that we typically don't experience strict physical simultaneity. Everything we experience lies in our past lightcone. Some of these events are simultaneous with one another (in one's local reference frame). Most of these won't sneak into the same experience for you because the signals are coming to you at different times. Nonetheless, we do experience some events as happening at the same time, whether or not these events are strictly simultaneous.

signals emanating from one and the same event arrive at the body over a large interval of time. Moreover, the body is a spatiotemporally extended object, so even where signal speed is otherwise the same, signals will reach different parts of the body at different times. Even within a modality this will be true, as sound will strike one ear before the other, heat one body part before another, and so on. This difference is not always negligible. For instance, a signal on the nose will arrive in the brain (cortex) approximately 30 ms before a signal on the toes. Once in the brain we still find great variance. Some pathways are longer than others. The finer features—such as the light intensity— can also affect the processing, as can whether you are paying attention. A priori, there is absolutely no guarantee that when the stimuli reach your consciousness they will bear any correspondence to the physical temporal relations among events. Yet if we're to successfully navigate the world, we had better have representations that correspond passably well to the objective temporal relations. We need to tell whether that rattling sound comes from the same place and time as that snake popping into view.

Remarkably, despite all these hurdles, we do get the approximate temporal sequence correct for salient macroscopic events. Take the binding of multisensory signals from a common target into a subjectively simultaneous whole. Let's concentrate on the audio-visual simultaneity window, for that is the combination about which the most is known, and also, given the goal of the next chapter, the one most relevant to the inter-subjective agreement about the now. Suppose a friend from across the room shouts "Now!" The light carrying the visual information is traveling at 300,000,000 m/s and the sound is lagging behind, traveling at merely 330 m/s. Note that this huge difference in speed is compensated, in part, by two processes that favor sound:

(a) the mechanical sound transduction by the hair cells of the inner ear is many times faster than the chemical phototransduction in the retina, and

(b) the neural transmission time from the visual cortex to the cerebral cortex is about 30 to 50 ms longer than that from the auditory cortex to the cerebral cortex.

The exact calculations of compensation will depend on additional features,[5] but it turns out that the horizons of simultaneity between light and sound in perception typically intersect at about 10 m from the subject. That huge difference in signal speed is more or less wiped out by (a) and (b) when you're 10 m away from your friend. Indeed, in a sense it almost does too good a job, for typically people report signals as simultaneous when the audio *lags* a bit behind the visual.

What's striking is that our brain is surprisingly tolerant of asynchronous information. When your friend speaks to you from up close ($\ll$ 10 m) or far away ($\gg$ 10 m), the visual impression of her lips matches the sounds from her throat. She never appears as

---

[5] The narrow range implied by (a) and (b) should not be considered fixed. The intensity and contrast of the stimuli can affect the response latencies of nerve cells, as can the place of stimulation upon the retina in the case of sight. In the case of sound the signal speed from the target also varies and matters.

an actor does in one of those old poorly dubbed Japanese-to-English *Godzilla* movies. From experiment (Dixon and Spitz, 1980) we know that your friend could delay the emission of "Now!" by as much as 250 ms after moving his or her lips before you would notice the discrepancy. The brain *gets it wrong*—the sound came much later—to *get it right*, for typically those sorts of signals do originate from the same event.

There are limits: the phenomenon of thunder and lightning is perhaps the most conspicuous case, where we hear the former later than the latter. My children are afraid of thunder, but I always tell them hearing it is a good thing – it means you were fortunate enough to survive the lightning strike! The same kind of thing happens with the image and sound of a car door shutting far up the street. And if the event is up close, we can react quicker to an auditory source than to a visual one, so for quick reactions the brain might not wait for the visual information.

How the brain manages to get the objective temporal sequence more or less correct despite all the discordant signals and pathways is an active subject of study. Psychologists debate whether a "brain time" or "event time" theory is correct Holcombe (2013). According to the former, the experienced order in time is structurally isomorphic to the sequence of correlated brain processes (Köhler, 1947, p. 62). You experience events in the order in which they make it to the locations of your brain responsible for consciousness. According to the latter, the experienced temporal order depends not merely on the arrival of the signals, but on some features carried in the signals themselves. The first theory explains the above asynchrony tolerance by simply pointing out that we are insensitive to the difference between certain lags—and that's all there is to it. On this view, there is a window of temporal integration. If the amount of time required to bind together cross modal stimuli is large, then we simply wouldn't notice small asynchronies between the light and sound. (See Fig. 9.1.) This is surely part of the right answer. The question is whether this is the whole story, or are there adaptive processes helping one get the timing right?

A useful analogy is with "smart" and "dumb" keyboards. Many phones and other computing devices employ so-called smart electronic keyboards: the menu of options on the keyboard varies with the computing application in use. When emailing, the "@" symbol displays, when online, the ".com" button appears, and so on. What keyboard we see depends on the type of application. Perhaps our cognitive systems are similar. "Dumb" compensation is perceiving signals that arrive asynchronously as synchronous just because we don't notice the difference. Of course, this compensation might be very clever even if "dumb"; after all, our sensitivity has been shaped by millions of years of evolution. But we can also imagine "smart" compensation, compensation that adapts the size, shape, and behavior of the integration window or even bends signal streams to fit some windows rather than others. Smart compensation is perceiving signals that arrive asynchronously as synchronous due to some adaptation of the window or the signal stream because of the type of information carried by the signal. For example, perhaps the brain judges that the sound signal and light signal probably originated in the same spatial location and uses that as a hint that they ought to be bound together.

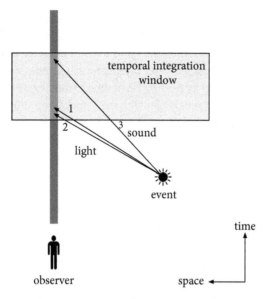

Figure 9.1 Temporal integration window

While research is ongoing, there are many hints that we do employ some "smart" tricks in dealing with time.

9.4.1 Temporal ventriloquism

Ventriloquism is the ancient art of making a sound appear in a spatial location that is not its source. As we've all seen, the sight of the puppet's mouth captures the sound emanating from the puppeteer. Since the puppeteer's mouth is not visibly moving the brain matches the sound's location in space to the mouth that is visibly moving. The *modality appropriateness hypothesis* (Welch et al., 1986) is the thesis that the sensory modality that provides the most accurate information will dominate the percept created by the brain. Vision is considered the most accurate source of information about spatial relations, and spatial ventriloquism provides some evidence for this hypothesis. The sight of the moving mouth dominates (i.e., attracts) the auditory information. Hearing, by contrast, is judged the most accurate modality regarding temporal relations. Are there cases, then, of sound altering one's visual impressions in time like those of sight altering one's auditory impressions in space? Does temporal ventriloquism exist?

Researchers have recently discovered many such cases. In Fendrich and Corballis (2001) subjects were asked to judge when a flash occurred by stating the clock position of a rotating marker. *Preceded* by a temporally proximate audible click, the flash was seen earlier; *followed* by a temporally proximate audible click the flash was seen later. The one stimulus "captured" the other, bringing them closer together in time. Many researchers have found similar phenomena (see e.g., Scheier et al., 1999; Morein-Zamir et al., 2003; Spence and Squire, 2003). In general, if the auditory and visual stimuli are within 200 ms apart, then the visual event is attracted toward the auditory stimuli. (These measurements are typically made using temporal order judgment (TOJ)—see below—or simply by reading clock time.) Similar results have been found between tactile and visual stimuli, again with the visual being shifted. In general, the phenomenon has been replicated many times and seems quite robust.

Temporal ventriloquism is one quite dramatic example whereby the brain can maintain a perception of synchrony even with quite asynchronous inputs. It's especially interesting to us because it's a concrete example wherein a "wide" window of integration doesn't appear to be the whole story. Here one stream of information is being bent toward another stream of information. Why does this happen? With the modality appropriateness hypothesis in mind, we can note that there is evidence that the effect optimizes the chances of matching stimuli from the same event (Alais and Burr, 2004). The effect clearly reduces the lag between senses, and as such, possibly helps produce a coherent temporal organization of the world.

9.4.2 Temporal recalibration

Another type of mechanism the brain may employ is not the capture of one input by another, but the recalibration of the simultaneity windows themselves. Although there may be connections between ventriloquism and recalibration, the key difference is that recalibration is an effect that happens to temporal integration over an extended period of repetition.

The idea common to many of these studies is to present observers with a succession of stimuli with different lag times between them. For each subject a lag time that maximizes reports of perceived subjective simultaneity (point of subjective simultaneity (PSS), see Section 9.5) is calculated. Then subjects are exposed to a *fixed* lag for a few minutes. Do the subjects, after experiencing these fixed lags, eventually adapt and "erase" the lags, perceiving the events as synchronous? The answer, discovered by Fujisaki et al. (2004) and Vroomen and de Gelder (2004), is that they do. The PSS is shifted, to a degree dependent on the lag size introduced, in the adapted direction. This effect has been found in every combination of signals from audition, vision, and touch.[6]

Consider the audiovisual case. Subjects are exposed to *beeps* before *flashes*. Through repetition subjects gradually adapt to this lag, their PSSs gradually shifting. That means they became more likely to judge the two stimuli as simultaneous than before. These shifts in PSS were sometimes greater than 50 ms. Fujisaki et al. (2004) make

---

[6] Recently temporal recalibration has been shown to exist even within one sensory modality. If a subject sees an object that changes its color (red to green, and vice versa) and also its direction of motion (left to right, and vice versa), one can ask whether the change of direction is synchronous with the change of color. As in the multimodal cases, temporal recalibration was observed (Arnold and Yarrow, 2011).

plausible that this "lag adaptation" is an adaptation in sensory processing rather than cognitive processing by showing that the lag adaptation changed the temporal tuning of a hearing-influenced visual illusion. They claim that their findings "suggest that the brain attempts to adjust subjective simultaneity across different modalities by detecting and reducing time lags between inputs that likely arise from the same physical events" (p. 773).

How this recalibration works in the "wild" is not yet known. Apart from American teenagers, human beings do not spend their days sitting in front of computer screens looking at single targets. We live in a cluttered environment, bombarded by multiple events at any given time, some requiring different recalibration than others. Early indications are that the brain can manage this adjustment with audiovisual stimuli from multiple physical events (Roseboom and Arnold, 2011). For audiovisual pairs coming from one event, it may recalibrate in one direction, and yet for audiovisual pairs coming from other events, it may recalibrate in another. Furthermore, recalibration between two stimuli from a common source can be affected by introducing additional audio or visual events (Roseboom et al., 2009). Recalibration between any pair of stimuli may hang on the presence of a third and that may in turn be independent of recalibration between another pair of stimuli.

Recalibration can produce very surprising effects. Suppose we induce a recalibration by repeated exposure to a lag of length L. Now we "surprise" the subject by presenting the two stimuli with the lag L completely removed. One would predict that one would then see the two events as asynchronous for a while. That is indeed what happens, but it is especially striking in cases where motor control and vision are the pairs recalibrated. In one experiment (Cunningham et al., 2001), subjects moved a mouse that caused a spot on a computer screen to move. Gradually a lag between the movement of the mouse and the resulting effect on the screen was introduced. Subjects informally reported that soon their actions and effects were subjectively simultaneous again. The experimenters then suddenly shut off the lag, and the subjects reported that the effects on the screen occurred *before* they moved the mouse! Haggard et al. (2002) set about testing directly whether a subject's intentions affected the experience itself of what things happen simultaneously. They confirmed that it did. See Eagleman and Holcome (2002) and Stetson et al. (2006) for more work and discussion.

Perhaps the most controversial instance of simultaneity recalibration is an example where the brain seems to take into account *target distance*. In Sugita and Suzuki (2003) subjects were presented with bursts of white noise (10 ms duration) through headphones to simulate external sound from the frontal direction. Brief light flashes were produced by a uniformly spaced array of 5 green LEDs at different distances (1 m to 50 m). The intensity of light was altered so as to produce consistent intensity at the eye. Subjects were then asked to imagine that the LEDs were the source of the light and sound, while listening to sound directly from the source. To estimate subjective simultaneity, observers judged what came first, light or sound. The surprising result is that subjective simultaneity increased by about 3 ms with each 1 m increase in distance up to about 40 m. Now, as it happens, sound travels at roughly 1/3 m/ms at sea level and room temperature. Coincidence? Sugita and Suzuki think not. They claim that the "results show that the brain probably takes sound velocity into account when judging simultaneity" (p. 911).

This fascinating suggestion has drawn plenty of opposition. Variations on this experiment have not always reproduced simultaneity recalibration as a function of target distance. Yet some have. As I write, we must wait on new data and theory to sort out the disparity among experiments. There is some type of recalibration going on, but the cause is probably not simply target distance. Taking into account target distance would seem to be a computationally complex task, so there are probably various proxies for target distance employed by the brain. The question is then what heuristics the brain uses to tell whether inputs are likely from the same source. Zampini et al. (2005) show that one very natural strategy seems to be employed, namely, that subjects are more likely to report stimuli as simultaneous when they originate from the same spatial location than when they come from different spatial positions. In any case, this effect, like the lag exposure effects described above, suggests the existence of malleable windows of simultaneity.

9.4.3 Comments

The phenomena of ventriloquism and recalibration challenge the "brain time" theory of experienced temporal order. How much of a challenge they pose depends on how widespread they are, and this is too controversial to call at the time I write. Impressed by how well we compensate for all the signals lags, processing lags, and so on, Kopinska and Harris (2004) suggest that we should in the light of the evidence conceive of ourselves as possessing a mechanism of *simultaneity constancy*, comparable to other mechanisms of perceptual constancy. Examples of perceptual constancy include our ability to represent a penny as circle-shaped throughout all our viewing angles or our ability to represent an object as a single color despite all the variation in shading, contrast, and so on.[7] In general, the idea of a perceptual constancy is that we represent something as invariant even when the information arriving at our senses is not. In the case of time, simultaneity constancy is the claim that we represent events as simultaneous despite the huge variations in signal speeds, processing, and so on that we have discussed (Harris et al., 2010).[8]

However, as striking as these two challenges are, we shouldn't be led to believe that the brain is *always* striving to obtain the correct temporal sequence through adaptation. The adaptive processes just described, if correct, are processes working

---

[7] See Cohen (2012) for discussion of perceptual constancy.

[8] Perhaps, as Harris et al. suggest, there is a three-stage process. First, stimuli are fitted into various temporal windows. The inputs in these windows are the candidates for recalibration. Second, unfamiliar stimuli are delayed according to fixed rules (say, 40 ms delay for sound to be bound with light plausibly from the same source). Third, for familiar stimuli often experienced from the same target, a more fine-grained delay is used. This process might try to erase a lag as two stimuli are experienced together more and more frequently.

on the millisecond scale, not the scale of minutes, days, and years (Holcombe, 2013). Nor is it clear that the brain is in fact striving for the correct temporal sequence even on short time scales. That's an open question. Sometimes it seems to be striving for other goals at the expense of getting the order right. Over a hundred years ago Benussi pointed out that particular sound sequences will result in perceived illusions of temporal order (Benussi, 1913; Albertazzi, 1999). If one is presented with a low tone, a noise burst, and then a high tone, one will typically hear instead a low tone, high tone and then a noise burst.[9] A reasonable supposition is that one perceives the low–high tones together because they form a natural package or gestalt. Getting the objective order may not always be of paramount importance.

To be clear, the mechanisms we've been discussing typically operate at an unconscious level. The intervals of time we've been talking about are very short. How do these present patches connect up with consciousness? Answer: I wish that I knew. Although some cognitive factors may be at play, typically all that window bending, integration, and so on is happening before anything gets conscious. My claim is only that this work is the *foundation* of any conscious present. Subjective experiences that underlie reports of subjective simultaneity depend upon this integration. How exactly the products of these integration mechanisms get ratcheted up to consciousness is not something about which I'm making a claim.

9.5 Interlude: Measuring Subjective Simultaneity

You witness a brief flash and hear a quick beep. Did you experience them at the same time? Cognitive scientists study your answer to this question and others corresponding to other sensory modalities and stimuli primarily with two different types of measurement, *simultaneity judgments* (SJ) or *temporal order judgments* (TOJ), both of which ask you to make reports about perceived timing. (Modifications of these tasks, plus entirely different ones, also exist, but TOJs and SJs are the canonical types.) They also study this question via the binding together of stimuli, such as whether you hear one or two tones when two short tones are played closely together in time.

Take timing reports first.

In the first type of measurement, SJ, one presents two stimuli separated by different temporal gaps ("stimulus onset asynchronies") and simply asks subjects whether the two stimuli are synchronous or not. Then one plots, as in Fig. 9.2, the number of reports of synchrony (the y-axis) as a function of these gaps (the x-axis). For very large gaps people do not report synchrony and for very small ones they do, as one expects. In general, therefore, the function plotted is typically a bell-shaped Gaussian. Two quantities are derived from this function. One is the *point of subjective simultaneity*, or PSS, which is defined as the peak of this function. This represents the gap at which the subject is most likely to report "synchronous." You might, for instance, say that the

---

[9] See Koenderink et al. (2012) for a convincing online demonstration of this effect.

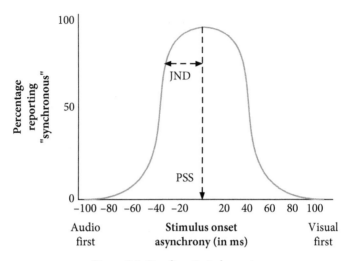

Figure 9.2 Simultaneity judgments

beep and flash occurred simultaneously most often when the beep follows the flash by 80 ms. Don't make the mistake of thinking that we perceive the click and flash as synchronous when they objectively are! Generally that is not the case. The other quantity is the *just noticeable difference*. This is the average interval at which the subject reports "synchronous" at least 75% of the time. You may answer "synchronous" 75% of the time when the gap is all the way up to 110ms, for instance. The just noticeable difference is sometimes regarded as revealing one's "window of simultaneity."

In the second type of measurement, TOJ, one again introduces stimuli separated by a variety of temporal gaps. Now instead of asking whether two stimuli are synchronous or not, ask subjects which stimuli came first. The flash or the click? Plot the "light first" responses against the temporal gaps (see Fig. 9.3). This time we will typically obtain an S-shaped curve. In this scenario the PSS is defined as the 50% crossover point in lag time, e.g., as that temporal gap between stimuli where there are as many "light first" as "sound first" reports. Intuitively, it's the gap at which you *just can't tell* which came first. For the same temporal gap, half the time you say that the flash happened first and half the time that the click did. Presumably you're guessing, and you're doing that because the two stimuli seem like they happen at the same time. The just noticeable difference in this case is given by half the difference between the 25% "light-first" point and the 75% "light-first" point.

Both measures, SJ and TOJ, can obviously be generalized for testing synchrony between stimuli from any pair of sensory modalities, e.g., auditory and tactile senses. They can also be used within a single modality, e.g., two flashes at different locations. These measures are both based on *judgments* of timing.

Turn now to sensory integration. It is also possible to measure when different stimuli are bound together. These tasks are also self reports, but they seem to be getting more

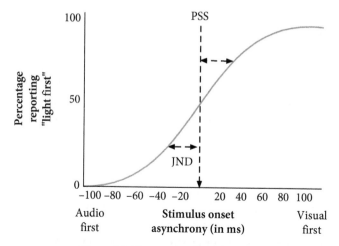

Figure 9.3 Temporal order judgments

at the experience itself because typically the integration either happens or doesn't. For instance, within one sensory modality, subjects will fuse two events together as one if they are presented closely enough to one another. Put headphones on a subject and let her listen to clicks lasting for 1 ms. If the left and right ears are stimulated simultaneously, then the subject hears not two clicks but one fused tone. One can then stimulate the two ears non-simultaneously but very close together at 2 ms apart, and the two acoustical stimuli will still be fused together. But if one stimulates the two ears much further apart, say at 3 ms or 4 ms, we pass the fusion threshold and suddenly hear two clicks (see the classic experiments by Hirsh and Sherrick (1961), followed by Fraisse (1984) and Pöppell (1988)). The experiment simply asks how many clicks you heard, one or two. Similar experiments and fusion occurs for visual and tactile stimuli (I'm not aware of tests with olfactory or taste senses).

The fusion thresholds for these three senses differ. Although they vary with individuals, to get a rough sense of the numbers, note the following typical fusion thresholds:

- Auditory: ~2 ms to 3 ms
- Tactile: ~10 ms
- Visual: ~20 ms.

Hence the auditory modality is our most discriminating.

Interestingly, it is a striking and robust finding that in each of these modalities there exists a temporal interval in which the two stimuli remain unfused and yet—surprisingly—their *order* nonetheless *cannot* be perceived. After the fusion threshold, there is what is sometimes called an "order threshold," the threshold at which order is perceived. If we do the same experiments as above but wait a few more milliseconds past the above thresholds, then a subject can tell that there are two clicks, yet she cannot tell which one came first. The order threshold, the minimal interval between two stimuli at which one can discern which came first, comes much later. If you see brief flashes separated by only 25 ms, you might report that you saw two flashes but not be able to say which happened first. One would have thought that a condition of individuation for sensing two events *as two* is that one could tell their relative temporal order, that their spacetime location is what individuates them. Not so. Just as there exists a fusion threshold, so too does there exist an order threshold, the point by which order is reliably determined. This threshold varies less by modality than fusion does. When stimuli are separated by about 30 ms, be they auditory, visual, or tactile, they are generally perceived as having a discernible temporal order. The not-fused-but-unordered window thus tends to be largest in the auditory modality.

Of course we also integrate multisensory stimuli. Compared to the within-modal integration, here we find a generally much larger window of integration, or put the other way around, a much lower temporal resolution. Typically for maximum integration in the audiovisual case, the sound will lag the visual cue by as much as 100 ms. A particularly interesting way to measure integration is to measure it in the context of a perceptual illusion. The Stream–Bounce illusion is a popular audiovisual example (Fig. 9.4). Here two identical "balls" intersect each others' path on a computer screen. When the intersection is accompanied by a sound, subjects report that the balls bounce off one another; when no sound occurs, the two balls instead seem to stream through one another. A temporal gap can be introduced between the visual interaction and the auditory event and one can then determine the gap necessary for maximum chances of seeing a bounce rather than no bounce, i.e., maximum integration of click with visual intersection.

Hearing affects vision in this case, but there are similar illusions between other modalities. For instance, in the striking McGurk effect, visual impressions affect heard

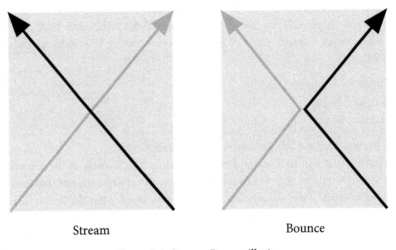

Stream   Bounce

Figure 9.4 Stream–Bounce illusion

speech. The canonical example is of a video of someone repeating the phoneme /ga-ga/ dubbed with a sound recording of someone saying /ba-ba/. The lip movements of /ga-ga/ together with the incoming sound of /ba-ba/ confuse the brain and what we hear is /da-da/. Close your eyes and the illusion immediately vanishes, leaving the listener hearing /ba-ba/. As with the Stream–Bounce illusion, one can now introduce a gap between the signals, here the lip movements and phoneme sounds, and then determine what gap maximizes the illusion.

9.6 Exploding the Now

Now that we have an idea of how the brain accomplishes the amazing feat of unscrambling the temporal sequence from a confusing mass of signals, we can return to our question of whether we perceive synchrony. Granted, we certainly can and do report that some stimuli appear simultaneous and others do not, but that by itself doesn't imply that we directly experience these stimuli as synchronous. In psychology it is notoriously difficult to tease apart what is cognitive from what is sensory. When two flashes occur at different locations, do you genuinely experience their synchrony, or is that put together later? We'll soon consider experiments meant to address how "automatic" synchrony is. For the moment, assume we do experience the synchrony or asynchrony of the flashes. Typically more is going on, of course, and we can ask whether the different senses of synchrony hang together as we think they do.

Introspectively it seems clear that they do. We're presented with what seems to be a tidy unified percept of what is happening at any one time. Yet if there is anything we've learned in cognitive science, it's that initial appearances are deceiving. Investigation of the spatial binding problem, for instance, shows that spatially coincident color, shape, orientation, and motion can all come apart. Take a field of dots moving back and forth, colored black when moving rightward and colored white while moving leftward. Even though we can tell that the color and direction are both changing, at some rates the color becomes completely "unbound" with the direction (Moutoussis and Zeki, 1997). Our experienced "here" is not as unified and tidy as we think. Nor is our experienced now.

Consider the two ways of testing synchrony discussed in Section 9.5, testing via PSS values, and testing via multimodal integration. Take PSS first. Based on introspection, one would expect the PSS as measured by SJ tests to approximate the PSS from TOJ tests. One might reason as follows. If one reports two stimuli as "synchronous" in a SJ task, then one doesn't notice a temporal gap between them. Therefore, when judging temporal order via a TOJ task, one should be guessing and performing at chance levels. And vice versa: if the temporal order isn't stably noticeable, then one should expect reports of "synchronous" in SJ tasks.

A similar argument would link the PSS in either sense with sensory integration. Consider the Stream–Bounce illusion again. Surely one thinks that the bounce happens because the sound is bound to the visual intersection of balls, thus making it appear that they collide. Hence one would expect that we would obtain maximum bounce in the Steam-Bounce illusion at an asynchronous gap approximately equal to the PSS calculated from one of the synchrony tasks. That is, you would see the bounce when the two stimuli are simultaneous for you. One might reason similarly with the McGurk effect.

Although the above expectations are perfectly natural, it turns out that for the short time spans we're discussing neither result obtains:

- *No correlation.* In the small time scales of interest, it's well known that there is fairly poor agreement between the PSS derived from SJ tasks and the PSS derived from TOJ tasks. In fact, there is no correlation between them (see, e.g., Eijk et al., 2008).
- *Negative correlation.* Recently Freeman et al. (2013) report that the point of maximum bounce is significantly *negatively* correlated with the PSS (via TOJ) calculated from asking which happened first, visual intersection or click. Similarly for the McGurk effect: maximum McGurk illusion is negatively correlated with one's PSS via TOJ).

Both results are surprising. The first one means that you might reliably notice, say, an audio signal arriving before a visual one even if you classify the two as synchronous; alternatively, you might not be able to reliably tell which happens first and nonetheless classify them as asynchronous. The second result is perhaps even weirder. The intersection and click seemed to happen simultaneously (in the sense of TOJ) when the intersection temporally *led* the click, but the bounce phenomenon was maximized when the intersection *lagged* the click. The same also occurred for the maximum McGurk illusion for speech. (In the next chapter we'll meet a subject referred to as PH for whom the point of maximizing audiovisual integration and PSS values differs by a whopping 450ms (longer than it takes a baseball to go from the pitcher to the catcher).) One wants to ask: why did we see a bounce if we didn't think the click happened during the visual intersection?

The explanations of the above divergences are still open questions. The divergence between the two PSS values might be due simply to the experimental questions prompting different default behaviors, that the first test's question ("simultaneous or not?") inclines the subject to think that events are simultaneous whereas the second ("which came first?") inclines the subject to think they never are. In addition, we know that there is a range of onset asynchronies where it is possible to tell that two events happened but not reliably tell which happened first. It may not be clear to subjects confronted with such stimuli whether they ought to answer "simultaneous" in an SOJ task. The divergence in PSS values may not be evidence that one's *experience* of synchrony is in any way disunified. Still, the lack of a correlation is unexpected. If one hoped for evidence from PSS tests confirming that we do have a strong unified percept of simultaneity, one doesn't find it here. The negative correlation between PSS and the point of maximum integration, by contrast, seems less likely to be due to questions prompting different default reports. One either sees a bounce or hears a "da-da" or not, and one cautiously expects that subjects can report this reliably. While TOJ tests have their limitations, certainly it's astonishing that one doesn't get maximum bounce when one can't tell which happened first, the click or the intersection. This case pushes Freeman et al. to ditch the so-called "unity hypothesis," the idea that the brain has mechanisms to enforce a unified percept of external events.

Now put these results in context. We already learned that each sensory modality has its own fusion thresholds. Two sounds might be bound together at a different temporal resolution than two visual stimuli. The same goes for the different types of multimodal integration. We also know that the brain is "filling in" or "leaving out" a lot of what we experience for physiological reasons. To build up a good visual model, the eyes rapidly dart around (saccade) approximately three times per second (on average). We don't notice the chaotic blur that would result from this. Instead saccadic suppression turns off visual processing during these moments; that is, visual processing suppresses the motion blur that would otherwise be distracting. For example, when one grabs a continuously moving second hand with one's attention, it seems to pause just as one grabs it. Only the second hand pauses, not everything else, and we never notice all the portion of the day wherein we are effectively blind due to saccadic suppression.

Now imagine an ecologically valid scenario. People are talking to you, asking you the time perhaps. Meanwhile objects are colliding, making noise, and altering their shapes. Someone snaps their fingers. Music from the radio arrives at the ears. What are you experiencing *right now*? It's hard to say. It's not so clear that our conscious experience has a single temporal resolution.

From the above results it would be hasty to conclude that we don't have an immediate perception of synchrony. We have not explored all alternatives yet. However, we are inching toward the startling conclusion that the more we probe the subjective now the more it falls apart. Is there any positive evidence of an immediate experience of synchrony?

9.7 Does Synchrony Pop Out?

The way cognitive scientists approach the question of whether we have immediate experience of synchrony is by asking whether the detection of synchrony is *automatic*. A process that is not automatic is one that is typically slow and serial, one that demands attention; a process that is automatic, by contrast, is one that requires minimal attention and is otherwise very fast. Obviously automatic and non-automatic represent ends on a spectrum. The idea is that the automatic detections are lower level than the non-automatic ones. While it's hard to tease apart what is directly experienced from what is not, it's at least possible to separate fast low-level processes from slow higher-level ones.

One often determines how automatic a process is with a test for "pop out" (Fig. 9.5). These tests can be implemented with a standard search diagnostic (Beck,

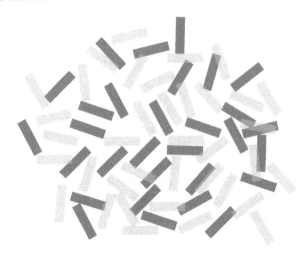

Figure 9.5 Pop out

1966; Treisman, 1982). The idea is that if the phenomenon is immediate, it should become salient to the observer. And if so, the reasoning goes, then the subject should be able to use that pop out effect to help in a search for a particular type of object. If I ask you to search for tilted lines against a background of vertical lines, the orientation will quickly pop out and you'll subsequently find the tilted lines very quickly. But as we know from word jumble games, it's often very difficult to find words when immersed in nonsense words. The conclusion is that orientation is signaled early in visual processing, whereas word recognition happens much later.

Is our experience of synchrony automatic? Unfortunately the evidence as I write is unclear. There are some experiments that suggest that perception of audiovisual synchrony is a slow and serial process, that one needs to compare temporal information from within-modal signals in order to obtain audiovisual synchrony (Fujisaki et al., 2006). Yet as Keetels and Vroomen (2012) point out, this study and others pointing to a similar conclusion offer stimuli that are presented very quickly or lack sharp transitions. It might be that for more salient stimuli synchrony pops out.

And indeed, in some interesting experiments by Van der Burg (2008, 2009), that seems to be the case. In the 2008 experiment, a bunch of lines oriented in different directions are presented. The lines change color at random times, from green to red or red to green. Your job is to search for the vertical or horizontal lines amongst all these distractors. The authors found that the search time dramatically improves when one presents a sound—a pip—synchronized with the target line changing color. This assistance was barely affected by adding more distractors. And a variety of experiments were done to tease out the "warning" effect that such a pip might present, as in "warning, the target is about to change direction" message that would clearly affect performance. Here it's important to note that the pip was presented simultaneously with the color change or even just after. The authors conclude that synchrony does pop out at you fairly automatically. If it didn't, the reasoning goes, then the pips—which give no hint of spatial location of the target—wouldn't be much help. Just as a line's orientation can jump out at you against certain "backgrounds," so too does synchrony help one find the present targets. In the 2009 experiment synchrony again popped out, this time with tactile stimuli simultaneously presented with the target. If the interpretation of these experiments is correct, these are the first demonstrations that synchrony, like spatial orientation, can pop out at us.

9.8 Conclusion

On the beach someone behind me was blowing bubbles. Due to the offshore breeze, the bubbles, like the surfers and small children, had trajectories taking them to the ocean. I was vaguely aware of two or three of the bubbles. One bubble hesitated dangerously near the sand before heading out to sea and popping. Two popped before making it to water, the big one first and then the small one. I'm aware of the temporal features of each bubble's worldline—that the first bubble hesitated before going out to sea—and I can compare temporal features—if asked, I can say that the big one passed by me first and the small one second. But one can additionally ask me whether two of the bubbles popped at the same time. Certainly I can give an answer, but is that answer inferred from comparison of the worldlines and my modeling of the situation, or is the synchrony something I directly sensed? Did their popping pop out?

We began the chapter asking whether we sensed a special (tensed) property of presentness missed by physical science. We found precious little evidence for this assertion. In fact, all we found were some linguistic features associated with temporal indexicals, features shared by all indexicals. This question then took us to a more modest one, whether we experience not a tensed property of presentness but rather synchrony between some (past timelike) events. In the van der Burg experiments we did find synchrony pop out. So it does seem that a low-level automatic detection of synchrony may sometimes occur. Laboratory tasks with few and short duration stimuli may pop out, but that is a far cry from a general pop out in ecologically valid (i.e., normal life) situations.

Did the bubbles popping at the same time pop out at me? Perhaps. But it's very unlikely that synchrony popped out between those events plus all the other events going on at the beach at that moment, e.g., the child yelling, the surfer ducking under a wave, the feeling of sand on my toes. Experiment and theory don't yet reveal the answer. Yet one cannot help but feel that the more we probe the subjective "now" the more it falls apart. It appears that "no single underlying representation is responsible for all judgments of time" (Durgin and Sternberg, 2002), and in particular, that there may not be a single task independent representation of synchrony at all. The metaphysician's present seems undermined by physics, and the experience on which it is based gets fragmented by cognitive science.

10

Stuck in the Common Now

Manifest time attributes to the world an objective global now, one that divides the past from the future. Manifest time of course attributes more to this now than that it is global and objective, it also attributes to this now the idea that it flows and also that the world is on a particular now right, well, now. The latter two features seem to be bound up with the idea of a flow of time, and therefore, I'll postpone discussion of them to Section 11.2. Here I'm concerned with explaining why we think the now is global and objective. The last chapter called into question the idea that we even perceive a unified now. And as we'll soon see, this perceived or judged now differs quite a bit from person to person. Nonetheless, the now seems to be a real entity in the world. Why?

A good way to start is by comparing and contrasting the temporal *now* with the spatial *here*. The two are similar in many respects, so isolating where they aren't is a useful way of concentrating on what we want to explain.

Begin with space. We commonly represent space in at least two quite distinct ways, allocentrically and egocentrically. The distinction hangs on whether a particular spatial viewpoint and corresponding frame is required or not: egocentric spatial representation, as the name suggests, makes reference to the spatial frame my viewpoint picks out; allocentric representation does not. Here is Evans on egocentric space:

> The subject conceives himself to be in the centre of space (at its point of origin), with its co-ordinates given by the concepts "up" and "down," "left" and "right," and "in front" and "behind." "We may call this "egocentric space," and we may call thinking about spatial position in this framework centring on the subject's body "thinking egocentrically about space." (1982, pp. 153–4)

There are contexts where the viewpoint may not be the subject's body. What's important for us is that a certain egocentric viewpoint defines the space. Examples of egocentric spatial relationships include:

- The coffee cart is here.
- The fork is on the left, the spoon on the right.

The first claim makes implicit reference to my current location in the use of "here." If the coffee cart has just been located, it is here relative to where I currently stand. It's not here for you if you're far away from me. The division of the world into {here, there} is egocentric. We may disagree about these determinations: what's here for me may be there for you. Same goes for the left and right and front and behind relations. The fork may be on my left and the spoon on my right, but if you're sitting across from me, then the fork is on your right and the spoon on your left.

Examples of allocentric spatial relationships include:

- The distance between San Diego and Providence is 2366 miles.
- Kansas City is between (longitude-wise) San Diego and Providence.

The first refers to a metrical relationship between the two cities, one that holds wherever I am or you are. The second refers to a non-metrical ordering relation, but again one that holds without reference to me or you. A listener perfectly well understands these claims without the speaker including the egocentric space associated with his or her viewpoint. As these examples and others show, we can make egocentric and allocentric representations with any of the usual properties thought to be enjoyed by space; i.e., its topological, ordering, and metrical properties.

Egocentric spatial relationships are, naturally, egocentric, and hence there isn't much intersubjective agreement about them. I say the fork is on the left and you say it's on the right. We're not going to agree unless we insert the implicit reference into our assertions and turn the statements into allocentric ones.[1] Then you can agree that, relative to my location the fork is on the left, and I can agree that, relative to your location the fork is on the right. Hence we commonly adjust and speak of *your right* and *your left*. Doing so is a way of translating egocentric spatial claims, about which we commonly disagree, into allocentric spatial claims, about which we commonly agree.

Science, which prizes intersubjective agreement, naturally builds models of space using allocentric spatial representations. If we look to science as a guide to what's objective, we'll conclude that the allocentrically represented spatial relationships are objective and the egocentric ones are not. The world has objective spatial topological features. The jelly of a jelly donut is inside or bounded by the donut, regardless of where you are or how you orient yourself. The world also has objective metrical features. The planet Jupiter, at its closest approach to the Earth, is about 365 million miles away from us, regardless of where you are. Physical space doesn't include an objective Here, There, Right, Left, Front, Back, and so on. The world satisfies those discriminations only in virtue of us, and more particularly, our spatial Heres. Notice, crucially, that the manifest image of space largely agrees. No one is tempted to think his or her Here is objective. The intersubjective disagreement is too apparent, confronting you every time you sit at a dinner table.

Temporal representation also admits of egocentric and allocentric distinctions. In this case the distinction hangs on whether a particular temporal viewpoint and corresponding frame is required or not. Egocentric temporal representation picks out such a viewpoint and allocentric doesn't. This time egocentric representation refers to

---

[1] One must be careful here, however, for as Caponigro and Cohen (2011) show, one often can achieve agreement about egocentric reports despite differences in the egocentric elements.

one's *now*. Just as in the spatial case, the division can be made with all kinds of temporal features. Examples of allocentric temporal relationships include:

- The duration between the 1986 visit of Halley's comet to the inner Solar System and its 2061 visit is 75 years.
- Halley's comet's 2061 visit is after its 1986 visit.
- My class graduated simultaneously with Halley's comet visiting.

The first claim refers to a metrical notion, i.e., the duration between two events, and the second and third refer to species of ordering relations. Examples of egocentric temporal relationships include:

- Halley's comet visited in the past.
- Halley's comet is not visiting now.
- Halley's comet will visit again in the future.
- Halley's comet will visit in 75 years.

All four claims are true or false depending on *when* they are said, that is, on which *now* implicitly picks out the time frame (e.g., past, present, future, in 75 years). The first claim is true if said any time later than any comet visit, but its meaning depends on the now that determines what is past. The second is true if said anytime not simultaneous with a visit by Halley's comet, but its meaning explicitly depends on what time is now. Ditto for the third, so long as it's said between the first and last visits. And the fourth, which gives tensed metrical information, is only true if said during a year simultaneous with Halley's comet visiting (and strictly, only those years where the period is going to be 75 years, not 76 years), but its meaning depends on the now that acts as the standard from which 75 years is measured. There are many other egocentric temporal expressions, some metrical (e.g., yesterday, next week, in one year) and some not (e.g., then[2]).

It might be better to think of the egocentric/allocentric division in terms of spatiotemporal categories. Notice that for spatial egocentric representation we implicitly invoke time too. The fact is that we move around spatially in time, and so our *here* is a function of time. Gourmet food trucks move about, and the excited claim that one is here is as much picking out the now as the spatial here. By contrast the claim that Chubby's Burger is at the corner of 101 and J Street on August 3, 2012 is something always true from no particular spatiotemporal viewpoint—and exciting only if you happen to be there then (and eat burgers).

While it's true that we can make the egocentric/allocentric division in terms of spatiotemporal categories, we shouldn't let this obscure the deep division between time and space here. The linguistic material that gets time into the truth conditions, in

---

[2] "Then" might be metrically tinged when used indexically to pick out some some specific duration away, but it seems it can also have a demonstrative use whereby it picks out an arbitrary temporal location without any specific metrical relation to another time.

English, lies in the aspect and tense of the verb. Chubby's Burger *is* here, as opposed to *was* or *will* be here. Other languages also convey this temporal information, although not always via the verb. So spatial egocentric representation tells us about time. But notice, interestingly, that temporal egocentric representation doesn't tend to implicitly invoke space. Asserting that Napoleon's death is past doesn't require any indexing to space. It doesn't matter where you are when I say this, so long as you're listening now. There is no spatial counterpart to what the verb does for time. And that's not a peculiar feature of English: most or all natural languages build egocentric temporal indexing into sentences but not egocentric spatial indexing into languages. This is a striking and important asymmetry between spatial and temporal representation. It shows how deep this asymmetry penetrates into our language.[3]

This asymmetry is a symptom of a much deeper asymmetry, one that motivates this chapter. Let's dub it the "Representational Asymmetry":

Representational Asymmetry: we think, speak, and act in ways that treat the egocentric temporal categories as objective, but we do not think, speak, and act in ways that treat the egocentric spatial categories as objective.

We tend to think of ourselves as sharing a common mind- and frame-independent now but aren't tempted by such a claim about the here. We think the now is objective when we talk about the present being what's truly real, what's truly happening, and so on. We don't think like the time-traveling Doctor Who when he says:

You're going. You've gone for ages, you've already gone, you're still here, just arrived, haven't even met you yet. It all depends on who you are and how you look at it. Strange business, time.   (*Doctor Who*, "Dragonfire: Part 1", 1987)

Here the Doctor is treating the now like the here, but this talk strikes us as odd. (Compare also with the Yogi Berra quip mentioned in Chapter 1.) We seem endowed with some very powerful intuitions supporting the view that reality is divided into past, present, and future, and that this is so not merely relative to one's current perspective. The strength of such intuitions is evinced by the existence of philosophy of time itself, with so many philosophers arguing for a metaphysically distinguished present, as well as the reaction one finds in students when teaching the relativity of simultaneity. Part of the shock of relativity is its conflict with the idea of a special common now. The relativity of co-location, by contrast, garners mild interest. We find further evidence of this alleged objectivity whenever we treat as objective temporal discriminations that themselves rely on the now. The past and future are judged objective. Events fall irredeemably into the past. Pastness is something that happens to an event; an event is

---

[3] Here is an interesting tidbit brought to my attention by Alex Holcombe: whereas compound words like *anywhere* and *everywhere* now prevail, their temporal counterparts, *anywhen* and *everywhen*, which briefly made an appearance in the early twentieth century, died out. You can see the trend here: http://books.google.com/ngrams/graph?content=anywhen%2Ceverywhen&year_start=1800&year_end=2000&corpus=15&smoothing=3&share=.

judged either past or not, future or not, full stop. One doesn't need to say that they are past or future only with relative to some other event. Yet note: since the past and future are past or future only with respect to the now, if they are objective then so is the now. In general, as emphasized in Chapter 1, the idea that the egocentric temporal categories are objective is deeply ingrained in us, indeed, a crucial part of the way we live our lives.

A corollary of this (or maybe a cause?) is that we commonly agree about egocentric temporal representations in a way in which we do not with respect to egocentric spatial expressions. We (happily) disagree about the spatial here, but agree about *the* (not *a*) now. However, as stressed in the Introduction and defended later, physics does not make use of egocentric temporal representations—at least, explicitly. We have a gap between the time of our lives and the time of physics.

Phrased in terms of the first chapter, we can state our problem very simply. "Manifest space" agrees with "physical space" about the egocentric spatial discriminations: none are objective. "Manifest time" and "physical time", by contrast, disagree about the egocentric temporal discriminations: the first treats them as objective but the second doesn't. Assuming physics is at bottom correct about this, *why do we feel so strongly that the egocentric temporal relationships mark something objective whereas we don't have such corresponding feelings about the egocentric spatial relationships?*

10.1 Disagreement and the Case of PH

Answering our question is made more difficult by the fact that perceived synchrony, as understood at the end of the previous chapter, seems to differ markedly from person to person. Relativity theory, with its relativity of simultaneity, presents a well-publicized threat to the common now. What is not generally appreciated is that psychology presents a similar kind of threat. In this section I'll lay out this challenge. Then, invoking physics, our environments, and more, I'll suggest an answer to the question of why we objectify the now but not the here in later sections.

Despite the onslaught of time-varying information from internal and external events, we somehow manage to get things more or less right. But it's a mistake to think that we all agree. In the summer of 2013 news broke about a 67-year-old retired pilot living in England with a very strange condition: he hears people before they speak (Freeman et al., 2013). Referred to as PH, he is a very high functioning individual, scoring above average on a battery of standard psychological tests, yet lesions left in his brain due to an operation seemed to have caused visual information to consciously lag auditory signals. Sight and sound are out of synch. He noticed this while watching television at his daughter's house one day, telling his daughter that she had two televisions needing fixing due to the lack of audiovisual synchrony. The television shows seemed like they were poorly dubbed. The problem wasn't with the televisions, but rather with the fact that for him auditory signals consciously lead visual signals.

Scientists rigorously tested these claims and verified his subjective reports. To establish subjective synchrony, they had to artificially lag voices relative to visible lip movements by 210 ms. In other words, his PSS—as measured by TOJ tasks—is 210 ms, as compared to mostly lower PSSs in controls. Stranger still, to get the McGurk illusion, a kind of binding between sound and lip-movements, one needed the *visual* signal to lag the *auditory* signal by 240 ms! His point of maximizing audiovisual integration (in the McGurk effect) is negatively correlated with his PSS, the two differing by a whopping 450 ms. Although some other cases like this have been mentioned in the literature, PH is the first subject rigorously studied who consciously notices differences between audio and visual signals emitted from common events in his immediate surroundings. PH does with your speech what you do with thunder and lightning, only on a much faster scale and with the auditory and visual signals lagged. PH, of course, has his own subjective now, but it's a mistake to think he binds together the same events into his now as you do yours.

As surprising as PH's numbers on PSS and audiovisual integration were, what seems to have made an equal impression on the researchers is that this finding *wasn't* that unusual compared to the controls, subjects who do not notice audio-visual discrepancies in timing. The large gap between maximum integration in the McGurk effect and PSS stood out in PH, but not the negative correlation between the two, nor even the very high PSS. These findings may lead you to wonder whether some of the healthy subjects also could notice auditory or visual lead, just as PH. And indeed, 10/37 healthy subjects consistently reported seeing auditory or visual lead when the stimuli presented were objectively simultaneous (Freeman et al., 2013). Like PH, they reliably experience synchronous stimuli as asynchronous too. These people don't report problems, presumably because they only notice such discrepancies in laboratory conditions. With these findings, it's hard to resist the conclusion that many of us are binding different events together into our subjective presents; however, in ecologically valid situations, we tend not to notice.

Disagreement is the norm in people's PSS values too. Experiments by Stone et al. (2001) found that the point of subjective simultaneity (PSS) in the SJ task is remarkably stable within an individual, as we would expect given its importance in motor control and other mechanisms important for functioning. However, the interpersonal differ-ences were sometimes astonishingly large; for example, in their study, one observer needed sounds to precede flashes by 21 ms whereas another needed flashes to precede sounds by 150 ms for maximum synchrony. We see the same kind of disagreement as in the PH study and many others.

In the last chapter I discussed fusion thresholds, the asynchronous gap tolerated by subjects before they perceive, say, two tones as two tones instead of one. What I didn't mention is that these thresholds, in each modality, vary to some degree from person to person. For audition, for instance, the fusion threshold can vary from about 2 ms to 5 ms. It also varies with age, older people fusing more events than younger people, and with training, up to a point. In each person, the threshold of simultaneity cannot be shrunk beyond a minimal amount. This is the same for the other sensory modalities. In each of these, there will be a minimum window of fusing events together as subjectively present, but this will vary from person to person.

We don't usually notice these differences, yet they can become apparent when making very precise measurements—as in an infamous episode at the Greenwich observatory in 1796 (see Mollon and Perkins, 1996). There the Royal Astronomer Maskelyne fired his assistant Kinnebrook for systematically deviating from him by a whopping 0.8 s in observations of stellar transits. The observer's job was to note the time, given by audible clicks, at which a star was viewed to pass a hung vertical wire. Hence the task relied on multisensory integration. Given the regular pattern of discrepancy, however, it may be that Kinnebrook's so-called "personal equation" simply differed from that of Maskelyne's. The personal equation is the concept later used in astronomy to describe the fact that perceptual and reaction times differ amongst individuals. Astronomers learned that even trained experts' senses of simultaneity might differ by up to a second. Kinnebrook's discrepancy may have arisen through no fault of his own. That said, he may have been fired for altogether different reasons.[4]

All of these cases, plus others I could give, paint a very clear picture. We think that we all perceive the same now, but really we are walking around binding different stimuli together as simultaneous and judging different stimuli to be synchronous. These activities don't converge even in a single individual, as we saw last chapter, so it isn't that surprising that they also each differ quite a bit from person to person. If PH sat in a philosophy of time class and the instructor said that the moment of his or her finger snap—SNAP—is real, PH might ask, "which moment?" Disagreement with PH might be dismissed because he is a bit of an outlier. But an abundance of evidence suggests that we're not really so different from him. All of us, if tested rigorously, would find that we disagree with our peers about the subjective now too.

In a way, this disagreement data is worse than the disagreement about what is simultaneous found in relativity theory. There we witnessed that what events are simultaneous for an observer depends on his or her path through spacetime. Different paths meant different events were simultaneous, even for observers intersecting at an event. That is the basis of Putnam's famous argument against manifest time. But notice that that disagreement is merely theoretical. Two relativistic observers will calculate that the events simultaneous with some common event differ. That difference isn't observed, for relativity doesn't make use of simultaneity—it's not an invariant and therefore not real. Here, by contrast, the difference is one that can in principle be observed. So we can imagine a kind of psychological counterpart to Putnam's challenge to manifest time. You experienced events e_1 and e_2 as happening now but your friend experienced e_2 as happening after e_1. Assuming neither of you is special, we have a challenge to the idea that your experienced now is distinguished in some way.

---

[4] He had recently turned down a marital match proposed by Maskelyne—another "personal equation" that differed from Maskelyne's.

10.2 Manufacturing the Now: Signals, Speed, and Stamps

Despite the relativity of simultaneity, we are able to explain why we think simultaneity is absolute. The answer is well-known. Relative to the speed of light, we move at very low relative velocities with respect to one another. Expressed in terms of Minkowski coordinates, when the relative velocity is low the Lorentz transformations approximate the Galilean transformations, the defining symmetry of classical spacetime, a spacetime where all observers (in principle) agree on the time. Put more intuitively, notice that when expressed in natural units, such as miles, and focused on terrestrial distances, the lightcone emanating from an event is for most intents and purposes instantaneous, i.e., everywhere at once. The spacelike region from any event seems huge in a Minkowski representation of spacetime. In terms of human affairs, it is very tiny. Reach your arms out as far as you can and try to snap your fingers at the same time. Can you make the two snaps spacelike related to one another? To do so you must snap your fingers within one billionth of a second or better of each other (relative to your inertial frame)! It's possible, but not very likely in any given try. And anyway, by the time you notice the two events they are in your past lightcone. As "big" as the spacelike region is relative to an event on your worldline, you probably can't get the two snaps in it. Since events need to be in these regions to get relativistic disagreement, the relativity of simultaneity won't matter in ordinary life.

That, very roughly put, is the reason why we think simultaneity is absolute despite the relativity of simultaneity. Can we give a similar argument to explain why we agree on the experienced now? One of my favorite papers in philosophy of time, Butterfield (1984), does precisely this. I'll present my take on his theory in the next subsection and later offer some refinements to it.

10.2.1 Time stamps not needed

The idea begins by pointing out that our signaling to each other is very quick. Light is obviously super fast, and as we've seen, so too is audio processing. But quick relative to what? To answer this, we should first point out the scale differences between the microscopic and macroscopic world. The microscopic world is changing very rapidly in comparison to the macroworld.

In front of my desk there lies a clump of electrons, protons, and neutrons, roughly 1.2×10^{29} of each of them. They are moving very rapidly, some at almost two million meters per second. These particles have spins that are rotating and an assortment of additional changing properties. They are colliding at something of the order of once every 10^{-10} seconds. There are also other particles involved, continual absorptions and emissions, spontaneous decays and fluctuations. Yet despite all the billions and

Figure 10.1 Odie

billions of particles and interactions, all whirring by at almost unimaginable speeds, together these subatomic particles comprise Odie, my dog, who has not moved since breakfast (Fig. 10.1).[5]

The emergence of a stable macroworld is one of the more amazing facts of physics. Quantum mechanics, chemistry, solid state physics, statistical mechanics, and more combine to tell us that by macroscopic scales systems will tend to be electrically neutral, chemically stable, unlikely to spontaneously fluctuate to new macrostates, and so on. Whatever the detailed reason, while subatomic particles and fields vary their properties at an incredible pace, macroscopic properties change relatively slowly and infrequently. Odie, for instance, lays in a certain location and sports a brown mop of fur . . . and still . . . and still. Even when deftly stealing my shoes, Odie still sports a brown mop of fur, still has four legs, and so on. His location is changing very slowly compared to the pace of locational change in the microworld. Appropriately motivated by a cat, Odie can hit 25 mph, tops.

Suppose I form a belief about Odie's location. Light reflecting off Odie travels at approximately 300,000,000 m/s into my eyes. Since Odie sits only a few feet from me, the travel time is only about 3 nanoseconds. Processing the light in the visual system takes much more time, both in the chemical transduction on the retina and in the neural transmission, but still the total is roughly 0.5 s. This is about the time lag it takes between the state of the object at time t and the formation of our belief (tacit or not) about it at time t^*. The interesting thing about this time lag is that, given that objects change their macroscopic properties relatively slowly (typically), the time lag does not end up falsifying the relevant belief. Photons reflected from Odie are absorbed in my

---

[5] Okay, the numbers shouldn't be taken seriously. For instance, I assumed for convenience that Odie was made of carbon-12.

eye and processed in a complicated mechanism. The result of this process is a belief at t^* that the large object 1 m away at t is dog-shaped, furry, and brown—and at t^* it (typically) still is dog-shaped, furry, and brown! The lag $t^* - t$ typically doesn't make the belief about Odie or other local macroscopic objects false.[6]

The same is true when I communicate this belief to you. If I sign "Odie is brown" or "Odie is in front of my desk" to you, there will again be a lag. This time, the delay arises from the lag from my visual processing system, the lag arising from my signing, and then the lag of your visual processing. But this total lag is still very small, just over a second, and typically it won't be enough to falsify the information. Odie's brownness and approximate location will survive the lag. We can take the statement to be true at the moment of utterance and the moment of reception. Naturally this feature will be crucially important when trying to coordinate action. The same story can be repeated for sound and touch. Sound, of course, travels much slower than light. However, the auditory system, relying on mechanical transduction, is much faster than the visual system. The tactile sense is the same way.

What these facts imply is that we can reliably form beliefs and communicate in typical environments about macroscopic objects without including *time stamps* in our beliefs or communications (Fig. 10.2). A time stamp is a dating provided by calendar or clock time. We can say "the bird is in the tree" or "the bird is now in the tree" and manage to impart useful information. We don't need to say "the bird is in the

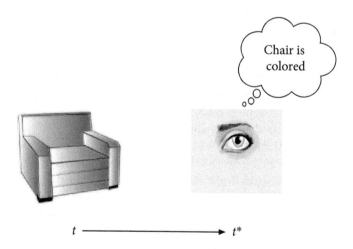

Figure 10.2 No time stamp needed

---

[6] One therefore can easily see why for most of history experts and non-experts agreed with what Jammer (2006) calls the *visual simultaneity hypothesis*, namely, the idea that everything you see is simultaneous with your seeing it. Authorities such as Aristotle, Descartes, and Kepler argued that this must be the case, and the ordinary person would have no experiences suggesting otherwise. Throughout history there were dissenters to this view, including Empedocles, Avicenna, Alhazen, Roger Bacon, and Francis Bacon, but I believe they always represented a minority opinion.

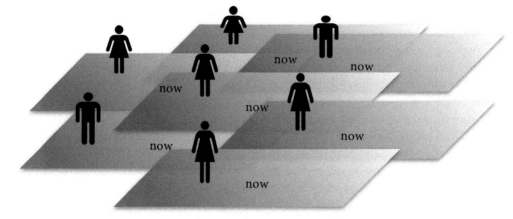

Figure 10.3 Our common now

tree at 12.58.03 GMT" to someone standing near enough to hear. We can drop the "B-time." Because the "disagreement data" is masked, we can therefore treat the now as a monadic property and omit mention of its relata. For this reason there is a strong temptation to conflate our egocentric temporal representations with allocentric ones, and hence, to regard the now as objective. It explains why we have intersubjective agreement about a global now that does not fundamentally exist.

This framework lends itself to the following picture. Define a *now patch* as a spatiotemporal region over which typical observers in typical environments do *not* require a time stamp in order to reliably navigate their environments. A global common now is then built by sewing together these local now patches into a much larger entity and then extrapolating (Fig. 10.3). The inter-subjective agreement leads to the idea that this "global now" is objective.

Notice two facts about this explanation.

First, we couldn't tell the same story if we communicated by means of smell or post. Burnt toast can linger in a house all day. Its smell does not reliably stamp when it was burnt. Similarly, a letter in the mail without a date on it can often be useless. In both cases the lag makes a difference. Smell as a consequence of its lag is often unreliable; the post is reliable but only if the calendar date is stamped on the letter. With no date provided, a letter from your bank might be worthless. Fortunately, in our world, we don't need to rely on smell or the post. As a result, given the stability of macroscopic properties and the short $t^* - t$ lag for modes of information transfer, we can reliably gain and communicate information about the world without the use of a time stamp.

Second, although the "time stamp" picture explains our ability to treat the now as objective, like all other species of what is sometimes called the "mind-dependent theory of the present" (Grünbaum, 1968), it curiously leaves the mind out. The lead character in the story is missing. One can't rest content with time stamps because we aren't simply forming reliable beliefs in certain physical regimes and communicating these beliefs effectively. Presumably we're also perceiving and experiencing events together as simultaneous. I'm experiencing various visual and non-visual impressions together as happening at the same time. I hear a dog bark while seeing a red coat. Facts like these shouldn't be left out; rather they are the basis of the patches that get sewn together to create a common now. The time stamp theory offers a kind of "behaviorist" picture, one ignoring the underlying cognitive states, the states that ultimately explain why we perceive the same nows. But we can simply fill this gap with our picture of how our brains actually construct subjective nows, and we can go further and even explain this construction by appealing to evolution. That's what we'll do in the next section.

10.2.2 The common now

The first point I wish to make is that the foundation of our agreement lay in the fact that *for the events that are salient to us* the differing PSSs, points of maximum integration, and more, generally don't matter too much in ecologically valid scenarios. Earlier I gave the example of the finger snap. The snap sound is associated with the finger movement, but not with any particular portion of it unless we're told to pay attention to it. Events in ecologically valid scenarios are coarse-grained like this. We can and do notice differences in the lab, with flashes and beeps and taps. Fundamentally we're not so unlike PH, but in this respect we are. If I snap my fingers and say "right now!" this picks out the same coarse-grained moment for everyone in the room.

The differences among us aren't enough to undermine the idea that we share a common now; rather, the similarities among us and the hazy boundaries and lengths of events enforce a shared present. With integration windows as wide as 250 ms, light so fast, auditory processing so speedy, and macroscopic objects changing properties relatively slowly, these differences—which are experimentally detectable—won't often be noticed in daily life. This point finds some confirmation in a recent experiment by Boenke et al. (2009). Here subjects had their audiovisual integration tested through the standard temporal order judgment (TOJ) choice of "light first" versus "sound first," but in this case the intensity and duration of the stimuli were varied. As usual, for stimuli lasting 9 ms long, individual's PSSs varied considerably, with some negative (sound first) and some positive (light first). When the duration of the stimuli was increased, however, first from 9 ms to 40 ms, and then from 40 ms to 500 ms, there is a remarkable convergence of PSSs. If your PSS was negative at 9 ms, it was shifted in the positive direction when duration increased; if positive at 9 ms, it was shifted negative when duration increased. Ditto for greater times. Moreover, the amount of shift depended on the duration difference, plus how far negative or positive one was originally. For a given visual intensity, a certain PSS value would act as an *attractor*, erasing the PSS variability found at small durations. As the stimuli duration grew, the probability that two such stimuli were perceived as present also grew.

Of course, this result is to be expected eventually at long stimuli durations. No one sees the ten o'clock sitcom while listening to the eleven o'clock news. Nonetheless, the experiment shows that at even very small durations there is a tendency for people's integration windows to catch the same information. In some exceptional circumstances—such as the case of the astronomer watching dots of light pass vertical strings at Greenwich, the occasional offside call in soccer, and in PH—our individual differences do show up. But in the kinds of conditions in which we normally interact—macroscopic events of duration of (say) a second or more—these differences rarely matter. As a result, we can expect with high probability that the events that are present for you are present for me whenever we interact. We don't literally see the same present, but we'll agree on enough of it to not notice the difference. This likely agreement then survives communication with one another as Butterfield observes.

Why do we tend to agree? The answer, I think, ultimately must come from the fact that we all have a shared evolutionary history. Evolution has shaped our windows of simultaneity. The survivors thus don't differ so much because they have all been produced by the same mechanism; and moreover, since successful communication is crucial to our survival, it is no accident that we're able to communicate as we do.

An attractive way to think about what is going on is to model it with a simple decision setup (Colonius and Diederich, 2010). The outside world is a messy jumble of signals, but there really is an objective temporal sequence of events (among timelike related events). Sometimes it pays to find that objective sequence (you avoid the tiger), sometimes it doesn't (expend too much energy). Focus on audiovisual information. Given a visual stimulus and an auditory stimulus, they were either generated by a common source or not ($C = 1$, a common source; $C = 2$, unrelated). And you, the creature involved, will either integrate the two stimuli together or not ($I = 1$ or $I = 0$, respectively). This setup therefore can be described with a simple 2×2 matrix with the four outcomes allowed (Fig. 10.4).

U_{11} and U_{20} describe cases where we get the timing right, integrating information from common sources and not integrating from unrelated sources, respectively. U_{10} and U_{21} correspond to getting it wrong, with either type I or type II errors, integrating what isn't common or not integrating what is. Each possibility has a particular value to us, and naturally, they needn't be the same value. Since integration must be energetically taxing, there will often be reason to make type I and type II errors. You can imagine an organism developing the apparatus to determine that the noise generated by a finger snap is generated by the end of your finger's motion, not the

Gain/Cost	Integration ($I = 1$)	No Integration ($I = 0$)
Common source ($C = 1$)	U_{11}	U_{10}
Separate Sources ($C = 2$)	U_{21}	U_{20}

Figure 10.4 Payoff matrix

whole snap motion. Yet what benefit obtains from tying the sound to that part and not the whole snap? Given the costs of exact match, it might make sense for an organism to attach the sound to the whole snap. We shouldn't be tempted into thinking that getting time right is always best. It's best to integrate that low rumbling noise with the sight of the tiger, but maybe it's better not to waste time on integrating when you hear a roar right next to your ear. Alternatively, you can imagine an organism that receives a light signal from lightning far away, but doesn't include it in a percept until it hears the thunder; that is, the organism binds thunder and lightning the way we do heard speech and lip movement. But the costs of waiting around in a kind of fugue state are obviously potentially high: such organisms would tend to get eaten when thunderstorms pass by! Here it's best to separate the signals.

Given some values for the utilities, one can calculate an idealized *optimum window of integration*. This is the window of integration that will maximize your expected utility. Let $t = |V - A|$ be the window, where V and A represent the arrival times of the visual and auditory signals, respectively. Then a little calculation reveals that if you adopt the following decision rule

$$\text{If } \frac{P(C = 1/t)}{P(c = 2/t)} > \frac{U_{20} - U_{21}}{U_{11} - U_{10}} \text{ then integrate; otherwise, don't integrate}$$

then you will maximize your expected utility. We can obviously generalize this simple toy model to different types of PSS values, to different fusion and order thresholds, and to many other sensory modalities. This method of reasoning isn't restricted to the determination of $t = |V - A|$. Making assumptions about the U's, one can then plot curves that describe the optimal window widths as a function of probability of common sources. In fact, by making a ton of assumptions[7] and idealizations, some groups run experiments to check whether we are using anything like the optimal temporal width. Many groups try to see whether we are, in particular, Bayesian optimizers, when it comes to multimodal integration (for a review, see Ernst, 2005).

I won't take a stand on whether we use an optimal width or not. It's clear that we don't *have* to be optimizers, as evolution is hardly always an optimizer. And the real world's payoff matrix is unimaginably complicated. We'll never know the correct values for U's in this toy matrix, never mind the real one with indefinitely many rows and columns, plus connections to other developmental options and their costs and benefits. I mention this line of research only because it offers a vivid picture of how I'm thinking about our temporal integration. There are costs and benefits to be balanced, and whether two aspects of an event get bound together hang on this balancing (optimal or not). Noticing this fact allows us to see how sensitive our nows

---

[7] The assumptions are claims such as that, all else being equal, it's generally better to integrate common flashes and beeps than not in normal macroscopic conditions, that the times to arrival at cross-modal integration for auditory and visual stimuli have certain values, that the probability of common sources is such-and-such, that it's good for creatures like us to bind together beeps and flashes from the same locations, and so on.

are to the environment. In this toy example, if we make a few assumptions, we can infer that:

- increasing the prior probability of a common cause $(C=1)$ will make the optimum window larger
- keeping costs U_{10} and U_{21} of the payoff matrix (Fig. 10.4) fixed, an increase in U_{11} leads to a window increase
- keeping costs U_{10} and U_{21} of the payoff matrix (Fig. 10.4) fixed, a decrease in U_{20} leads to narrowing of the window.

Here one can see in a very clear way how our subjective nows are dependent upon the environment. The values of the U's will depend upon predator and prey populations, the weather, and much else.

We can also imagine how our simultaneity windows evolved. Imagine the payoffs are given in terms of fitness and not utility. Then we could populate a toy world with creatures having such a fitness matrix and endowed with a range of values of t. Running a replicator dynamics or some other similar dynamics, we could watch in our model the value of dominant t evolve with time given certain environments. Although indefinitely more complex, presumably in outline form something like this shaped our subjective nows. In any case, that we are all the products of this common evolutionary history explains why we tend to agree on what's now.

Furthermore, I suspect that one might test this theory of the development of subjective nows in various ways. As mentioned, it already is tested on human beings, but these tests require large assumptions about our actual payoff matrix. I'm conceiving instead of tests based on more modest assumptions and directed at the animal kingdom.

Animals have evolved in a diverse variety of habitats and they also must solve the problem of setting the widths of their subjective nows. Some of these habitats are dark, others humid, some underwater, others underground. Some animals communicate via signals quite different from the ones we use. Based on a creature's dominant environment—the types of predators it has, the means and importance of signaling, and so on—one might be able to generate coarse predictions about the widths of their simultaneity windows.

For example, sound travels almost five times faster through saltwater than through air. Do ocean dwellers who navigate the world via audiovisual signals have significantly different "nows" than we do? Sound also varies significantly with altitude, humidity, and temperature. Do critters in cold, dry, high-altitude environments differ in their integration from those in hot, wet, and low environments? Animals also respond to the types of dominant predator around. Do animals eaten by snail-like predators have different sized windows than those fleeing from cheetah-like predators, all else being roughly equal? The former may have more freedom to allow wider width windows than the latter. Animals of course also differ physiologically. Most obviously, some are big and some are small. If we consider tactile sense, note that the distance in a giraffe from a tendon in its foot to its brain is almost 3 m! It's already known that this distance results in relatively long sensorimotor delays and more limited sensorimotor resolution than in smaller mammals (More et al., 2013). Does it mean the giraffe has a "long now" in addition to its long neck?

The nice thing about the contingency of the subjective now is that it makes the now subject to theoretical and empirical study.

How should we conceive the presents being generated? Since it's not clear that there is a single representation underlying judgments of synchrony, it's not certain that there is *one* patch for any individual at any time. A bird tweets just as it alights on a branch 20 m away. What's in your subjective now? The answer depends on many factors. There are many different times scales possible. A "now" only 1 ms long permits sound localization but not music recognition. Birdsong is not in this now. But it could be part of a 1s "now." Your chances of experiencing the call and the landing as happening at the same time will depend not merely on its spatiotemporal location but perhaps also how intensely colored the bird is. For this reason the subjectives presents are probably not best conceived as spatiotemporal patches.[8]

Rather than a spatiotemporal patch, it's better to think of us as each carrying around a subjective probabilistic disposition to bind stimuli together living in an abstract stimuli space. Given any two stimuli in this space, there is an associated probability of them being integrated together. Despite the differences in these dispositions, the probability that I'll bind together signals as present that you do is very high, given certain spatial and temporal scales and other physical features. We therefore won't typically notice that our local "patches" aren't global. Since everywhere we go we meet people who agree with us on the present, we have strong reasons for regarding the present as objective and global.

10.3 Wiggling in Time vs Wiggling in Space

If the foregoing is correct, we have explained why we might mistake egocentric temporal discriminations with allocentric ones. But we have not explained why there is a difference between space and time in this respect. Why do we have "disagreement" about the spatial discriminations but not the temporal ones? After all, presumably we have a similar story to tell about the construction of the spatial here.[9] By itself, the theory I've developed doesn't explain the Representational Asymmetry.

---

[8] Even if we were to take the local patches seriously and try to sew them together, probably we would fall far short of simultaneity as it's usually understood. Simultaneity is typically regarded as having many properties, the most central one of which is that it is supposed to be an equivalence relation. Equivalence relations are reflexive, symmetric, and transitive. But it's hardly obvious that the thinnest present patches would be transitive (see Hansteen, 1968, 1971; cf. Sternberg and Knoll, 1973).

[9] The spatial counterpart of a temporal integration mechanism is a process that welds together inputs arriving at different spatial locations into one unified percept. The reader may be tempted to think the counterpart of simultaneity constancy is the phenomenon known as "location constancy." Location

We might seek some psychological features that are part of the answer. Certainly one can imagine science fiction universes wherein creatures are rigged so as to be conscious not just of recently past events, but also of past and future events—and other strange wirings. But given the way we are rigged, I think the source of any answer will ultimately bottom out in some basic physical features. The physical facts, taken together, imply that there is a sense in which we are—unlike Billy Pilgrim in Kurt Vonnegut's *Slaughterhouse-Five*—"stuck" in time in a way that we aren't "stuck" in space. The physical facts are the suite of temporal features discussed in Chapter 6. Of particular importance is the mobility asymmetry, i.e., that there is no time travel, or weaker for present purposes, that there is no exploitable time travel. The background facts that we and everything we interact with are representable by continuous non-spacelike worldlines, plus the one-dimensionality of the timelike directions, are crucial too. But they are only crucial insofar as they contribute to the fact that we have a mobility asymmetry.

Think of how intersubjective disagreement is noticed in the spatial case: essentially, one can move forward and back, up and down, left and right, and turn around—and any combination of these acts. However, the one-dimensionality of the timelike directions implies that one cannot "turn around" in time by a continuous transformation. An arrow, pointed along the past or future direction, cannot be rotated to an opposite direction. An arrow pointing east, by contrast, can be rotated continuously so that it is pointing west. This central piece of physics makes one way of seeing that the "here" is egocentric unavailable in the temporal case: one can't spin around and see other nows as one can other heres. Ditto for ups and downs and lefts and rights. Moreover, in the spatial case, along any given axis, we can move relatively freely back-and-forth along that axis. If we can walk west, we can usually walk back east. This is not true with respect to time. The mobility asymmetry (plus the condition on worldlines) imply that nothing we consider a genuine object has ever done this, so far as we know; and in any case, what's relevant here is what *we* can do, and we can't do that. In sum, the most common motions that we take for granted in space—e.g., walking west to see the sunset, then spinning around and returning to your car—are impossible in time. This basic physical difference between space and time has huge ramifications for our lives, one of which is that two natural ways of generating spatial egocentric disagreement simply do not have temporal counterparts.

---

constancy refers to the fact that objects generally seem to stay in the same place as you move your head around. However, simultaneity constancy is a mechanism that welds together two inputs emitted at the same time from the target event which arrive at distinct times; so a spatial counterpart is not location constancy but a mechanism that welds together two inputs emitted from the same location but that arrive at different locations. In Fig. 9.1 it is a mechanism that welds together inputs 1 and 2 instead of 1 and 3. Clearly we employ such mechanisms, even if only in the forming of one percept from information received at two eyes or two ears. We also have mechanisms that help us figure out spatial order, spatial distance, and so on. These mechanisms are different than the temporal ones, but I'm not claiming this difference is the whole difference.

True, everyone thinks that where they are is here. However, spatially, one knows the egocentric frame is not the allocentric frame. The idea that my here is metaphysically distinguished meets nothing but disconfirming evidence everyone I go. I walk 30 feet and it's changed again. I talk to you and learn that it's the same way with you; and moreover, that your here is not necessarily my here. "Here in California," I say on the phone, "it's warm and pleasant." You respond, "Here in New England it's dark and cold." We say "wish you were here" but not "wish you were now." Of course, when "here" is opened up to its widest extent, there may be a sense in which I regard everyone I see as here; but that's certainly not the case when restricted to contexts where "here" means "in this room" or somewhere finer. It's clear that often we're not speaking of the same heres. And of course, the non-objectivity of egocentric spatial determinations is equally apparent when we turn to properties such as right, left, up, down, and so on. All of these, one quickly learns, vary with my and your locations and orientations. There is no temptation to confuse my egocentric here with allocentric spatial determinations. There is too much intra- and inter-personal disagreement for that.

For the now, by contrast, that is not the case. When I see a dog and a cat both running, we agree that now both events are happening for all the reasons already discussed. Furthermore, physics prohibits either of us from rotating or traveling in any way that will generate disagreement. It's simply not possible to execute the necessary motions, try as we might. Consequently, everyone you meet thinks that the time is now, and not merely in some wide sense, e.g., the same century, but also in the narrowest sense, e.g., this very second. We can't jump into time machines that whisk us away to other nows—which is exactly what we do spatially with no trouble at all. For all of these reasons, there is massive intersubjective agreement about the now. And this leads, I claim, to a reification of the now, to a conflation of the egocentric temporal properties with the allocentric ones. The conflation is not so much a mistake or illusion so much as it is a natural response to the fact that we really are objectively stuck in time.

10.4 Conclusion

Let me conclude with some observations about how our discussion connects to work in philosophy of time. Philosophy of time for much of its history has focused on whether the fundamental temporal properties of events are monadic or relational. Are properties such as presentness, pastness, and futurity monadic, or are they instead cashed out via relations to particular times? Caesar lived in the past, but does the truth of that claim make reference to the fact that it is said in the twenty-first century? The central debate in philosophy of time is framed as between tensers, those who hold that some monadic temporal properties are irreducible to relations, and detensers, who deny this claim. Many philosophers of time feel that experience weighs in this debate, but I've never understood how it could do so. Chapter 13 can be read as a denial of the claim that it could do so.

Shoemaker's example of the property "being heavy to lift" can be used to illustrate the point of this chapter. Imagine someone who thinks being heavy to lift is a monadic property. The property partitions all the objects in the world into two, those heavy to lift and those not. One day this person goes into Gold's Gym, however, and there confronts a lot of *disagreement data*. Some weights that are heavy to lift for this person turn out to be easy to lift for the others! The property is relational after all. But then why did he think it was monadic? Well, firstly, it's not surprising that he didn't explicitly represent himself in the content of his experience, as Shoemaker says. If we point out that he lived alone on an island beforehand we could explain why he didn't previously notice the disagreement data. We would then have an explanation of why he thought being heavy to lift was a monadic property. We have done essentially the same with the present. Manifest time portrays the present as a monadic property. But then one day we're presented with disagreement data: we meet PH, we learn of the different individual PSSs, we learn that people integrate sensory stimuli differently, and so on. We leave the "island." So why did we think the present was monadic? The answer relied on many contingent factors: there aren't too many PHs around, events are coarse-grained, signals travel very quickly compared to typical macroscopic changes, and more. We have given the psychological counterpart of the explanation in physics of why simultaneity seems absolute even when it is not.

Philosophers of time who favor a tensed metaphysics may feel safe in ignoring this whole discussion. I have not attempted to show that the scientific explanations that I've invoked can be purged of tensed presents. Hence it is open to those who would add a fundamental Now to our basic description of the world to adopt this explanation wholesale without feeling that it impugns their view.

The tenser should not feel too comfortable with this reply. Traditionally it is thought that the tensed theory's best case came from experience and that the detenser's explanation of experience looked comparatively hollow. Now, the "best case" is exposed as empty. The present theory predicts and explains various ways in which we can intervene and manipulate lots of variables regarding the perceived present. This data is compatible with a tensed world, but tenses *do nothing* to help explain it.

It appears that with regard to experience, the roles are now reversed. The argument for tenses now relies instead on a half-hearted hope that cognitive science explanations can't be written except with tensed concepts, a far cry from the usual argument that the now is experienced. With less work for a tensed now to accomplish, there is less reason to posit it. Using these present "patches," we now enjoy the beginnings of a theory that accounts for the persistent belief that we inhabit a metaphysically distinguished present but not metaphysically distinguished here. The massive subjective and inter-subjective agreement on what happens now but not what happens here explains why we're tempted to conflate the egocentric and allocentric representations of time but not space. With this theory we have enough to block the hypostatization of the felt present.

When Einstein lamented the lack of a now in physics, Carnap replied that it could be explained with psychology. Many others have dismissed the now as a psychological illusion. This dismissal is wrong, as there is a phenomenon here, properly understood, and why the now seems objective is susceptible to explanation. While the previous two chapters hardly offer the final word on the story, they do—I hope—deliver on Carnap's promise. The trick to providing an answer was to clear away distracting features of temporal indexicals, focus on the Representational Asymmetry, and then go interdisciplinary. Our answer crucially required physics (Section 8.3), but it also needed serious doses of evolution, psychophysics, cognitive science, linguistics, and philosophy. Physics by itself can't explain the now, but science can.

Now, let's get this present moving.

11

The Flow of Time: Stitching the World Together

> Well, we think that time passes, flows past us, but what if it is we who move forward, from past to future, always discovering the new? It would be a little like reading a book, you see. The book is all there, all at once, between its covers. But if you want to read the story and understand it, you must begin with the first page, and go forward, always in order. So the universe would be a very great book, and we would be very small readers.
>
> Ursula K. Le Guin, *The Dispossessed*

Of all the ways time is distinguished from space, perhaps the idea that time flows but space does not is among the most significant and pervasive. Time's passage or flow is firmly entrenched in the manifest conception of time. People commonly speak of the whoosh of time as it goes or flies by. It's hard *not* to think of time as unfolding in some way. When you imagine your life, even when you think of it as stretched out four-dimensionally, it's natural to think that there is a moment it's *on* right now and that this moment is changing. Nothing is more obvious, for instance, than that we (fortunately) haven't reached our deaths yet. But what does "reach" here mean, if not that time hasn't flowed there yet?

Not only do people speak, think, and act as if time is dynamic, but many claim that they *experience* the passage of time. Such is the strength of this experience—or at least the pull of the manifest theory leading them to claim this—that many thinkers assume that the positing of a fundamental flow of time is the only action fully respecting it. Thus the flow of time is accorded a central place in many researchers' temporal metaphysics. Distinguished physicists, philosophers, and philosophers of physics all make room in their metaphysics for objective time flow.[1] The typical idea underwriting temporal flow is that time passes or, more generally, is dynamic in some way. The now, viewed as an objective temporal feature, moves or changes. As Broad puts it, "Along [the order of events], and in a fixed direction, . . . the characteristic of presentness [is]

---

[1] See, for example, thinkers as different as Bergson (1922), Broad (1923), Čapek (1991), Ellis (2006), Maudlin (2002), Prigogine (1980), Shimony (1993), Smolin (2013), Whitehead (1929), Zimmerman (2007).

moving, somewhat like the spot of light from a policeman's bull's-eye traversing the fronts of the houses in a street. What is illuminated is the present, what has been illuminated is the past, and what has not yet been illuminated is the future" (Broad, 1923, p. 59). Or as Santayana memorably writes, echoing Heraclitius, "the essence of time runs like fire along the fuse of time" (1942, p. 491).

Popular though it may be, the idea that time flows faces many challenges. As we've seen, in modern physics time does not flow, or at least, not in the senses typically intended. Even allowing for the possibility that physics is incomplete on this score, making sense of temporal flow is no trivial feat. Running like fire along a fuse is a gorgeous metaphor, but for what exactly? Smart (1966)'s innocent-sounding query, "if time flows, how fast does it flow?" leads to an investigation of whether it's even *coherent* to say that time passes at 1 sec per 1 sec. Even if coherent, still many metaphysical models beg related questions: at what time do four-dimensional branches "drop off" branching spacetime? do "moving spotlights" crawl up worldlines in terms of a second time? and so on. Difficult questions dog virtually every metaphysical model of temporal passage. That is just the beginning of the challenge, however. Coherence is a low bar. The literature often seems to forget that once coherency is achieved the metaphysical models must pass the further bar of being *explanatorily powerful*. Comparatively little has been done to show that these models earn their keep in the way we expect of other scientific models. In Chapter 13 we'll return to this question.

For these reasons we have plenty of motivation to check whether the phenomena inspiring belief in a flow of time is explainable without departing from physical time. Not much work has been done in this regard. Physicists typically label the flow of time an "illusion." Doing so removes the burden of explanation from the desk of the physicist and places it on the desk of the psychologist, freeing the physicist to work on something else. Meanwhile, psychologists don't know the problem has been placed on their desks! The problem of passage consequently falls into the gap between these two fields. Philosophers sometimes occupy this gap, yet with most of the focus on metaphysical models of time, interdisciplinary explanations have languished. All of this neglect is regrettable, for this "illusion," if it is one, is a central and deep feature of human life. We conceive of ourselves as "flowing" up our worldlines and organize our lives accordingly. Convictions about free will, causation, the self, and much more are all connected to this "flow." To leave it mysterious is to ignore one of the most basic and perplexing features of our existence.

In what follows I'll outline the beginnings of a theory of time flow. While physical time does not itself flow, we can explain why creatures like us embedded in a world like this one would nonetheless claim that it does. I fully acknowledge that this sketch needs development. To me, however, it is vastly more plausible as an explanation of time flow than any existing model relying on exotic temporal metaphysics. The sketch appeals throughout to known and independently posited physical constraints and psychological mechanisms. The net effect of these mechanisms and constraints is sufficient, I submit, to undermine the confidence that such a theory cannot be right and that that there must instead be genuine and fundamental temporal flow.

11.1 Sharpening Focus

The alleged experience of the passage of time is striking: people *know* that they experience it, yet few can speak sensibly about it. All hands agree that it is like fire running along a fuse, but when asked to supply details, people tend to disagree on what the target phenomenon is. There is not a single experience or intuition representative of passage. Hence we must canvas alternatives. My tour will be rather brisk, as a careful unpacking of passage will not be particularly exciting to the reader. I just want to gather enough of a sense of passage so that later we can provide a check that our later explanation is in the right vicinity.

Let's narrow our focus. I want to concentrate on what is sometimes called the "argument from experience" for temporal passage (described below) and variations on this argument. Hence I will bracket arguments for passage that rely on other routes. Maudlin (2007), for instance, argues that passage is needed to explain the thermodynamic arrow of time. No human beings are necessary to get this argument going. Although I don't find his argument decisive, we're ignoring it and others like it. Here I wish to concentrate on the far more pervasive idea that some distinctive aspects of experience or immediate intuition supports the positing of flow.

The argument from experience can be phrased as an inference to the best explanation. Coarsely put, the argument has two premises. The first is that we have experiences as of passage. Note that I don't say "experiences of passage." As we witnessed in our discussion of the present, many philosophers read theory into the data when describing what is explained. The experience of passage is characterized as that appearance that is triggered by objective passage, or something close to this.[2] That's not helpful, especially when this appearance is then cited as evidence for objective passage. The second premise is the claim that if we have such experiences, then an objective fundamental flow of time is the best explanation of these experiences. One then concludes that objective fundamental passage exists.

The problem with the argument from experience is that few advocates of objective passage spell out the experiences alleged to support it in any detail. It's a curious fact that the richest descriptions one can find of passage phenomenology typically arise in discussions seeking to debunk it. Rather than restrict attention to contested phenomenology, I propose that we also deal with the deep intuitions that support passage too, whether or not they are truly part of experience. So we need to canvas not just experiences but also some of the deep structure of manifest time. Later,

---

[2] For example: "Let me begin this inquiry with the simple but fundamental fact that the flow of time, or passage, as it is known, is given in experience, that it is as indubitable an *aspect of our perception of the world as the sights and sounds that come in upon us*" (Schuster, 1986, p. 695).

I will question whether "tensed" metaphysical models really offer good explanations of experience or these intuitions (see Chapter 13). Here I want to engage in the positive project of explaining what needs to be explained.

What experiences and strong intuitions lie underneath the idea that manifest time flows?

Phenomenology of passage? Many commentators claim to directly experience the "whoosh and whiz" (Falk, 2003) of passage. When not spelled out in question-begging terms, invariably this experience is characterized in terms of some aspect of perceptual experiences of movement or change. Le Poidevin (2007, p. 76) writes:

> We *just see* time passing in front of us, in the movement of a second hand around a clock, or the falling of sand through an hourglass, or indeed any motion or change at all,

while Norton (2010, p. 33) concentrates on the following:

> Even if we are in an environment that it totally static—an empty, noiseless doctor's waiting room—we still perceive our own bodily functions changing, such as our breathing and heartbeat, and even the process of our thought.

Schuster (1986) and Skow (2012) suggest that the experience is *analogous* to our experience of motion and change. Described as above, there is little question that the features mentioned are properly phenomenological, e.g., seeing a second hand, perceiving our stream of thought. Many questions naturally arise—especially why seeing a second hand provides evidence of time itself having some dynamic feature. No matter. It will turn out that we can explain the experience of change in a manner perfectly consistent with physical time in any case.

In the life sciences and also ordinary life, when people speak about the "flow of time" they often mean subjective or felt duration. When you say time flies by when you're having fun, you mean that the duration of the event feels shorter than it otherwise would if it weren't fun.[3] Your judgment is of (timelike) *interval length*. An event feels like it takes a certain amount of time. As such, it is typically assumed to be part of phenomenology. No philosophers that I am aware of use the experience of duration to support an objective passage of time, even if many end up talking about duration experience when defending their views. Whether surprising or not, we'll see that the experience of duration provides no support for objective passage.

Temporal phenomenology provides us with many exciting lines of research. Yet, I'll argue, these turn out to be challenges for psychology and not metaphysics.

---

[3] More accurately, just because time flies when you're having fun doesn't automatically mean that you'll judge that period of time as shorter. In Wearden (2005) some subjects were asked to watch nine minutes of the exciting meteor-destroying film, *Armageddon*, while others were asked to sit for nine minutes in a boring waiting room. The first group reported that time flew by while the second complained of it dragging; yet surprisingly, the first group judged the time interval as ten percent longer than the second group. Retrospectively, there were more salient events to remember and take account of in *Armageddon* than in the waiting room.

Novelty. C. D. Broad writes that "the *continual supersession of what was the latest phase by a new phase*, which will in turn be superseded by another new one . . . seems to me to be the rock-bottom peculiarity of time" (Broad, 1959, p. 766). Broad's statement is about temporal reality and not experience or intuition, but some have sought experiential counterparts:

> Temporal becoming . . . is an indubitable feature of temporal reality; it is whatever meta-physical phenomenon underlies: (1) the *phenomenological novelty* of each moment, that is, the *continuous sense that one has not encountered this state of the world* (despite gross qualitative similarities it may share with many others) and (2) the fact that successive uses of temporally indexical expressions (such as 'now') are possible and have different referents.
>
> (Fiocco, 2007, pp. 4–5)

From our previous discussion in Chapter 9 we will not be impressed with the datum described by (2) in the quote. Yet this gives us a clue. McTaggart insisted that *real* change involves events having and losing A-properties (pastness, presentness, futurity). Putting it like this forms a tidy little circle if one tries to argue from real change to the plausibility of a tensed model of time, as some have. But we can understand the idea in a non-question-begging sense too, namely, as saying that *what's present* changes, where we don't yet commit to what the present is, metaphysically. In other words, flow means that the tensed tripartite division of reality itself changes with time. What's continually novel are the new presents. The now is changing. In Chapter 9 I argued that presentness is not a phenomenological feature of the world, contrary to some claims. If correct, then its updating is also not presented to us phenomenologically—even if the subjective psychological nows' changing content is in some sense phenomenological. This datum is about manifest time, not phenomenology. What we need to explain is why creatures like us take the apparent tripartite structure of reality to be updating, not why we actually sense this.

Growing Nearer. Night approaches. The dawn is receding away into the past. We're moving through time. Related to the above novelty is the idea not merely that the tripartite structure is changing, but that something that persists through this change is moving in some way. The now—or us—has identity through time that allows us to speak of advancement through time. The tripartite structure is not just changing; rather, *something* is moving or surviving through it. That is why the now is like fire running along a fuse or a moving spotlight. As a kind of corollary on novelty, this is also a feature of manifest time and arguably not part of phenomenology.

Asymmetric Power. This next cluster of ideas is probably best conceived as connected to the past/future asymmetry more than flow. For the latter, see Paul (2014). Yet the two are so intimately connected that I hope the reader will bear with me in raising it here. In the now we make decisions, we cause things to happen . . . we act. We enjoy a *sense of agency*. This sense is tied to the present, and like the present, it is continually refreshing itself. It is also notably temporally asymmetric. As Torre (2011, p. 361) puts it, "We take ourselves *to have power* over the future, yet lack power over the past." This cluster of features is connected to temporally asymmetric production. Some feel that time flows in the sense that one state of the world *produces* or generates the "next" state of the world. However, the idea here is more personal and direct, as it is based on our power to act. While we deliberate, the future seems open to us and the past closed or fixed. When deciding between two flavors of ice cream, one has an irresistible sense that one's options are open. But when contemplating past events or actions, we have a vivid sense of powerlessness. We may regard these feelings as deep and virtually unavoidable intuitions of the mutability of the future and fixity of the past. Flow in this cluster is related to our agency. What we have power over is continually shifting with each new now.

Surveying the field, therefore, one can isolate a number of distinct clusters that one may wish to explain:

- *Perceived* change, or more generally, perceived temporal properties (e.g., duration)
- Convictions about *real* change (i.e., updating novel presents)
- Convictions about something moving into the future
- Convictions about and perhaps phenomenology of producing the future.

Not all aspects of these clusters can be properly separated, nor can they always be formulated with great analytical rigor. That should not be surprising or disturbing. Manifest time, after all, is the product of messy real-life psychology, biology, and social interaction, not the product of the refined world of PhD dissertations. Fuzziness is expected.

I want to explain why creatures like us might take time to flow even when it doesn't. The above inference to the best explanation from experience, and the nearby one from conviction, benefits from having few competitors. Against an impoverished field it's easy for a poor explanation to be among the the best. The aim in what follows is to provide the objective passage theory some healthy competition.

11.2 Meet IGUS

We ought to eliminate this flow idea from the real picture [of the world], but before we can eliminate it we ought to understand how it arises. We should understand that there can be a self-consistent set of rules that would give a beast this kind of phony picture of time.

Thomas Gold (1967, p. 182)

Not all physicists have been happy to dismiss temporal flow as an illusion and leave it at that. From Richard Feynman's biographer James Gleick we learn that in 1963 the philosopher of science and time, Adolph Grünbaum, told physicists and mathematicians at a workshop at Cornell that the flow of time was merely an illusion, a byproduct of consciousness, and as such, it was nothing for them to worry about. Feynman, identified only as "Mr X" in the published transcript, would have none of this. Gleick reports:

> It seemed to Feynman that a robust conception of "now" ought not to depend on murky notions of mentalism. The minds of humans are manifestations of physical law, too, he pointed out. Whatever hidden brain machinery created Grünbaum's coming into being must have to do with a correlation between events in two regions of space—the one inside the cranium and the other elsewhere "on the spacetime diagram." In theory one should be able to create a feeling of nowness in a sufficiently elaborate machine, said Mr. X.
>
> (Gleick, 2011)

In the same spirit, in the quote that begins this section, the cosmologist Thomas Gold likewise feels that we understand flow only when we know the "rules" that "beasts" like us obey.

To my knowledge neither Feynman nor Gold pursued this project of understanding such a beast or machine, but more recently another physicist, James Hartle, has begun such an attempt in a paper entitled "The physics of now" (2005). Hartle seeks to explain some features of our temporal experience by building a model IGUS (Fig. 11.1), that is, an information gathering and utilizing system (Gell-Mann, 1994) and looking at how it experiences time. IGUSs are a general class of creatures, including humans, robots, and many biological organisms. Hartle's paper focuses on a particularly simple model robot. Let's build this robot, understand the rules governing it, and then see how it experiences time. My claim is that Hartle's IGUS explains some but not all of the above features of passage; nonetheless, by adding various components to his IGUS we can more or less get the rest. The main novelty of my view is that eventually we will add to the IGUS an evolving self-conception and provide some evidence to connect this to time flow.

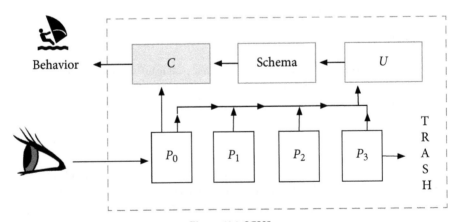

Figure 11.1 IGUS

Although an IGUS is really an abstract species, I trust that there will be no confusion if we name this particular robot we're building "IGUS."

IGUS is exceedingly simple, a kind of glorified camera. The robot has a perceptual apparatus that gathers information about the external environment. In Hartle's robot, the apparatus is entirely visual, although we could easily remove this restriction. After capture by the visual system, information is stored in $n + 1$ memory registers $P_0, P_1, \ldots, P_n$. Each register is a snapshot of a different time. IGUS takes new pictures at fixed intervals $\Delta\tau$. When a new picture is taken, it is stored in P_0 and the previous images are pushed down the memory registers, i.e., P_n is erased and replaced by the image in P_{n-1}, P_{n-1} is erased and replaced by the image in P_{n-2}, and so on. Thus, as Hartle says, IGUS has a coarse-grained record of its environment over a time $(n + 1)\Delta\tau$.

Even if IGUS's eyes are poor, a lot of information is stored in the memory registers. IGUS doesn't use all of it. Instead, via a process dubbed U, IGUS creates a representation of the world, what Hartle calls a *schema*. The schema is a useful model of the world, containing the locations of food, mates, enemies, plus generalizations inferred from the information in $P_0, \ldots, P_n$. When it comes time to act, IGUS uses a distinct process, C. Process C takes as input only the newest image, that stored in P_0, and the schema. Using these it computes the most likely future and decides what to do based on some cost–benefit analysis. Given the model of the world in the schema and the hungry tiger portrayed in P_0, it may compute: run!

IGUS is a poor imitation of us. We enjoy (or suffer) hundreds of features missing from the IGUS, even at this gross level of discussion. Our memories are notoriously reconstructive, for example, and they aren't arranged in linear registers. Sometimes memories are directly accessible even if old. The suggestion that U is our unconscious and C our conscious computations implies that these are utterly separate processes, when that is far from the case. In addition, our brains and visual systems have all sorts of mechanisms that make us very different from cameras.

All of this is admitted by Hartle and me. Nonetheless, some aspects of the IGUS make it a suggestive toy model of us. Like us, the IGUS forms and updates percepts of the world at distinct time intervals, it extracts information from memories to build models of the world, and it makes decisions and acts based on new information and its model of the universe. In particular, as regards time, by construction the IGUS is conscious only of the present, the information in P_0 extends over an interval of time, and yet it has awareness of changing schemas too, making predictions based on these schemas. Its virtue is that because it is simple we can isolate the features that contribute to IGUS feeling that time flows.

Hartle believes that even this simple robot can be said to "experience" the present, "remember" the past, and also "feel" a flow of time. I agree that in some sense this is correct. But—except when talking about perceived change—I'm going to avoid phenomenological talk. I also think that a few more features need to be added to IGUS before we can say it is modeling time as flowing in its schema.

11.3 Getting IGUS Stuck in Time

Let's begin with some very basic facts about IGUS. IGUS has identity through time; that is, one temporal stage of IGUS is genidentical with another stage. Earlier we thought of the identity of objects as a kind of thread tracing out a curve in the manifold of events. We assume that IGUS is composed of matter with mass. Hence our creature inherits all the rights and restrictions of objects represented by *always timelike curves* in physics. As we've seen in previous chapters, the features that accrue to always timelike objects are non-trivial. Recall, it means that IGUS is constrained by:

- the mobility asymmetry
- one-dimensional movement in the timelike directions

due to the matter–energy conditions and metric of our world. Thus the IGUS begins to be "stuck in time" in the sense of the previous chapter. But the IGUS also reaps the benefits of being timelike, in the sense that

- timelike curves possess an invariant temporal order

and thus its "experiences" (occupants of C-registers) possess an invariant temporal order.

To be "stuck in time" IGUS's world needs more structure. Before putting this in, notice that even at this early stage, we have a deep difference between space and time in terms of IGUS's possible beliefs. Note something about its "experiences" that is so obvious in us that it may escape attention. At any given place, IGUS has different mental states at different (sufficiently distant) temporal parts—the contents of box C changes with time—yet at any given time, IGUS doesn't have different mental states in different spatial parts—there is only one C box. Momentarily switching to people, what I mean is that in you right now, your head doesn't believe that hoppy beers are best and your right leg believe that wheat beers are best. Except in split brain cases, we ascribe mental states to whole organisms at a time. But at a place, we ascribe multiple mental states across time.

Why does this matter? As Mellor (1981) points out, this means that there is no *spatial* variation of *spatially* perspectival beliefs at a given *time*, yet there is *temporal* variation of *temporally* perspectival beliefs in a given *place*. IGUS doesn't, for example, have a pathway from one eye to one C box and a separate pathway from another eye to another C' box. Maybe the IGUS can blink or turn around and thereby obtain variation of spatially perspectival beliefs? Perhaps, but that is not spatially varied because blinking and turning around take time. By contrast, IGUS while just sitting still gets temporal variation of temporally perspectival beliefs.

Populate the IGUS's world with macroscopically stable objects, fast signals, and other IGUSs with whom to communicate. Endow IGUS with:

- the means to emit signals
- the ability to move relative to one another
- intra- and inter-modal temporal integration mechanisms

but inhibit them so that

- their relative velocities with respect to each other is very low relative to c
- the signals and processing are fast relative to typical macroscopic changes of salient properties
- their observations are restricted to "macroscopic" time scales

plus whatever other features are understood as required from the previous chapter. Then, if Chapter 10 is right, it will make sense for IGUS to form a model of the world that attributes a global shared now to it. In Hartle's terms, the present for IGUS is the time on its worldline over which it keeps one content in its P_0 register. What events on another IGUS, IGUS*, happen "at the same time" as the set of events comprising P_0 over the proper time $\Delta\tau$ along IGUS's worldline? If IGUS* is far away from IGUS, then a potentially indefinite amount of IGUS*'s lifespan may happen at the same time as P_0, as pictured in Fig. 11.2. IGUS*, we would expect, would "experience" many different contents in its registers over the time IGUS "experiences" P_0.

But if IGUS* is up close to IGUS, not moving very quickly with respect to it, and is endowed with the abilities and limitations mentioned above, then often what will happen "at the same time" as IGUS's P_0 will be a span of events with small proper time on IGUS*, small enough so that the content *doesn't* change. This can be visualized nicely by switching to natural units instead of adopting the usual convention of letting $c = 1$. Then the lightcones effectively flatten out for hundreds of kilometers when the time interval is a second, as in Fig. 11.3.

In this sense they will share a common albeit local present. This present will be useful in many aspects of IGUS's life and so it will be represented in its schema of the world. The robot hence divides its world into three regions: past, present, and future.

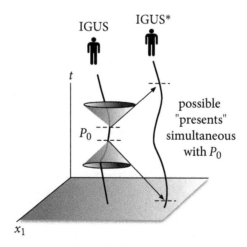

Figure 11.2 IGUS and IGUS*

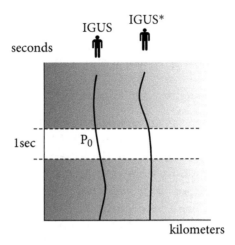

Figure 11.3 IGUS near IGUS*

Getting this division of the world is very important. We can now zoom in to a point on the creature's worldline and view the world from that perspective. There we see the world not in terms of solely earlier and later relations, but instead from a temporal perspective that sees a future and past.

In addition, these nows will update themselves regularly. Think back to the genesis story told in the previous chapters. The rationale for a now wouldn't hold up if the now persisted unchanged through great durations. True, our IGUS is wildly simplistic in updating its experience all at once at fixed intervals. In us there is perhaps no all-at-once periodic updating—a small field in cognitive science is now developing that looks into the question of the speed and mannner of update in human beings (e.g., Holcombe, 2009). We can update our experience remarkably quickly. You hear a noise and turn your head. In a split second you take in the new coherent scene. The temporal resolution is amazing. We can refresh our nows many times per second if need be. For this reason, we assume that the contents of IGUS's C-box update rapidly and that the memories shift boxes accordingly.

Although we have only begun, we can perhaps already discharge a worry about flow. Norton (2010) claims that the experience of flow is related to updating presents and sensing motion. Yet he seems to think that the absence of distortions from serial order is evidence of genuine passage:

> Most significantly, the delivery of the doses is perfect. There are no revealing dislocations of serial order of the moments. While there may be minor dislocations, there are none of the types that would definitely establish the illusory character of passage. We do not, for example, suddenly have an experience of next year thrown in with our experience of today; and then one of last year; and then another from the present.   (p. 27)

With all the tricks at their disposal, why can't an inventive researcher induce dislocations of the order of a day or a year? But if the passage of time is an objective fact independent of our neural circuitry, that failure is no surprise. (p. 29)

As I read this, Norton is asking why, if passage isn't an objective metaphysical fact, we don't sometimes experience things in the wrong order? This challenge is easily answered. To begin, as Norton knows, on short time scales, we *can* induce dislocations of temporal order. Many experiments confirm this. By abruptly halting a lag time, Stetson et al. (2006) reported that subjects perceived the lights they controlled on the computer to jump before they moved the mouse, not after. Given the different signal speeds, processing times, and temporal order thresholds, it is clear that we can induce, say, auditory experiences before visual experiences in some regimes even where the visual event is after the auditory event. But even within one modality, by shifting a subject's attention between two stimuli, one can induce order reversals (Stelmach and Herdman, 1991). Our IGUS doesn't have the mechanisms to do this, but we do.

On long time scales Norton is of course correct that we don't experience dislocations of day or year. Is *this* so hard to explain? It's no more mysterious than the fact that an IGUS taking successive pictures sees the pictures successively. The perceptual system can only weld together percepts based on stimuli that have arrived. The stimuli come in succession, the successive updating then happens after a fixed interval, and the order of all these events is invariant because IGUS is an entirely timelike worldline. Given the "laws" of IGUS, no stimuli from Thursday can make it into any Tuesday window. Modulo worries about small time intervals, how could events which succeed one another by large intervals appear as anything but successive processes? Unless one thinks the experiences are utterly detached from our cognitive architecture, it's hard to see how huge disorder dislocations could happen. They simply *can't* in IGUS. Nor are they *likely* in us over long intervals given our "neural circuitry."

Norton is right that it *could* happen in us. But the fact that it *does* over small time intervals—as predicted by neural circuitry—and *likely doesn't* over longer ones—as predicted by neural circuitry, physics, and our environments—added to the fact that the alternative theory says nothing about either shapes up into confirming evidence of the mind-dependence of flow, not disconfirming evidence.

11.4 Outfitting IGUS

One benefit of having children is that they provide cover for you to purchase items you otherwise couldn't justify buying. My (their) Lego *Mindstorms* robot is a great example. It is an IGUS whose modular structure allows you to add bricks, wheels, motors, and more to create a robot designed to implement specific tasks. Best of all, the Lego robot comes with ultrasonic, touch, light, and color sensors, allowing it to explore much of its environment. Want your robot to "feel" rotation and know its orientation? Just snap on the gyro sensor and update the software. Want proximity detection? Snap on the infrared sensor. And so on.

Similarly, if we want our IGUS to "experience" time at all like we do, at least in certain crucial respects, we're going to need to snap on some sensors and other circuitry, plus update its software. I don't think any of these additions really gets to the heart of the flowing now conception. However, when canvassing what one means by the flow we saw a diversity of views, so it may well be that some of these additions matter for some conceptions of flow.

11.4.1 Sensing motion and change

Many authors—especially authors seeking to debunk objective passage—take change to be evidence of metaphysical flow. Here I am not talking about McTaggartian *real* change. The appeal to *that* in support of flow is plainly question-begging, and anyway, I think the above updating present explains its non-question-begging counterpart. I am instead referring to the experience we have when we perceive something changing: the second hand moving around a clock, the notes changing when listening to music, and so on. At a perceptual moment, we seem to experience motion and change as opposed to a set of "stills." Here is Broad:

> [I]t is a notorious fact that we do not merely notice that something has moved or otherwise changed; we also often see something moving or changing.   (1923, p. 351)

Because movement and change require at least temporal order and (if speed or velocity are noticed) duration, then it seems that we are experiencing temporal properties at a perceptual moment. Is this experience of temporal properties evidence of flow?

If this is all the experience of passage amounts to, then it doesn't present much difficulty for physical time. Focus on the paradigmatic example of watching a second hand move on an analogue clock. We know from cognitive science, perceptual physiology, and much else, more or less how this works. Human beings are change detectors. We look at the world and notice not just static configurations but also all the "vectors" indicating motion. Motion triggers in us certain change or motion qualia.

There even exists some neurobiological evidence for how motion detection works. It's commonly thought, for instance, that our motion detection may arise from Reichardt-like motion detectors (Reichardt, 1969; Adelson and Bergen, 1985; van Santen and Sperling, 1985). These are simple circuits inspired by modeling the neural circuits of flies. A toy example of such a detector is pictured in Fig. 11.4.

Imagine IGUS has two detectors, O_1 and O_2, each sensitive to an edge. The edge is moving from left to right. It first triggers the left detector, but that information is sent into a delay filter before reaching our given set of neurons. If the edge continues at the right speed and direction, then it will trigger the right detector which sends its information directly to the set of neurons. The set will fire if the information from both sources arrive at the same time. In this way IGUS can have mental contents

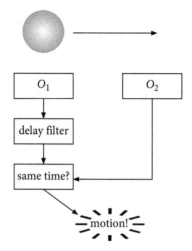

Figure 11.4 Reichardt-like motion detector

containing information about speed. These sorts of detectors are already implemented in computer chips, so there is no question that IGUS can be so equipped. In us, of course, our motion detection is vastly more complicated. Insofar as the perception of velocity is a problem, however, it seems easily solved.

Motion doesn't always trigger in us perceived change, nor do change qualia always require motion. The famous waterfall illusion makes one see change (in particular, upward motion) where there is none. The cells in your retina get fatigued by the downward motion of the waterfall, so that when you turn away and look at something not changing, you see it moving upwards. There are also many experiments that determine the spatial locations and time scales that will induce in us a sensation of something moving, as in the famous color phi experiments. In these experiments, an impression of motion is produced by still images. The classic case is of a picture of a blue dot on the left followed rapidly by a picture of a red dot on the right. At the proper speeds and distances, subjects will see the dot on the left move to the right, changing color in the process. Our motion detection equipment, like the rest of our perceptual mechanisms, can misfire. From this kind of misfiring, we learn much about how these mechanisms work.

Do we learn more? Le Poidevin (2007) and Paul (2010) claim that color phi is crucial to explaining flow. Although I would love to agree, I confess that I don't understand the reasoning. Paul writes, "[I]f the brain can create the illusion of flow in cases of apparent motion, then it can create the illusion of flow in cases of experiences as of passage." If I read her correctly, the idea is that our change qualia *animate* the static "block." Color phi shows that we can generate motion qualia from static inputs. The block also gives static inputs; but we use our motion detectors to generate change qualia here too.

Color phi phenomena teaches us that we can't infer from motion qualia that something actually moves. But I don't see it doing more. After all, all sorts of objects on the block *really* move and change. Is Paul saying that it's all static? Compatible with a "block" universe, there are objects that really change (e.g., a ball rolling down a hill) and those that do not (e.g., the color phi dots). Add experiencing subjects to the world. Sensations sometimes trigger motion qualia in them. Type I and type II errors can happen: sometimes motion qualia happen without real motion, and sometimes no motion qualia happen in the presence of real motion. To think it's all like color phi obscures the difference between the illusory and non-illusory cases. For more (similar) discussion, see Deng (2013) and Hoerl (2014).

Moving on, perhaps the question is instead that we experience the second hand moving in a different way than we do the hour hand? The second hand whooshes and whizzes by while the hour hand does not. As Hoerl (2014) points out, however, that difference is easily explained in terms of structural aspects of the phenomenology of motion perception. We do literally *see* the second hand move, whereas the sense of experiencing the hour hand is quite a different sense of experience produced by quite different mechanisms. The difference is not that both are cases of perceived succession, yet the second hand enjoys an additional experiential ingredient (whoosh and whiz) that the hour hand lacks. The two differ because one is the product of immediate visual perception and the other one is not.

11.4.2 Specious present

There exists a tremendous puzzle over how best to understand how change gets embedded in an experienced present. I refer to the problem of the so-called *specious present*. To appreciate the initial puzzle, model our experiences at times as snapshots, like the IGUS's snap stored in P_0 and made conscious in C. How do we have experiences of any temporal property apart from simultaneity? Some cameras now possess "burst" operations: the ability to take ten pictures or more in less than a few seconds. Train your camera on a basketball bouncing and hold down the burst button. When printed, the still photos will show the basketball at a great range of locations. Even if ordered properly, however, not one of them shows any motion. True, flipping these photos very quickly in order will make it look like an animated basketball, just as animations convey motion. But this just makes the point: by hypothesis the snapshots are the experiences, so something outside of the snapshot—the flipper—is responsible for the experienced motion. Our IGUS, with its snapshots in C, won't experience temporal properties apart from simultaneity. Yet it seems that we do, at a perceptual moment, experience temporal properties besides simultaneity.

There are many responses available (see, e.g., Dainton, 2001; Grush, 2007; Phillips, 2013; Andersen and Grush, 2009).

First, one can deny the background assumption of the above argument. The *vehicle* of experiences may be snapshots, but the *content* of those snapshots needn't be stills as in a photograph. The content could be of motion, change, succession, duration, and more. In general, the temporal features of the represented content needn't be isomorphic to the temporal features of the world itself nor in particular of the vehicles bearing this content. As psychologists are wont to say, brain time needn't be experienced time. Nothing in our setup precludes the content of the experience in IGUS's C box from having temporal properties.

Second, the vehicles of experienced content needn't be snapshots. Of course, since experiences are processes, the snapshot view must be wrong sometime before the limit of $\Delta t \to 0$; but maybe it is *seriously* wrong and the vehicles aren't tiny sensory atoms but instead temporally structured in various ways over a large interval of time. If they are like short films instead of snapshots, then the objective temporal relations among parts of the film may *themselves* serve to convey temporal information. On this picture, we can regard C as like a short film. Temporal experience then turns out to be special insofar as there is a close match between the temporal properties of the content of experience and of the experience itself (for defense, see Phillips, 2013; for criticism, see Watzl, 2012).

Third, one could deny both assumptions. The vehicles needn't be snapshots nor the contents depict stills.

Fourth, perhaps the experience itself is a snapshot, but succession, duration, and so on are experienced due to the experience's proximity to the recent memory of other experienced snapshots. Making a resurgence now, this idea can be traced all the way back to Whewell. On this picture, we could add an "arrow" in the IGUS from the recent memory registers back to P_0 and/or C to reflect this mechanism; alternatively, perhaps the memories work via the schema's affect on C and nothing "new" is added.

Finally, some are skeptical that we directly experience succession and other temporal properties at all.

Inspired by this puzzle and also various temporal illusions, e.g., representational momenta and the flash lag effect, models exist for many of the above options. The trajectory estimation model of Grush (2007) and the fixed-lag smoother model of Rao et al. (2001) are examples of the first option. The trajectory estimation model, as the name suggests, has the organism create an anticipatory model of what will immediately follow and then update this model as information comes in. On the fixed lag smoother model, by contrast, the organism waits a specified time period, say 80 ms, before creating its percept; it thus has access to an extended interval before "deciding" on the percept's content.

Incidentally, *all* of these proposed models are temporally asymmetric in crucial ways. The memory models obviously inherit the asymmetry of memory. The smoothing models wait 80 ms *and then* form the percept. And the trajectory estimation model gives us in experience the *last* non-overwritten estimate.

Work here is admittedly ongoing. The important point for us is that there is no reason to think that change perception requires anything more of time than physical time (Dorato 2015). Disputes over vehicle size, extent of time represented in content, injection or not of memory, and so on can all be accommodated by "snapping" the appropriate mechanisms to IGUS. None of these models requires of time anything more than physical time.

11.4.3 Felt duration

Although psychology often takes "the flow of time" to be given by felt duration, rarely is this the meaning of flow in philosophy. Typically duration is not taken as evidence for objective passage. Presumably the reason for this is that felt duration varies so widely. The movie that lasted two hours according to physics seemed longer than that to you but shorter than that to your friend. Hence there is little temptation to identify metaphysical passage with felt duration. Nonetheless, it's worth pausing for a moment to think about how our IGUS might feel duration.

Scores and scores of experiments have been done in psychology and cognitive science on felt duration. The mechanisms underlying these judgments are still an open area of research, yet much is known. Cognitive scientists often distinguish between the mechanisms responsible for encoding duration over short periods, say, of less than a second, and the mechanisms responsible for encoding information about longer durations, such as a second or even much longer, such as a week. Though the difference isn't hard and fast, nor sharp, the thinking is that judgments about short durations are in some sense automatic whereas judgments about longer times are cognitive. It is widely thought that for different time scales there are different mechanisms at work (Rammsayer, 1999). "How long were you on the phone with mom?" triggers different mechanisms, plausibly, than "is the left flash longer or shorter than the right flash?"

Not surprisingly—given the variance we find in ordinary life—experiments show that cognitive duration judgments vary with . . . well, virtually everything. Stimulus intensity, attention, color, gender, age, cognitive load, task demands, sensory modality, predictability, mood, task description, caffeine intake, and method of eliciting the judgment are just some of the variables known to affect judgments of felt duration. Especially important proves to be whether the task is prospective or retrospective. Knowing beforehand that you will be asked how long an event lasts can dramatically alter how long you judge it to be. Perhaps more surprising is that scientists are increasingly finding that even automatic judgments vary with many common environmental variables, e.g., novelty of stimulus. For instance, in the "oddball effect" a stimulus is repeated successively but interrupted by a surprise stimulus injected into this stream (Tse et al., 2004). Although the surprise stimulus lasts as long as the others presented in the sequence, subjects report that the new stimulus, the oddball, lasts longer than the others.

To keep track of interval duration, we must have various biological "clocks" available to us, both for short and long temporal distances (Block, 1990). These clocks are affected by all sorts of internal and external variables, so there are interesting variances among people in their duration estimates. The real story is not how badly these clocks perform but how well they do. Physics provides the genuine invariant duration of timelike paths through spacetime. Organisms must estimate these lengths to successfully navigate the world. The environment and our make-up will determine the cost–benefit calculation involved in our temporal discrimination. No organism will spend the energy to manufacture a clock measuring radiation going back and forth between two hyperfine levels of the ground state of the cesium 133 atoms (as it is measured in physics). Neither will many fail to invest in a clock that can't tell the difference between a second and a day. We'll probably never know whether nature has optimized us for duration detection, but we do know that we're pretty darn good at it, especially with practice (Montare, 1988).

What the biological underpinnings of these duration-detectors are is an open question. There are almost as many models of biological clocks as types of clock, or so it seems. Some are periodic devices like a pendulum, others accumulator devices like a sand hourglass. Some "classic" models are Hoagland's (1933) "chemical clock," Treisman's (1963) "internal clock," Thomas and Weaver's (1975) "attentional model," and Ornstein's (1969) "memory storage model," and so on. Ornstein's model, for instance, is based on a computational metaphor. The core idea is that remembered duration is determined by the storage size taken up by the encoded and retrievable stimulus information. This clock, like the others mentioned, is testable and known to face certain recalcitrant data. We don't know the correct model, never mind its neurological implementation.

Yet we know enough to see that adding clocks to IGUS won't pose problems. With enough ingenuity IGUS could be fitted with any of the above models. Following Ornstein, for instance, and to a crude first approximation, the IGUS could evaluate duration by measuring how much storage is used up in P_0-P_n. Essentially, all that is needed is a set of circuits whose ticks approximate—at a level appropriate for our IGUS given its environment and goals—the objective physical duration. If the flow of time is felt duration, IGUS can experience the flow of time.

11.5 Memories and Flow

If IGUS's world is like ours, then it is immersed in a temporally asymmetric environment. Heat spontaneously goes from hot to cold but never the reverse. Outgoing mechanical waves and radiation emerge from concentrated sources but seem never to converge on localized sinks. As we imagine our creature's worldline, we must imagine it as bathed in asymmetric processes. At any given point on its worldline, IGUS is receiving signals only from past events. If wavelike, these signals are not converging on that reception point but diverging from a concentrated source in its past. Entropy is going up everywhere, including in our IGUS itself. Surely this powerful and general wave of asymmetry affects IGUS in a deep way?

It must. In particular, it might be thought that the thermodynamic arrow of time implies the existence of what is sometimes called the *knowledge asymmetry*. Loosely put (for a better characterization, see Chapter 12), the knowledge asymmetry is the fact that we know more about the past than the future. Yesterday's DOW is easily discovered—I just look in a newspaper—but if I knew tomorrow's DOW already I wouldn't be writing this book. Narrowing further, one instance of the knowledge asymmetry is the memory asymmetry, the fact that we form memories of the past but not future. Simply remembering something doesn't mean that you *know* it. Permit me, however, that abuse of the term for the sake of presentation. Drop IGUS into this thermodynamic bath. Does it thereby suffer a memory asymmetry?

While the answer may seem to be obviously positive, it's a lamentable fact that so far philosophy and physics have yet to seal tight the connection between these arrows. The general outline of a connection seems pretty clear. Memory, to be memory, must satisfy certain conditions, especially a kind of counterfactual stability. Any macroscopic recording system must have had other options and be such that it would still create the record it does when the microstate is slightly perturbed (Mlodinow and Brun 2014). The thermodynamic arrow prevents memories of the future because slight perturbations in the microstates of future events will, when backward time evolved, lead to a very different present state (unlike the case in the forward direction of time). So it seems we can only have memories of the past in a thermodynamic world. See Albert (2000) for more.

Unfortunately, many important questions bedevil such accounts and the details are still being developed.[4] It would have been nice to say: embed IGUS in a thermodynamic world and automatically it remembers the past and not the future. Instead the best we can do is say that we have many tantalizing hints that the thermodynamic arrow (or whatever is responsible for it) also aligns the memory arrow. The exact details elude us a bit, but it remains extremely plausible that physical and biological facts dictate that we and IGUS suffer from a memory asymmetry.[5]

IGUS's memories are temporally asymmetric in a few respects (Fig. 11.5). First, at any event it can only gather information from events within the past and not future lightcone. Thus P_0 contains information about events that happen in the past lightcone of the most future point of interval $\Delta\tau_0$, P_1 information about events in the past lightcone of the most future point of interval $\Delta\tau_1$, and so on. This is true of cameras in our world and it is also true of IGUS's. IGUS's macroscopic states therefore have correlations with some macroscopic states of the universe, and these states are screened off by past interactions but not future ones. Second, the information flow within IGUS is asymmetric: the "image" in P_0 goes to P_1, then P_2, and so on, until P_n, and not the temporal reverse. Third, the act of erasure is asymmetric. We haven't said

---

[4] The idea of tracing the memory arrow to thermodynamics has a long and distinguished history in both philosophy and physics. Reichenbach (1956) gave it the first serious try, but he had to resort to a metaphor in the notion of "quasi-entropy" to attach thermodynamics to the so-called fork asymmetry. Albert (2000) takes a more sophisticated indirect path, arguing that a constraint on initial conditions implies both the thermodynamic arrow and knowledge asymmetry. Callender (2011c), Frisch (2010), and others point out worries. In physics, Landauer's Principle seeks to demonstrate that information erasure requires energy dissipation; if correct, it would connect records to thermodynamics too. But there are also problems here— see Earman and Norton (1998, 1999); Norton (2005); Maroney (2005, 2010).

[5] This asymmetry doesn't necessitate a metaphysical temporal asymmetry. With a fun toy model Norton (2000) shows that memory doesn't imply that the physical laws need be time irreversible or stochastic or that causation be a time asymmetric primitive. His argument also supports the conclusion that an asymmetry of time itself isn't needed to have memories.

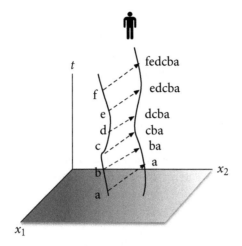

Figure 11.5 IGUS's memories

anything about the type of information IGUS can capture, but it seems unproblematic to allow it to capture and store spatiotemporal context. The IGUS doesn't yet have an "I" so it won't have properly autobiographical memories, nor will it reconstruct memories as we do. But there is no impediment to IGUS having a "layered cake" model stored in its schema, as the schema box gathers information from all memory registers and organizes the information temporally. If the sum of all these memory arrows serve to pick out "earlier" for the IGUS, then we can say with Hartle that the IGUS "experiences" a flow in its conscious register updating itself in the earlier to later direction.

Our own memory system is vastly more complicated than IGUS's. Our memories can be procedural, unconscious memories of skills such as navigating the kitchen at night, or declarative, memories of facts and events that can be consciously retrieved. Tulving (1972) famously divided the latter into episodic and semantic types. Semantic memories are those of particular facts and items that don't depend on spatiotemporal context. You might recall all sorts of mathematical, geographical, and sports history this way. I remember that Bill Walton came off the bench for the 1986 Boston Celtics, helping to create arguably the greatest basketball team in history. Yet I recall nothing about the causal history or experience that created this memory. By contrast, I remember catching a striped bass when 14 years old—not only *that* I caught the fish but also the pressure in my forearms, the bass pulling the small boat around, the sound of the line buzzing out, and more. I recall catching the fish "from the inside," as Shoemaker (1963) puts it. The "where, when, who" comes with the memory, whereas no such context comes with my Bill Walton memory. When the memory is "from the inside" and about oneself, some authors distinguish it as an *autobiographical* memory, a special type of semantic memory.

Some of our memories include memories of memories. Just a moment ago I remembered catching a striped bass and now I remember that act of remembering. I remembered that memory—or some similar reconstructed version of it—many times before too, almost every time I come across a picture of me with that fish at my father's house. The same goes for indefinitely many other memories too. Our autobiographical memories live in a thick multi-layered cake of memories. Arguably this cake is crucial to the formation of a conception of a real past with a linear structure (see Campbell, 1997). Our IGUS is already outfitted with the means of creating memories. In its schema we could also add an ability to abstract out a linear structure from nested memories. Memory is absolutely critical to our experience of time.

Some authors feel that it is the key to understanding the flow of time (see especially Mellor, 1998, pp. 122–3). It's easy to see why. Here is Barry Dainton:

> Even if our future experiences are just as real as our past ones . . . at any given point during our lives we have detailed knowledge only of the earlier parts of our life; indeed, for all we know we might be about to drop dead, and so not have a future at all. It is thus easy to believe that our lives don't extend beyond the present, even if in fact they do. This belief leads to another: that any continuation of our lives involves the addition of something new to the already completed past. Since we remember having these same beliefs at past stages of our lives, and anticipate having them in the future, we always have the impression of living at the outermost limit of an ongoing future-directed process. (2001, p. 31)

Dainton's first point is that memories help explain why the past seems fixed and the future open. We know what happened yesterday, but not tomorrow. In this ignorance lies a certain kind of epistemic openness that may slide into a belief in ontological openness. After all, everyone suffers this same asymmetry, so evidence disconfirming the belief that the future is open will be hard (or impossible) to come by. Dainton's second point is about passage: the growth of memories and the lack of any about the future makes it seem as if what exists is increasing. C. D. Broad, at one point an advocate of absolute becoming, writes that "[n]othing has happened to the present by becoming past except that fresh slices of existence have been added to the total history of the world. ... The sum total of existence is always increasing" (1923, pp. 66–7). However, one is tempted to ask Broad how he can distinguish fresh slices of existence being added to history from fresh *memories* being added to the total *remembered* history of the world. The phenomenology, it would seem, would be the same.

With Dainton and Mellor, I share the view that the memory asymmetry is absolutely crucial to the explanation of why we believe in passage. Dainton, Mellor, and others have put their fingers on important points, both about passage and the fixity of the past. If necessary, we can modify our IGUS so its memory is more like ours. Like a digital camera that includes GPS information about the image in the picture, we can allow our IGUS to form episodic memories. With enough ingenuity, we can also rig it to take pictures of pictures, and so on, and use these when building its schema of the world.

Is this enough for passage? I think it probably isn't. Ultimately with IGUS we have a robot instantiating various asymmetric processes and maintaining an asymmetric macroscopic correlation structure with the outside world. These asymmetries shouldn't be underestimated, especially when put in context with all the rest of manifest time we've already recovered. Nonetheless, I see at least two places where IGUS might do better.

First, it will be complained that we don't have movement yet, the *whoosh* and the *whiz*. Nothing seems to "crawl up" IGUS's worldline, thereby making time flow. I think that there is something to this criticism. I hasten to add that I don't believe anything *literally* moves up a worldline. We do, however, have the *conception* of something moving up the worldline, and I'm confessing that IGUS may not yet think "it" moves up its worldline. The memory asymmetry doesn't provide us with our desired feature of *something moving through time*.

Second, the sense in which the future is open for the IGUS is limited. We've recovered an epistemic sense of openness. The IGUS in some sense knows more about the past than future. Ultimately the knowledge asymmetry, or what causes it, may be responsible for the conviction that the past is fixed and future open. However, the conviction that we have power over the future and not past is so deep that it would be nice to say a little more if we can. The memory theory of flow, on this score, isn't wrong, per se. It's just that if we can do better we should.

11.6 The Enduring Witness

Time presupposes a view of time. It is, therefore, not like a river, not a flowing substance. The fact that the metaphor based on this comparison has persisted from the time of Heraclitus to our own day is explained by our surreptitiously putting into the river a witness of its course.

(Merleau-Ponty, 1945, p. 411)

The extra ingredient needed to get flow, I submit, is *the self*. This idea isn't new to me. It can be found in contexts as diverse as ancient Eastern philosophy, continental philosophy, analytic philosophy, science fiction, and developmental philosophy. The reason it is ubiquitous, I maintain, is that there is an important insight here: the self, although intimately connected to our memories, provides the something that flows, the object with identity through time missing from the memory account of flow. The self furnishes the whoosh and whiz associated with something moving through time, the idea that you are here at this time now, and it also makes intimate the conviction that the future is open.

The nature of the self is an enormously contested topic. I cannot do full justice to the literature here, nor need I to make my point. To begin, note that we believe in selves. We take the self to be the seat of our experiences and the cause of our decisions and actions at a time. There are plenty of good reasons for organisms to have a conception of a self. For example, Dennett (1989) sees the conception of the self as developing so that organisms can know what to defend and what to plan for. We need a self/not-self division. The lobster, Dennett says, has to know not to eat itself when hungry. It needs to know what to plan for and what to protect. We need a boundary for "what you control and care for." Same goes for virtually all organisms. Even amoebae require some primitive momentary encoding of a self/not-self division. Organisms represent this division, and in this basic, minimal sense, enjoy a representation of a self.

Human beings demand a much richer notion of self. Not only do we act in the moment, but we plan, both short-term and long-term. At dawn, some of our ancestors thought, I will meet you to hunt for mammoths. Next summer, I now say to you on email, I will meet you for a conference in Grindavik, Iceland, but in a moment I plan to go downstairs to get some pistachios. Often this planning involves complex social interactions. Through it all we are constantly representing ourselves in thought, language, and behavior, both to others and to ourselves. The spatial boundaries of the "momentary" basic selves need to be stretched diachronically to accommodate these plans and interactions. But it's crucial that this diachronic self is conceived as one object, for its whole point is to distinguish itself from others in planning and other interactions. This observation brings us to an important point for our purposes: selves *endure* through time. Today you are different from what you were yesterday and much different from what you were a year ago and ten years ago. Your hair may be grayer, you may be taller or shorter, and you might even be a fan of a different sports team. Nonetheless, you feel that you are numerically identical to all those organisms at all those times, one self-same entity surviving all these alterations.

Endurantism is a theory of persistence through time. Instead of viewing material objects as temporal parts of a four-dimensional worldline (perdurantism), endurantism holds that such objects are three-dimensional and wholly present when they exist (Hawley, 2015). I'm not a big fan of turning this conceptual distinction between endurantism and perdurantism into a metaphysical one, for I regard endurantism and perdurantism as two different ways of carving up the same exact world. No matter— we simply need the conceptual distinction. Certainly in the manifest image of the world we don't believe that our present self is merely a temporal part of a larger four-dimensional self. Instead we hold that our selves are wholly present at each time. Nothing is *missing* from you at the moment you read this. You're all there! Likewise, you were entirely present when you had your first kiss and when you had your fiftieth birthday. Our selves, we think, are complete at each time they exist. This may not be true, but it accurately characterizes a feature of our pre-reflective conception of the self. This fact will play a role momentarily.

The literature on personal identity offers many metaphysical theories of what grounds talk of a self. Maybe the self is a manifestation of a soul? Or memories? Or a relation of psychological continuity and connectedness? Or . . . ? Because I am interested in explaining why creatures like us might develop a model of time approximating manifest time and not what selves really are, I can punt on this hard metaphysical question. Our model of the world includes enduring selves. I just gestured toward a sketch of why creatures like us employ enduring selves in our modeling of the world. That's all I need. The metaphysics of selves is deeply fascinating but ultimately irrelevant to my project. I require epistemological and psychological features of this conceived self, not special metaphysics.

One feature I need is that we take the self to endure. The other is that this self is identified in narratives. The *narrative theory of the self* is popular across the humanities and social sciences. One finds studies in philosophy, cognitive science, neuroscience, developmental psychology, cultural anthropology, sociology, and elsewhere. Let me immediately pause to note that one common version of the narrative theory of the self takes itself to be answering the metaphysical problem of personal identity, that is, it offers a metaphysics of selves. The thesis is found in Ricoeur (1986), Dennett (1992), and elsewhere (see Schechtman, 2011 and Hardcastle, 2008, for review, references, and discussion).

In Dennett's hands the core idea is that the self is a kind of self-constituting fiction that unifies different stages of a bodily worldline. This fiction is the protagonist in the story told by the human organism. Like a theoretical construct such as "center of gravity," this fiction can play an important explanatory role in science. Since you act for the sake of this fiction, good explanations of your behavior will refer to it. But the fiction really is a fiction: it doesn't exist.

Weaker metaphysical versions exist whereby narration doesn't constitute selfhood but is instead a crucial component of selfhood. Although I am partial to a view in this neighborhood, please note that I don't require the metaphysical version to be true. Obviously I need an answer to the problem of personal identity that is compatible with physical time. But I don't need the hypothesis that narration *constitutes* selves. What I am committed to is that the narrative theory more or less gets the epistemology right— that is, that the enduring self we posit is the subject of our narration. What we identify as our self is the subject of our story-telling. This claim is less controversial than its metaphysical counterpart, but still contested (e.g., Strawson, 2004).

The narrative theory hasn't settled into a canonical formulation. Not even close, really. Given the diversity of disciplines discussing the theory this situation is hardly surprising. Even among philosophers inclined to agree that the self is a fiction, such as Dennett and Velleman, there exist disagreements over whether the fiction is really illusory or singular, with Dennett holding there to be a single illusory narrator and Velleman positing different non-illusory selves for different episodes in one's life. Plenty of other disagreements exist over how *thick* the narration must be. Must the story of our lives be one that we consciously tell and accept, or can it be implicit? Is narration involved in just getting a cup of coffee, or is it more involved, such as in the thought that I'm a decent father? And how are emotions and moral reasons involved? We won't take a stand on these fine details of the narration account of the self. I think a fairly minimal notion of narration is sufficient for the view here. If something more substantial turns out to be true, so much the better.

For a start, we can agree that

> at least two events must be depicted in a narrative and there must be some more or less loose, albeit non-logical relation between the events. Crucially, there is a temporal dimension in narrative.   (Lamarque, 2004, p. 394)

The key is to fill in the details of this relation. Minimally, our lives should

> comprise an unfolding, structured sequence of actions, events, thoughts, and feelings, related from the individual's point of view.   (Goldie, 2000, p. 4)

But I think we can also add, with Schechtman (2007), that the way the events are related is teleological. The narrative isn't merely a history of what has happened, it adds content to that history, giving reasons for what one does and has done, all from a first-person perspective. I didn't merely boil water, then add tea, then walk to the refrigerator to get milk, and so on. I boiled water (etc.) for tea because I was chilly and wanted to get warmer, needed the caffeine to stay sharp while grading, wanted the latter because I desire to be a good teacher, and so on. With at least a notion of narration like this, and possibly much more nuanced, let's now continue, knowing full well that much more could and should be said about narration.

Let's accept that there is something right about the narrative theory of the self, at least epistemologically. Recall the observation that we tend to be *endurantists* about this self. When these two components are fitted together, we arrive at a picture reminiscent of what we wanted when explaining time flow. Velleman articulates this idea originally associated with Buddhism extremely well:

> Whatever the future draws nearer to, or the past recedes from, must be something that can exist at different positions in time with its identity intact. And we have already found such a thing—or the illusion of one, at least—in the form of the enduring self... [I]f I am an enduring thing, then midnight and I get closer together, and not just in the sense that I extend temporal parts closer to it than my earlier parts. I don't just extend from a 9:00 pm stage to a 10:00 pm stage that is closer to midnight, as I extend from my feet to a head that is closer to the ceiling; I exist in my entirety within the stroke of 9:00, and I exist again within the stroke of 10:00—the selfsame entity twice, existing once further from midnight and then all over again, closer. Midnight occupies two different distances from my fully constituted self. From my perspective, then, midnight draws nearer.   (2006, p. 13)

You take yourself to be one and the same entity throughout your life. Yet we face all these changes. These changes must be changes with respect to something. The suggestion is that the self understands time to be changing in some way. Like Merleau-Ponty and the Buddhists, Velleman notes that one needn't believe in a metaphysically robust notion of the self in order to sustain this idea.[6] A narrative conception will suffice. Hence the memorable slogan:

---

[6] Although one may. Interestingly, Lowe (2009) argues that the passage of time is metaphysically dependent upon the change of simple enduring persons. The current view is a metaphysically deflated

the illusion of the enduring self is responsible for the illusion of the flow of time.

What crawls up the worldline is not a substantial metaphysical entity, e.g., Weyl's moving spotlight, but rather the character in a kind of story. A narrative is being built up the worldline. At each moment, the main character in this story is being created from the resources available at that time. You are always the leading edge of the story. As more resources become available, the story and self change. Always, however, the story and self are responsible for the assumed unity of the self through time. What is "crawling" up your worldline is a story that unfolds "up" the worldline, the story of me (and for you, you). With this understanding, we obtain a reason for our deep conviction that something is moving through time.

In a moment I'll try to bolster this claim. Right now, however, let's notice that the memory theory of time flow is not far off this "flowing self" theory. The reason is that memory is crucial to constructing selves.[7] Memories are the ingredients out of which selves are born. Without episodic memory, arguably there wouldn't *be* a sense of personal identity over time (Atance and O'Neill, 2005; Klein and Nichols, 2012). And as Ismael (2007) stresses, we rely on reflexive autobiographical memories in telling any narrative. These particular types of episodic memory are often viewed as ways of "re-experiencing" a particular episode from one's past. They are important building blocks for any narrative (Eakin, 2008; Fivush & Haden, 2003; Klein, 2001; Klein & Gangi, 2010). Without them, it seems possible to know one's past traits, i.e., have semantic memory competence, and yet still lack a sense of ownership over one's past. This disassociation is vividly seen in the case of the patient known as D.B., who had knowledge of his past traits but nevertheless seemed unaware that he had a past due to his inability to form episodic memories (Klein et al., 2004). Perhaps even further aspects of our memory system are required (Klein and Nichols, 2012).

With this stock of memories "from the inside" one imposes an order among the jumble of memories and memories of memories. I may use all the available information to do this: photographs on fireplaces, correlations between procedural memories and declarative ones, testimony from others, and more. I know some Russian words, have a memory of taking Russian in college, know someone who saw me there—the best story puts two and two together and identifies that person with me. Thanks largely to autobiographical memories, I construct a narrative that, accurate or not, is what I associate with my self, the hypothetical entity that survives through time.

---

version of this link. The conception of time as passing is dependent upon modeling ourselves (and perhaps others and objects too) as enduring objects that change.

[7] Even skeptics about personal identity can admit as much: "As memory alone acquaints us with the continuance and extent of this succession of perceptions, 'tis to be consider'd, upon that account chiefly, as the source of personal identity" (Hume, 2000, p. 170).

The memory arrow turns out to be crucially important. It's just not the whole story. Memories are the raw materials out of which selves are built. Since these materials are temporally asymmetric, the selves created from them are too: they crawl up the worldline in one direction, not the other. Asymmetric memories, however, aren't the full story. The self they help create is the missing ingredient, the hidden variable that provides additional movement, or at least, the conception of it.

11.7 From Flowing Selves to Animated Time

How do we get from this self-narrating self to the idea that time itself is dynamic? Easy. The answer is that we readily slide back and forth between the two ideas all the time. We can make two observations to support this claim.

First, we can bring to bear some research done in cognitive metaphor theory (Boroditsky, 2000; Núñez et al., 2006). There it is common to distinguish an *ego-moving perspective* from a *time-moving perspective*. These are representational schemas with which we frame the world. In both of which the deictic center, the ego, is fixed on the now of the series. As the name suggests, the first has an ego moving through time, where the time series is like a background landscape. We adopt this framing when we say that we're coming up on Christmas, we're approaching the deadline, and so on.

The time-moving perspective, by contrast, has time moving past the relevant self. Here the ego is static and the temporal landscape changes. We adopt this framing when we say that Christmas is coming up on us, that the deadline is approaching, and so on.

Advocates of cognitive metaphor theory experiment on ways in which spatial and linguistic priming will recruit one or the other perspective. A well-known experiment presents subjects the ambiguous sentence

Next Wednesday's meeting has been moved forward two days

and simply asks them whether it means the meeting is now on Monday or Friday of next week. Pause a moment and decide for yourself. If you answered "Friday" you are employing an ego-moving perspective. The reason is that you have interpreted "forward" with respect to the moving ego, so the meeting must now be on Friday. If you answered "Monday," by contrast, you employed the time-moving perspective. The reason is that you interpreted "forward" to mean that time as it moves is arriving earlier, so the meeting, stuck on its timeline, reaches you earlier. In conversations, we unconsciously keep track of these representational schemas. Research has shown that the language we speak, the actions we perform, and much more can prime one or the other answer. Mandarin speakers and English speakers, for instance, significantly differ in their answers (e.g., Fuhrman et al., 2011).

What lessons to draw from this research is hotly contested. I have no stake in this debate. All I want to claim is that the mere existence of these two schemas and the fact that we can unconsciously flip back and forth between them based on the tiniest priming suggest that once we get the ego moving it's a very small step to getting time itself moving too. We simply need the other perspective to develop or have developed. We can then slide between the two nearly effortlessly.[8] So if you grant me that endurantism gives one a sense that the ego is moving through time, then I claim that is tantamount to regarding time as dynamic.

The second reason for believing this is that the whole notion of time flow is unclear on precisely this point. Even the great philosopher of time J. M. E. McTaggart seemed flummoxed by a question as simple as whether the flow was toward the future or past:

> It is very usual to present Time under the metaphor of a spatial movement. But is it to be a movement from past to future, or from future to past? … If the events are taken as moving by a fixed point of presentness, the movement is from future to past, since the future events are those which have not yet passed the point, and the past are those which have. If presentness is taken as a moving point successively related to each of a series of events, the movement is from past to future. Thus we say that events come out of the future, but we say that we ourselves move towards the future.   (1908, p. 470).

Relative to the idea of a privileged now, the future events are streaming backward, like locations seen through a train window if we imagine the train still and the outside world moving. Relative to the series of events arranged successively, by contrast, it seems that it is the now moving, just as your train window is moving relative to the

Past   Future

Figure 11.6 The self in time

[8] I say "nearly" because measurable delays can happen when shifting schema mid-conversation. Experiments have shown that it's cognitively slightly more demanding to engage in conversations that adopt inconsistent schemas. If you've just landed in Chicago's O'Hare airport from London and someone asks you about time zones, it's harder to answer them if they switch schema mid-question (Gentner, 2001). But even under these adverse conditions, we can still do it.

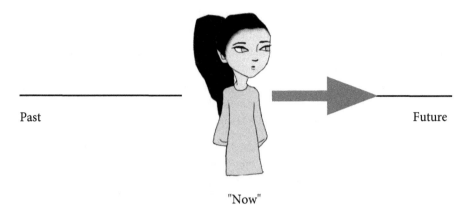

Past   Future

"Now"

Figure 11.7 Ego-moving perspective

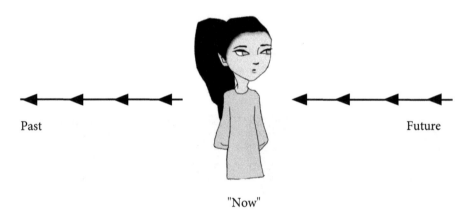

Past   Future

"Now"

Figure 11.8 Time-moving perspective

landscape. Clearly McTaggart has something like metaphysical versions of our ego-moving and time-moving schemas in mind.[9]

My argument is now simple. Velleman's enduring self is most naturally interpreted as a representation where we adopt an ego-moving schema. But given the way we effortlessly move back and forth between schemas, once you have the ego moving you in effect have time moving too. It's not a further step to explain why creatures endowed with these two representational schemas think time flows if they think the

[9] There is room for doubt that the essential divide is rightly drawn between the ego- and time-moving perspectives. Núñez et al. (2012), for instance, propose instead that the divide is between a perspective in which the present (not ego) moves and one in which time moves with respect to the present. That divide more accurately traces McTaggart's two possibilities. Given the tight connection between the self and the present—the self is wholly present, and the present, one might say, is the time when the self experiences, acts, and so on—either division will do the work we need here.

ego endures. I submit that to make our IGUS believe that time *whooshes* by we need it to represent itself in its model of the world as an enduring self. The model changes at each time step as P_0 is refreshed, but within this model there is a posited "I" that is represented as wholly present at each moment. In pictures, the argument is that the reality is represented in Fig. 11.6, yet pressures for the self to endure demand a self-conception represented as Fig. 11.7, which in turn we confuse with a picture of a "dynamic" time such as Fig. 11.8. This attractive idea adds the *whoosh* and animation that we earlier said our IGUS lacked.

11.8 Temporal Decentering and the Self

This theory of passage fits nicely with what we know in developmental psychology. In fact, I suggest that the present theory of passage proposes a non-accidental association between the development of the self and the development of our temporal representations.

How do children develop their understanding of time? Work on this topic is ongoing, but the broad outlines are becoming clear. As many have noticed, very young children seem to have a very different grasp of time than adults do. Past (future) events seem to be grouped together in an otherwise temporally undifferentiated bag of events. *Yesterday* refers not just to yesterday but to the entire past. And that is not simply a feature of the word use but it seems to be a feature of a small child's conception of the past. In a picture, it looks something like Fig. 11.9.[10]

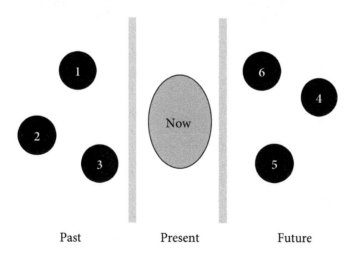

Past   Present   Future

Figure 11.9 Undifferentiated past and future

[10] The set of pictures is based on Povinelli (1999).

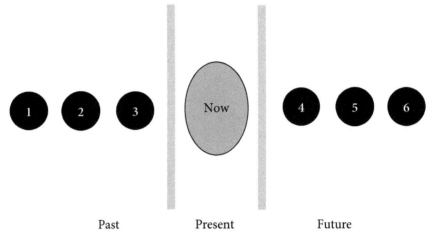

Past   Present   Future

Figure 11.10 Timeline

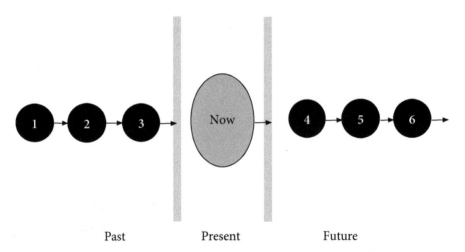

Past   Present   Future

Figure 11.11 Causal order

Children at this young age—2 or 3 years old—tend to attach times to particular events (e.g., birthday parties) and not really have a timeline in mind. But as they grow, they are eventually able to provide a rough order on events (Fig. 11.10), and even then regard the events at one time as arising due to the events at another (Fig. 11.11).

Children are able to understand the now as one moment in a temporal sequence of events. Past (future) events are not simply a temporally undifferentiated blur, but rather children begin to appreciate that the past (future) has temporal properties. The past event of Tommy eating the crayons was *before* his having an upset stomach,

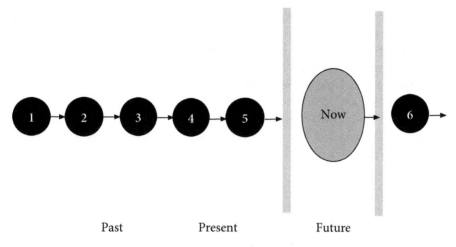

Figure 11.12 Temporal decentering

not simply in the past of the now. The upset stomach was due to the prior eating of crayons. Soon children can distinguish between the history of the world and the history of the self.

Crucially, by approximately age 4 or 5 years old, children are able to *temporally decenter*. As adults we are able to move back and forth mentally through different temporal perspectives. I know that when Socrates was drinking hemlock, his death was in the future, whereas while I write this now, it is in the past. Having the ability to temporally decenter oneself is a crucial component of our mature conception of time, as it allows us to understand the now as itself moving in a series of events (Fig. 11.12).

When we temporally decenter, we can lift up the now and place it forward or backward in a sequence of times, e.g., when Tommy ate the crayons, getting sick *was future*. At this stage children begin to have the resources necessary to form an A-series representation of time, a conception of time as flowing. McTaggart's A-series relies on an ordering (his C-series), and we obtain an A-series conception of time when we understand a now as moving along this ordering creating a past and present.

How and why do we learn to temporally decenter? We spatially decenter from our spatial perspective by simply turning our heads and taking in different perceptual information. Or we can get up and spin around, walk from here to there, and so on. Such actions allow us to see that my right may be your left, and so on. Disagreement among spatial perspective is right in our face—literally, if you consider the different perspectives adopted when looking through your left eye versus your right eye. Temporally, however, it is much harder to decenter. As we learned in Chapter 6, the mobility asymmetry and one-dimensionality of timelike directions will prevent us from decentering in time the way we do in space. The inability to do so is more or less the cause of our being stuck in the present. So how do we accomplish this feat? Enter selves. A plausible suggestion is that we temporally decenter by being aware that we are temporally extended, i.e., by understanding ourselves as having occupied different temporal perspectives.[11] Temporal decentering is "associated with a grasp of the fact that there were previous points in time at which one's temporal perspective on events was quite different but one's own nonetheless" (McCormack and Hoerl, 1999, p. 174).

Experiments show that the development of a concept of an enduring self is a very subtle thing. In many experiments, Povinelli and collaborators (Povinelli et al., 1996; Povinelli and Simon, 1998) look at the development of a self-conception in small children. In a temporal variant of the well-known "mark test" known as *delayed self recognition* (DSR), the scientists videotaped 2-, 3-, and 4-year-old children playing a game involving being patted on the head. Unknown to the children, the scientist would surreptitiously place a bright sticker on the child's head. A few minutes later the child is shown a video of the encounter, including the placing of the sticker upon the head. What would you do if shown such a video? You would immediately reach up and feel the top of your head looking for the sticker. Me-three-minutes-ago is me-now, one knows, so probably the sticker remains on the head.

Interestingly, this immediate and intuitive understanding of the situation seems to develop between ages 3 and 4. In this experiment and others like it, few if any 2- and 3-year-olds reach up to their head for the sticker, yet most 4-year-olds do, e.g., 25% of 3-year-olds versus 75% of 4-year-olds. Nevertheless, the younger children do recognize themselves in the videos. Paul can point out that it's Paul in the video. Yet something crucial is lacking. The younger children often disassociated themselves from their video counterparts, remarking that the scientist put the sticker on *his* or *her* head, not *my* head. In one case Povinelli reports that a 3-year-old named Jennifer answers the question of who has the sticker in the video with, "It's Jennifer . . . but why is she wearing my shirt?" (Povinelli, 2001, p. 81).[12]

If this is correct, then something like the following picture emerges. As our abilities to remember increase and develop, we are gradually able to put together a conception of a self. Autobiographical memory is crucial to the development of a self (Fivush, 2001). As the self develops one begins to be able to temporally decenter. And with temporal decentering, one begins to entertain an A-series conception of time as flowing.

Since having a self, autobiographical memories and temporal decentering are not all-or-nothing abilities, it's a bit messy to determine whether this picture is empirically supported. However, I predict that we'll see some non-accidental correlations among these three variables, autobiographical memory, self-conception, and temporal decentering. Right now we know that interesting developments occur in all three abilities at roughly age 4 years old. Once we accept some diagnostic task as representative of each

---

[11] See Campbell (1994), Moore et al. (2001) for discussion.

[12] Whether what is lacking is actually a conception of an enduring self or the ability to reason about the self is an open question. Some researchers who take the latter path suggest that we acquire our self-conception as early as 2 years old (Zelazo and Sommerville, 2001). Either way, all agree that by age 5 typically children have the bare bones of the mature conception of the self as enduring through time.

ability (which is the theoretically tricky part), one could test this theory by seeing if correlations exist among the various tests, just as one often does in developmental psychology. One might measure, for instance, performance on the DSR task with performance on various tests of memory or the ability to temporally decenter. Apart from this association, however, I make no further prediction. Which is first, temporal decentering or a full-functioning self-conception? That is a bit like a chicken and egg problem. Here I also caution that neither need be considered the cause of the other. It might be that children at a certain stage begin to be able to negotiate conflicting non-spatial representations and that this ability is the common cause of both temporal decentering and a self.

If temporal decentering is important to the conception of time as flowing, as I think it is, then in developmental psychology we find some support for the connection between time flow and the self. It is not the whole story. What I've just described depicts an A-series conception of time, but it does not pick out a preferred and objective flow. The narrating self gets us the preferred temporal perspective, the identification of the present with the leading edge of the story. And the material from Chapter 10 gets us a now that is apparently objective.

11.9 The Acting Self

With a directed whoosh in hand, we have regained almost everything found in manifest time. One major feature remains. We believe that the direction that hasn't yet whooshed is in some sense open. Why? Part of the answer must be that the kind of knowledge that we have about the past is of a different character than that we have about the future. Another part of the answer is that the world is causally asymmetric. Causes always seem to precede their effects. Or if one wishes to frame this asymmetry in terms of counterfactuals, we might say that the future depends counterfactually upon the present in a way the past does not. Future outcomes depend upon actions now whereas past outcomes do not. These two massive temporal asymmetries, the knowledge asymmetry and the causal-counterfactual asymmetry, have a powerful effect on creatures in our world. They entail that creatures like us will be very uncertain about later events on their worldlines even though they tend to have some measure of control over these events.

Like many others, I hold that these asymmetries are ultimately responsible for the past/future asymmetry. Here is David Lewis describing the general idea:

> In short, I suggest that the mysterious asymmetry between open future and fixed past is nothing else than the asymmetry of counterfactual dependence. The forking paths into the future—the actual one and all the rest—are the many alternative futures that would have come about under various counterfactual suppositions about the past. The one actual, fixed past is the one past that would remain actual under this same range of suppositions.
>
> (1979, p. 462)

To survive IGUS will need to anticipate what will happen. It does so, very roughly, by building a range of possible models over the course of future events, placing probabilities in front of these options, and updating accordingly. Because IGUS's deliberations and actions are only productive in one direction, it doesn't do this in the past direction. Already employing a model of the world that includes a distinguished present and a self moving through time, IGUS will understand these allocentric asymmetries as egocentric. Hence IGUS will have a tendency to paint the future as a whole open and the past as a whole fixed. At bottom, this is the reason why IGUS, and we, hold that the future is open and the past fixed.

In the next chapter we'll look at how these asymmetries affect our attitudes and beliefs in somewhat surprising ways. Here I would like to connect this asymmetry to our discussion of passage (see also Chapter 14 for more connections). This helps explain why the belief in an unfolding future is such a deep and intuitive part of manifest time.

Consider IGUSs like us. We don't simply calculate expectations over a range of models of the future and automatically go with what best maximizes our utility. We're agents who are constantly making conscious decisions. Maybe it is at a car dealership, when deciding whether to succumb to the dealer's relentless pressure to purchase a particular car, or perhaps it is in front of a crowd, wondering whether to actually go through with uttering "I do." Or maybe it is just deciding whether to scratch an itch. In each of these cases we possess a strong and irresistible sense of agency in one direction. We lead lives—we don't merely sit back and enjoy a temporal sequence of experiences.

For Velleman, our feeling of freedom and belief in an open future results from the belief that, within a certain range, whatever we decide to do, we will do. Many of our decisions are *self-fulfilling predictions*. Within limits, if I decide to do X then X happens. My snap decision when the waiter comes to get chocolate ice cream instead of strawberry means that I'll actually end up with chocolate ice cream. And at any moment prior—right up to the very moment of the decision—I could *bilk* anyone's prediction of what I will do (Ismael, 2011, p. 163).

That's not to say that we are independent of large social or physical forces or that we even have any genuine freedom, however that is understood. What I mean is that our experience gives us every reason to think that we *could* get up in disgust and walk away from the car dealership, we *could* create a massive spectacle at the wedding by not saying "I do," and we *could* resist the pull of the itch. All of our evidence is confirmatory of the idea that our decision is the causal trigger that leads to, or brings about, the event. Anything prior to that decision can be trumped by the decision itself. We feel like minor gods, not powerful enough to change the universe, but able, within certain limits, of bringing into being some events. Part of what it is to be an agent is to have this sense of freedom, a sense that other future options are in some sense live.[13]

---

[13] We don't have this sense aimed at the past. It's not just that the past is known—maybe it isn't—but that our decisions now aren't self-fulfilling predictions of the past. I can decide to have strawberry ice cream yesterday, but this decision won't be even close to self-fulfilling. Thanks to the causal asymmetry, such

And these options are live to me . . . and you:

> You experience your arm, hand, and fingers as being moved by *you yourself*—rather than as experiencing their motion either as fortuitously moving just as you want them to move, or passively experiencing them as being caused by your own mental states. You experience the bodily motion as generated by *yourself*. (Horgan, 2007, p. 187)

It is the self—the star player in our story of passage—that brings about future events. As the self is narrating its own story, much of what it uses is what it itself has wrought. I chose the chocolate ice cream, to write this book, to go surfing, and so on. So as the self "evolves" up its worldline, it not only passes through various events and feels time to be flowing; it feels it is part of a worldwide surge of "actualization" of one path over others. The sense of passage is intimately connected with the sense that the past is settled and the future open.

Hence, outfit our IGUS with a sensation of agency whenever it's in a position to make self-fulfilling predictions. There is controversy about whether the sense of agency is actually phenomenological (for discussion see Bayne, 2008). If it is, the output of this device will enter the IGUS's C box; if not, it can be part of its schema of the world. Either way, the IGUS will see itself as on the cutting edge between what is fixed and what is coming to be.

11.10 The Explanation of Passage

We've sketched the "rules" that might give beasts like us a sense of passage. We've outfitted our IGUS with a host of gadgets, including:

- detectors that perceive movement, velocity, succession, duration, and other objective temporal properties
- the means to emit signals, move, and integrate signals, intra and inter-modally, such that the signals and processing are fast relative to typical macroscopic changes of salient properties
- a self-conception—a "self"—based on patterns in its memory registers
- a sense of agency when in situations where it makes self-fulfilling predictions in its anticipated model of the next time step
- sensory modalities sensitive to "macroscopic" time and space scales.

We've also embedded IGUS in a world possessing:

- a sharp difference between the timelike and spacelike (and all that implies)
- physical limitations, including low relative velocities and observations restricted to macroscopic time scales
- large physical asymmetries, which either coexist with or ground the massive knowledge and causation asymmetries.

---

decisions seem to have no probabilistic effect on the past event (or at least, none detectable to macroscopic creatures like us—see Kutach, 2011).

With all of these features and constraints, I submit that IGUS will entertain in its schema of the world the idea that time flows. It is a natural reaction to the situation in which it finds itself. IGUS is a crude device. What about us? Our psychology and physiology is indefinitely more complicated than IGUS's, so we will disagree on many aspects of temporal experience. For instance, due to our brain's handling of eye saccades, we perceive the second hand of clocks to pause momentarily when they enter attention. Unless rigged with a brain like ours and similar visual detectors, there is no reason to suspect that IGUS will similarly experience paused second hands. IGUS is primitive, but that in a way is the point. These core ingredients seem to be *enough* to induce a modeling of passage. Since we overlap with IGUS in these respects, we should expect to employ models of time with passage too. We're still learning the developmental story of how exactly this model develops, but we already have a pretty clear grasp of the outlines.

With Hartle let's pause to note how contingent passage is. It didn't *have* to be this way. We can imagine "split screen" IGUSs with content from P_0 and (say) P_7 accessible to C. This IGUS would "see" the present and some point in the past at once. It thus might have two "presents." There might have been "always behind" IGUSs whose conscious registers C are filled with input not from the most recent "snapshot" but from an earlier one. Due to finite signal speeds, of course, we and all other IGUSs run "behind" a little bit. But the "always behind" robot runs unnecessarily behind by "experiencing" the report from, again say, P_7, instead of from P_0. This IGUS is thus experientially stuck in the past, as it regards the events encoded in P_7 as "present." It remembers events encoded in P_8-P_n and has reliable premonitions of the events encoded in P_0-P_6. Wired this way, it's not clear whether the IGUS would believe the events encoded in P_0-P_6 open or not; the "always behind" IGUS knows what happens in that period but hasn't yet experienced it. Conscious refreshing lags behind memory acquisition. Hartle speculates—and I concur—that these cases are perfectly acceptable to the laws of physics but likely would fare poorly biologically. Given the generally Markovian nature of the world, to catch a fly it suffices to know its location and speed from the time captured by P_0. Adding information from P_7 won't help. Worse, the additional energy needed to run a "split screen" won't be warranted. The same goes for the mis-wiring from P_7 instead of P_0 in the always behind IGUS: one wired from P_0 will catch many more flies. And probably the "dumb" IGUS won't catch any, for the fly will be long gone by the time that IGUS crunches all the information stored in its memory registers.

11.11 Conclusion

The theory of passage developed here requires elaboration and improvement. However, I believe that there are plenty of hints on how to do that. Indeed, some questions may turn out to be answered already and others subject to experimental tests. The theory of passage may turn out to be false or significantly underdeveloped, but it is a rich theory. In contrast to simply positing a primitive metaphysical flow and crossing one's fingers hoping that somehow we sense it, the present theory advances independently suggested mechanisms, makes a number of specific claims, unifies some types of phenomena and theory, and suggests fruitful lines of inquiry. By any reasonable standard of theory choice it is a better theory of passage than any currently on offer in metaphysics.

12

Explaining the Temporal Value Asymmetry

> There is another phenomenon of a like nature with the foregoing, viz. the superior effects of the same distance in futurity above that in the past. This difference with respect to the will is easily accounted for. As none of our actions can alter the past, 'tis not strange it shou'd never determine the will.
>
> Hume, *Treatise* 2.3.7.6

You wake up in a hospital bed, momentarily disoriented.[1] "Where am I?" What am I doing here?" you cry out. A nurse walking by hears you and enters your room. "Hmm," he says, "there are two charts, not one." He looks bewildered, but you demand that he read the charts. "Well," he says, "apparently you are either Patient A or Patient B." Patient A, he tells you, is about to be transported to the operating room where a very special and painful operation awaits. The operation is special because it requires that you be conscious and without any pain relief during the procedure, but it will be followed by a pill that causes amnesia just for that time period, so that no trace of the painful experience will persist. The operation itself is an hour long. "If you're Patient A, then momentarily we'll wheel you out to an hour of agony," the nurse reports. Patient B, by contrast, has already undergone this same operation, except that due to complications it required five hours of conscious agony, a hospital record for this kind of procedure. "If you're Patient B," the nurse states, "you've just been through hell but now can relax."

Who would you prefer to be, Patient A or Patient B?

That is the question the philosopher Parfit (1984, S64) asks after describing essentially the same scenario. Patient B suffers five hours of agony, whereas A suffers only an hour. Five is greater than one, so surely you prefer to be Patient A? No! If you are like most people, you will have a strong preference for being Patient B, despite the greater total pain. No matter how bad it was, Patient B's pain is *over and done*. Past pain gets discounted at a rate sufficient to make it preferable to be Patient B rather than A. The discounting of past pain is so steep that some might even prefer five hours of forgotten

---

[1] Much of this chapter was co-authored with Chris Suhler in Suhler and Callender (2012).

past pain to almost *any* future pain, no matter how small. It's almost as if that past pain happened to someone else.

This example vividly illustrates a general tendency. Our valuations of goods and harms vary in systematic ways with our temporal perspective on them. Past headaches mean little to us, but future ones are dreaded. We eagerly await future holidays, not past ones. We are not temporally neutral—indifferent about the timing of goods and harms—even if, as some have suggested (e.g., Sidgwick, 1907; Brink, 2011), there are reasons we should be. Let's call this asymmetry the

Past/Future Value Asymmetry (PF asymmetry):   all else being equal, we tend to prefer past pain (future pleasure) to future pain (past pleasure).

As we'll see, much more than our valuations of goods and harms varies with temporal perspective. Psychologists have recently discovered a treasure trove of ways in which attributions of intentionality, ethical norms, judgments of likelihood, and more depend upon temporal perspective.

The Past/Future Value Asymmetry ought to be distinguished from its more famous cousin, the

Proximal/Distant Discounting Asymmetry (PD asymmetry):   all else being equal, we tend to prefer distant future pain (proximal future pleasure) to proximal future pain (distant future pleasure).

Examples of the PD asymmetry might include skipping a flu shot to get a coffee or not saving adequately for retirement. This asymmetry is between two types of future events, whereas first asymmetry is between past and future events. Economists and psychologists interested in departures from dynamical consistency among preferences have studied the proximal/distant asymmetry extensively. This literature describes the asymmetry in rich mathematical and empirical detail and comments on its relevance to personal, social, and political decisions (e.g., drug addiction, global warming). The literature is also filled with models drawn from psychology, economics, and evolutionary theory that seek not only to characterize the phenomenon but also to provide a deeper explanation of it.

Figure 12.1 distinguishes the two asymmetries graphically. Here we weight our preferences against time—at one time.

The solid line displays the hyperbolic curve characteristic of diminishing impatience toward the future (the PD asymmetry) and the less studied drop-off in care once an event becomes past (the PF asymmetry) is sketched with the dotted line.

Interest in the PF asymmetry, in contrast to that in PD, has generally been confined to philosophers. Rather than seeking to describe or explain the temporal value asymmetry in any detail, philosophical discussions tend to take as their starting point simply that the asymmetry exists and attempt to work out its implications for issues of traditional philosophical interest. That is the case, for instance, with work focusing on the possible implications of the temporal value asymmetry for ethics and

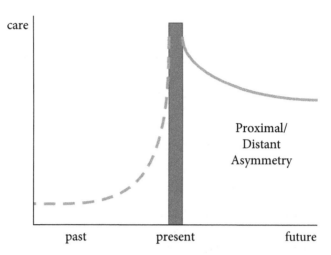

Figure 12.1 Care versus time

political philosophy (e.g., Parfit, 1984; Brink, 2011). The only literature that focuses on an explanation of the temporal value asymmetry is one based on the metaphysics of time, namely, Prior's (1959) famous "Thanks goodness that's over" argument. That argument is one of the two or three most central arguments in traditional philosophy of time, as it attempts to move from the PF asymmetry to the truth of a "tensed" metaphysics of time.

This chapter will critically scrutinize Prior's argument. But more than that, we'll take a step toward remedying the explanatory vacuum surrounding the PF asymmetry, a key piece of manifest time. The scientific literature on the PF asymmetry has recently grown. Study shows that the PF asymmetry is very deep and holds in surprising places. This data undercuts Prior's argument, I believe, for a uniform Priorian reply to it is untenable. This work also points to various psychological and evolutionary mechanisms that I believe provide the foundations of a detailed explanation of the PF asymmetry, one analogous to common explanations of the PD asymmetry. And this research also allows us to propose a much richer version of the philosophical answer to Prior developed in parts by Hume, Parfit, and Horwich. While in the spirit of these past sketches, the theory improves on them in significant respects and suggests clear avenues of future study. My hope is that work on the PF asymmetry will eventually attain the degree of rigor and explanatory power that the PD asymmetry currently enjoys, for like this latter asymmetry, the PF asymmetry has relevance to many practical issues in decision-making.

12.1 "Thank Goodness That's Over"

Here is one version of Prior's 1959 "Thank goodness that's over" argument:

I have a very good friend and colleague in Australia, Professor Smart of Adelaide, with whom I often have arguments about this. He's an advocate of the tapestry view of time, and says that when we say "X is now past," we just mean "The latest part of X is earlier than this utterance." But, when at the end of some ordeal I say "Thank goodness that's over," do I mean "Thank goodness the latest part of that is earlier than this utterance"? I certainly do not; I'm not thinking about the utterance at all, it's the overness, the now-endedness, the pastness of the thing that I'm thankful for, and nothing else.   (p. 106)

The argument is generally taken to be that the "tapestry view of time," i.e., eternalism, stands diminished in some way due to our practice of thanking goodness that something is over (and similar claims). Eternalists are supposed to have a special problem in explaining such beliefs, desires, and feelings, for their commitment to the reality of past, present, and future leaves all events of "equal significance" (Cockburn, 1998, p. 19). Put like this, however, the argument is a non-starter. That the eternalist believes past, present, and future events all exist does not imply she believes that they are all of equal significance any more than the physicalist who believes all events are material must believe some clumps of matter (e.g., my mom) are as significant as any other (e.g., my desk). What, then, is the argument?

No doubt part of the reason the argument survives is that it can be unpacked in various ways. I can imagine at least three quite distinct versions, each emphasizing different facets of the above claims, plus multiple finer instantiations of each version. Focusing on the three families, I think we've already answered two of the three challenges.

The first and probably historically most accurate and popular interpretation is about the *indispensability of tense*. On this reading, the challenge is to say what fact the eternalist is thanking goodness for. Tense is said to be indispensable for answering this question. If one suggests replacing "the headache is past" with "the headache is earlier than my utterance" then he or she fails to replicate what the speaker means. As Prior says, the speaker does not mean to thank goodness for an earlier than relation holding between a headache and an utterance. Prior is of course absolutely right about this. Note that on this version of the argument, the time asymmetry of the thanking is inessential. One could also make the same point by trying to replace "the headache is now" with "the headache is simultaneous with my utterance."

This version of the argument is historically important, but it is merely the "temporal knowledge argument" of Chapter 9. Indeed, sometimes Prior's case is explicitly formulated this way, as in Kiernan-Lewis (1991). This type of argument can be formulated in many guises, but it ultimately relies on the fact that beliefs, desires, and emotions generate non-extensional contexts. Hence, one can thank goodness that the pain is past without also thanking goodness that the pain is earlier than the utterance. Yet by itself this is no more mysterious than the fact that one may thank goodness that Superman arrived at the scene but not thank goodness that Clark Kent arrived (Dainton, 2001, p. 36).

The indispensability of tense is just the essentiality of indexicals in general, not something unique to time. We can thank goodness that the fire is not *here* or even *there* or even that [the person with the pie in their face] is not *me*! The right reaction to this predicament, it seems, is not to devise radical new metaphysical models of space and persons, but rather to add an indexical mode of presentation to our explanation. Such an addition seems vastly preferable to one adding to our ontology of persons and places. So, if one is already convinced that we need indexical modes of presentation in these cases, parity of reasoning suggests that we ought to adopt that response here too in the case of time. To do otherwise would be to ignore without reason a solution to an otherwise identical problem.

The second way to read this argument is to focus on the word *over*. When you say that the headache is over, you mean that it happened, that you won't face it again. There is a kind of preferred perspective picked out. Think of the four-dimensional worldline that is you. When you thank goodness the headache is over, you are saying that the seat of this perspective has already experienced the headache and will not again. Mellor's (1998) objection to Prior is that an explanation in terms of thanking goodness the headache is earlier is just as explanatory as thanking goodness the headache is past. That's true, so far as it goes. But it misses the sense in which the thanker regards the spot at which he sits as privileged, the cut between the open future and dead past. In other words, it misses the flow, discussed in the previous chapter.

On my picture, one models time as flowing due to the narration of one's self. If correct, this theory has the resources for answering this version of the headache worry. Before the headache, one cannot narrate a self that experiences the headache "from the inside." The "self-building" materials don't yet include autobiographical memories or other psychological states associated with this event. Hence the self hasn't crawled up to the headache yet; it hasn't "happened" yet. After the headache, one has plenty of material for narrating a self through the headache period. The headache has happened for that self, i.e., those experiences were used in building one's narrative up to that point and therefore later selves won't have to go "through them" again, for they are part of the story already. In that sense the headache is not merely earlier, but over and done.

A third way to interpret Prior's argument is as an explanatory challenge. One might want an explanation of our temporally asymmetric biases in belief, desire, and other attitudes. That is how I take Zimmerman's challenge:

> When I notice that a headache, or some other painful episode, has become part of the past, I am relieved that this is so; and when a pleasant experience becomes past, I am often disappointed. If a theory of time makes such changes in attitude utterly mysterious, we should have grave doubts about its adequacy. (2007, p. 214)

On this version, the claim is that non-eternalist models *better explain* these temporally asymmetric attitudes than do eternalist models.[2] The latter don't provide sufficient

---

[2] Zimmerman's challenge is to some A-theories he finds objectionable, but we can consider extending it to eternalism too for it posits no objective differences amongst the past, present, and future. See also Cockburn (1998) and Carroll and Markosian (2010, p. 170).

resources for explaining relief, disappointment, and so on, it is claimed, whereas the former do.

Clearly this challenge is only as strong as the assumption that non-eternalist models *do explain* this class of temporally asymmetric attitudes. What is the explanation? How could the nature of time by itself explain any attitudes at all, one might wonder. The presentist Craig answers, "past pain is non-existent pain, and so no pain at all!" (2000, p. 157, fn. 69). And Zimmerman writes, "Past headaches do not exist; consequently, they have no properties whatsoever, including being painful" (2007, p. 216). The claim therefore seems to be that the non-existence of past headaches is what explains our lack of concern.

That can't possibly be the whole story. Even granting a connection between existence and our attitudes (see Chapter 13 for serious reservations about this), there are big problems. First, according to presentism, future pains don't exist either. How can non-existence explain a psychological attitude in the one case but not the other? Also, we dread future headaches, but if future headaches aren't painful, then why do we dread them? The answer is that non-existence is a red herring. Rather, the answer is that future pain *will* be present pain and is thus a cause for concern (Craig, ibid). Non-existence doesn't do the explanatory lifting. What makes past pains unworthy of dread is instead that they will not be, unlike future pains. Once that point is made, Mellor's (1998, p. 42) *tu quoque* response beckons. How is this primitive linkage between attitudes and what will and won't be any better than stipulating that it's appropriate to feel relief when pains are earlier than the utterances or experiences? Furthermore, note that a primitive directionality is certainly compatible with an eternalist perspective, so at best the argument points toward the addition of a basic directionality, not non-eternalism.

How good is this explanation? To me it scarcely qualifies as one. Once again the advocate of a tensed metaphysics is caught claiming that they have an explanation when in fact all that they have is an appeal to a new primitive. Chapter 9 witnessed the claim that tensed metaphysics better explains the phenomenon of the present than eternalism. What this came to was the postulation of a brute phenomenon primitively linked to a novel property. Section 11.2 witnessed the claim that tensed metaphysics better explains the flow. How? By appeal to a primitive. And now again we see an appeal to a primitive link between tensed properties and some attitudes. While it may be true that often good scientific explanations appeal to primitives, the mark of a good explanation is that the unexplained explainers do a lot of heavy lifting. Here one sees theory read into the data, primitives galore, and explanation by stipulation again and again. The explanatory program bears all the hallmarks of a pseudo-scientific explanation.[3]

---

[3] One might be tempted to think that presentism is the problem, but that a different "tensed" theory, namely, becoming—according to which the past and present but not future are real—is the answer. I emphatically disagree. See Chapter 13 for my reasons for why this modification won't help.

Prior's argument, when understood the first way, was a powerful reminder that a simplistic understanding of tense won't work. Understood the second way, it forces us to come up with a theory of the apparent flow of time. The third version is less successful, I think; still, it raises an interesting question: what is the explanation of these temporally asymmetric tendencies in our attitudes (etc.)? There is no denying the fact that the curious asymmetry on which it is based is an important one in our lives. So far the philosophical literature has done little to explain it. Prior's argument, despite the shortcomings just noted, serves as a reminder of this robust pattern in our attitudes and preferences, a pattern for which there is so far little explanation.

12.2 The Proximal/Distant Asymmetry

To get an idea of what types of explanation might be on offer, let's briefly look at the PD asymmetry.[4] Doing so will help us see the contrast between the richness of work done on the PD asymmetry with the relatively impoverished attempts to understand the PF asymmetry. While the two asymmetries are indeed two—and so may have different causes, explanations, etc.—the hope is that the explanatory strategies used for the one might be usefully employed in the other.

A moment's reflection on your credit card spending and other practical financial decisions probably will reveal that you discount the values of distant future goods and harms. This pattern is ubiquitous: in general, humans (and other vertebrates) don't weight the preferences of their distant-future selves as much as the preferences of their present or proximal-future selves. People tend to display a kind of immediacy bias that privileges their current self's expected utility and discounts the value of satisfying their later self. For example, in a context where borrowing and lending occur at 5%, if given a choice between getting $100 this year and $105 next year versus getting $120 this year and $70 next year, you might prefer the second option to the first—despite the second's diminished total lifetime utility.

Empirical studies reveal a number of fascinating features of our discounting. For instance, we tend to discount smaller future amounts more than larger ones (the "magnitude effect"), we discount future losses less than future gains (the "sign effect"), and we discount for delaying more than for expediting an outcome (the "direction effect"). Many of these tendencies represent departures from what one would expect if people are interested in maximizing expected utility in a consistent manner.

Attention has focused in particular on the preference reversals that develop when our discounting changes with time. We do not employ a stationary (constant) discount rate. Rather, we make plans for the future and then don't stick to them. Our discounting is not "exponential" but instead "hyperbolic" in nature.[5] There is disagreement

---

[4] A good introduction to this asymmetry is Read (2008).

[5] A discount factor $r(d)$ for a delay of time d represents the proportion future utility is diminished when converted into present utility. An exponential discount factor is one that doesn't change with future times,

concerning how best to model this trait, but the behavior characteristic of it is that we underestimate how impulsive we will later be. Suppose subjects prefer $100 now to $200 next year. The wait of an extra year is deemed too long. However, when asked if they prefer $100 five years from now or $200 six years from now, they may prefer the $200. The extra year, if five years later, is *now* not too long to wait. From the perspective of one's discount rate, one is here *not* discounting the future *enough* when goods become distant. Individuals do not typically realize the obvious partiality revealed when they think year 5–6 worth the wait but year 0–1 not worth the wait for the same amount of money. Of course, it could be that one just wants to be richer in 2025 than 2020; although a systematic preference for 2025 is certainly strange, there is nothing necessarily inconsistent in this behavior from the perspective of expected utility.

Where one meets real inconsistency is in the way this preference structure updates itself. In 2025 a person will also have the valuation asymmetry described above, but now year 0–1 (i.e., 2025), for which a year's wait is worth $100, is the very same year that in 2020 was considered not worth the wait. A person bound by his 2020 preferences to wait that extra year will be frustrated in 2025 by the extra delay. Or as seen in an all too common example: thinking about a dinner party next week, today I greatly value fitness and decide that I will skip dessert; but a week later and after dinner, I'll prefer dessert to fitness.

The discounting asymmetry has been extensively studied empirically. Much of this work describes how we in fact do discount in various contexts. But there is also work that seeks to explain *why* we discount the way we do. This research suggests mechanisms and rationales at various levels that predict and explain some discounting behavior. Sometimes the discounting may turn out to be a rational strategy (e.g., given certain levels of uncertainty and a particular statistical distribution of hazards) and other times not.

Rational or not, mechanism-positing explanations for discounting are sought in fields including evolutionary biology, psychology, and behavioral economics. In psychology, for instance, experiments have demonstrated the existence of the phenomenon of future anhedonia—the belief that more distant future hedonic states will be less intense than present or near future hedonic states—and shown its contributions to our discounting of the future (e.g., Kassam et al., 2008). It is furthermore possible to link this phenomenon to what philosophers of time call the *knowledge asymmetry* (Horwich, 1987; Albert, 2000), the fact that we tend to know more about the past than future. Because of the knowledge asymmetry, uncertainty and time delay tend to be correlated; the more we must wait for an outcome, the less certain we can be of it obtaining. The correlation is not perfect, e.g., one might be more certain of the next arrival of Halley's comet than of the result of a coin toss one minute from now. But

---

i.e., $r(d) = (1 - r)^{-1}$, where r is the discount rate. A hyperbolic discount factor, by contrast, is one wherein the discount factor is modified by the time delay in the denominator, e.g., $r(d) = (1 - kd)^{-1}$, where k is a constant, so that $r(d)$ now changes depending upon how long the delay is.

for otherwise alike events, time delay is correlated with uncertainty. The hypothesis that the immediacy bias occurs due to implicit uncertainty arising from time delay was put forward by Keren and Roelofsma (1995) and Souzou (1998) and confirmed empirically by Weber and Chapman (2005). Weber and Chapman found that adding uncertainty to outcomes eliminated the immediacy effect bias (and also the other way around, namely, that adding time delay to outcomes—without adding uncertainty— eliminated the certainty effect bias).

At the evolutionary level of explanation, a rationale for the emotions and preferences associated with the discounting asymmetry is also readily available. The idea is nicely expressed by Maclaurin and Dyke:

> [O]ur emotional responses have developed to make us more concerned about the prox-
> imate future, because (for much of our evolutionary history) the costs of trying to predict
> and influence the distant future have far outweighed the benefits. . . . We care more about
> proximate future pain than distant future pain for the same reason that herbivores care
> more about proximate predators than they do about distant ones.   (2002, pp. 288, 290)

Further support for this view comes from the economics literature, where precise discounting preference models have been linked to fitness-enhancement in simple evolutionary populations (see, e.g., Robson and Samuelson, 2007).

A crucial and frequently overlooked point is that this asymmetry, properly under-stood, is *tensed*. The discounting asymmetry should not be understood in terms of a single demand versus time curve, but rather a set of curves, one for each "now" or present.[6] Emphasizing this point brings the two asymmetries closer in line with one another, for the temporal value asymmetry also is tensed in nature. It depends crucially upon the tripartite division of the world into past, present, and future, and it would be a mistake to view it as a preference for minimizing *later* pain at the expense of *earlier* pain.

12.3 The Humean Solution

Return to our topic, the PF asymmetry. In seeking to explain this asymmetry, I believe that one ought to look at many of the rich resources used in explaining the PD asymmetry. Instead of jumping immediately to the metaphysics of time as a source of explanation, we should look much closer to home, namely, at the psychology, environment, and evolutionary history of creatures like us. Let's call explanations in this spirit "causal accounts" of the temporal value asymmetry.

No sooner does one begin looking than a rather obvious version of a causal account suggests itself. The main ingredients are the prevailing causal asymmetry in the world and evolution by natural selection. Hume puts his finger on the causal asymmetry

---

[6] This understanding of the discounting asymmetry is supported by the economics literature, where researchers including Strotz (1956, p. 165) and Rasmusen (2008) note that what's important is the temporal distance or delay from the *present*, not the actual calendar date.

part of the story in this chapter's epigraph, Parfit (1984) mentions the role for natural selection in passing, and Horwich (1987, pp. 194–6) puts the two together.

Start with the causal asymmetry. Much can and has been said about the causal asymmetry and the related asymmetries of counterfactual dependence, influence, and mutability. Since Reichenbach, philosophers and physicists have tried to trace the origin of the causal arrow to the thermodynamic arrow of time. The connection, if there is one, is very subtle and recent attempts to draw it have met with many criticisms (see Loewer, 2007a and Frisch, 2010). My own view is that I find it almost impossible to believe that the causal asymmetry doesn't have *something* to do with the myriad physical asymmetries that prevail in our universe. We're bathed in outgoing radiation from earlier sources and subject to various thermodynamic asymmetries in all macroscopic physics. This vast physical asymmetry must have some important connection with why we can do what we do, even if the fine details of this connection presently escape us in their entirety. If this is right, then here we have another physical temporal "hook" onto which creatures like us might fasten. Just as the manifest now and flow arise in part due to physical constraints, so too may the causal asymmetries (see below) depend in some complicated manner on the (broadly construed) thermodynamic arrow of time.

Whatever its origin, there is no question that the asymmetry exists. It's a plain fact of life that *our actions are sometimes effective in bringing about future goals, but rarely or never effective in bringing about past goals* (see, e.g., Kutach, 2011). Wiggling something now may produce massive changes in the future, but it won't generally affect the past one bit. This prevailing asymmetry means that some desires will be more effective than others. If we think of causes as typically raising the probabilities of their effects, then we can say, with Horwich (1987), that "having the desire that future selfish desires be satisfied" increases the chances of those desires being satisfied, but "having the desire that past selfish desires were satisfied" does not. For example, one can desire that yesterday one played basketball, but that won't affect the chances of having played one bit; one's desire that tomorrow one play, by contrast, does typically affect its chances of happening.

Now couple to this asymmetry the idea that, generally speaking, the satisfaction of our desires is conducive to survival and reproductive success. This move yields the beginnings of an evolutionary argument for why we care more about future desires than past desires. A tendency to care strongly about the satisfaction of future desires would have been selected for because this would have led to behavior that increased the chances of these desires (e.g., finding food, mates, and shelter) being satisfied. By contrast, the tendency to care strongly about past desires would not have been selected for, since no amount of caring or effort after the fact will increase the chances of those desires' being satisfied.

Of course, all of the above is at a high level of generality. One can desire to kill oneself, to not have children, and so on. The satisfaction of those desires will not be conducive to survival and reproductive success. The argument assumes, very loosely, that typically we want to maximize our expected utility. It also assumes, again at a very coarse level, that improving our expected utility has some non-trivial association with increasing our fitness. Surprisingly little work has connected the two, expected utility and fitness, but there are reasons to believe that the two are associated. See Okasha (2011).

As appealing as it is, the full story needs to be much more nuanced and better motivated. At present, a critic might scoff at this account as another ' "just so' story of sociobiological mythology," in the words of the philosopher of time William Lane Craig (1999). This scoffing is less plausible in light of Maclaurin and Dyke's (2002) excellent defense of the evolutionary story, and in particular, their relating particular emotions and preferences to fitness increase. Still, much more can be said and a richer explanation of the PF asymmetry is possible.

12.4 The Knowledge Asymmetry

The Humean solution relies on the causal asymmetry. That is fine, as far as it goes, but it ignores another asymmetry that is arguably just as important, namely, the knowledge asymmetry.

The *knowledge asymmetry* is often glossed as the claim that we know more about the past than the future. We know what teams won each of the past NBA championships, but we have no idea who will win the next one, never mind one a few years hence. Although this informal gloss on the asymmetry serves its purpose, it's better to characterize the asymmetry without speaking of "amounts" of knowledge. We're certainly not counting and comparing known propositions, after all.

David Albert offers a better characterization (Albert, 2000). Consider the macroscopic information available to one in the present. Plug it in to the laws of nature (or if you don't like laws, projectable generalizations of whatever kind) and then evolve this information forward and backward in time to see what this information plus the laws jointly imply about past and future states. The knowledge asymmetry is then said to be the fact that while all of our knowledge about the future originates in something like this procedure, much of our knowledge about the past does not. It's not so much that we know *more* of the past than future than that the world allows a *way* of knowing the past different from how we know the future. Conditionalizing on the present macrostate, we can know that Halley's comet will return in 2061. Yet that same procedure won't tell us tomorrow's Dow Jones industrial average. Yesterday's Dow average, by contrast, we obtain by simply picking up a newspaper. Albert believes that a special initial condition of the universe is crucial to explaining the knowledge asymmetry, but we simply need the fact that it exists.

How does the knowledge asymmetry relate to the causal arrow? Knowledge and causation are two of the more loaded concepts of philosophy, so we really can't say definitely. There might be plenty of connections. For instance, on a causal theory of knowledge (Goldman, 1967), the relationship is quite intimate: S knows that p just in case p caused S's belief in the right way. Alternatively, the knowledge asymmetry may be the foundation of the causal asymmetry à la Horwich (1987). They might also have the same origin and explanation (Albert, 2000). Or perhaps they're utterly separate. Fortunately I can afford to be agnostic. The two asymmetries exist. How they relate to one another won't matter for our purposes.

What's important is that we have these asymmetries. When we look at the universe, two of its most salient facts for creatures like us are the causal and knowledge asymmetries. From any event, events to the future are potentially *controllable* thanks to the causal asymmetry. Hume, Parfit, and Horwich rightly stress the importance of this massive asymmetry. Equally important, however, is the knowledge asymmetry. When we look into the future we face a yawning darkness.[7] Despite our best efforts, a storm could destroy our well-tended crops, your partner could hate the present you purchased, or a meteor could drop out of the sky and hit you. Typically you just don't have a way to know. Consequently, one of the most basic facts of life is that we face great *uncertainty* in the future direction. Coupled with the causal asymmetry, this adds up to one of life's principal predicaments: we're uncertain about many of the events we potentially can control.

And that makes us nervous.

12.5 The Affect Asymmetry

The causal and knowledge asymmetries are widely pervasive asymmetries obtaining throughout most of the universe. They affect people, dogs, amoeba, and yes, even koalas (who don't seem worried by much). In their wake they leave a trail of asymmetries. The one I'm presently concerned with is the *affect asymmetry*. This is the fact that our emotions are often quite sensitive to whether an event is past or future. Suppose that I tell you that Coca-Cola invented a machine that made the price of a can of Coke proportional to the outside temperature: expensive when hot, cheap while cool. They're going to test this machine in a few months. How *angry* does this machine's test make you? Studies show that if I instead asked you the same question but of a past testing of the same machine, you would probably have not felt as angry (Caruso, 2010). The type and intensity of our emotions manifest many characteristic time asymmetries. In general, we feel more intense affect when thinking about future events (anticipation) than past events (retrospection); see Van Boven and Ashworth (2007).

There are strong reasons to believe that both the knowledge and causal asymmetries contribute to the affect asymmetry.

Take knowledge first. The claim is that uncertainty triggers heightened emotional responses, so the uncertainty asymmetry contributes to an affect asymmetry. The idea is intuitive and familiar. Haunted houses are scary because we're uncertain what will happen. And why do we prefer that birthday gifts are a surprise?

---

[7] Literally: no information-carrying light emitted from future events ever reaches us.

Several studies have identified a number of factors backing up this intuition that connects uncertainty with the intensification of emotional responses to events—both occurrent and anticipated. For example, Wilson et al. (2005) gave students sitting in public places on a university campus an unexpected reward (a dollar coin). The reward was accompanied by a card. Some subjects received a card containing text that made the reason for the reward easy to determine (the "certain" condition), while others received a card that did not make the reason clear (the "uncertain" condition). The researchers found that subjects in the "uncertain" condition had more prolonged experiences of positive affect than did subjects in the "certain" condition. In another study, Bar-Anan and colleagues (Bar-Anan et al., 2009) had subjects watch positive and negative film clips. Instead of inducing a feeling of uncertainty by manipulating the amount of information that subjects had about the clips, the researchers simply had one group of subjects repeat phrases associated with uncertainty while watching the clips, while another group of subjects repeated phrases associated with certainty. They found that subjects repeating the uncertainty-connoting phrases had more intense affective reactions, both positive and negative, to the clips. Not knowing tomorrow's Dow heightens your affective responses.[8]

Now consider causation. First, although I know of no study testing this, I expect that causation amplifies the effect of the knowledge asymmetry. The future is not only "dark" but it is what we can control. Not knowing the future is bad, but not knowing what we have some responsibility for changing is especially nerve-racking.

Second, there is a broad theoretical framework that links evolution, causation, and affect. Emotions are states that have evolved to organize and motivate organisms' behavioral responses in certain broad classes of ecologically important situations. The specifics of evolutionary accounts of the function of emotion are a matter of some debate, but the general notion that emotions play an important role in allowing organisms to respond to situations and prepare them for action is far less contentious,

---

[8] Probably other routes from the knowledge asymmetry to the affect asymmetry exist. There is, for instance, the phenomenon of *focalism* (Wilson and Gilbert, 2005; Wilson et al., 2000). This is the tendency of people to focus excessively on a given event (e.g., failing to get a job one wanted) and its effect on their emotional state, neglecting the effects that other events will have on their thoughts and feelings at that time. As a result of this focus on one event at the expense of others, people tend to overestimate the impact that the event in question will have on their affective state. Since we know more about the past than future, it's possible that imagining a future event (e.g., a headache) is more liable to focalism than thinking about past events. The past events are diluted by knowledge of all the other events that happened when that one event occurred. Another route might be via differences in the extensiveness of *mental simulation* of past and future events. In an experiment by van Boven and Ashworth (2007) subjects rated not only the emotion that they (currently) experienced upon thinking about the event, but also the degree to which they mentally simulated the events. They found that subjects gave significantly higher mental simulation ratings in the anticipation condition than in the retrospection condition, and that this difference in mental simulation mediated the reported differences in experienced affect. The causal asymmetry may also play a role here. Present simulation of anticipated future circumstances could aid in planning, preparation, and so forth, and this could in turn help to shape one's responses to those future circumstances. As a result of the causal asymmetry, however, present simulation of past circumstances would have no such ability to shape one's responses to those past situations.

having been championed by a wide range of researchers (e.g., Frijda, 1988; Frijda et al., 1989; Griffiths, 1997; Nesse, 1990; van Boven and Ashworth, 2007). Fear, for instance, is a response to perceived danger and is associated with a suite of physiological changes that facilitate and motivate an organism's fleeing or taking defensive action against a perceived threat (for overviews of the neurobiology of the fear response, see LeDoux, 1995; Panksepp 1998, Ch. 11).

Why are emotions preferentially activated in response to contemplation of future, as compared to past, events? The answer to this question follows the general outlines of the evolutionary explanation suggested by the developed Humean account. The advantages enjoyed by an organism with a tendency to experience (contextually appropriate) affective states upon imagining future events, relative to an otherwise similar conspecific not disposed to experience such states, should be clear. These future-triggered affective states—and their associated physiological changes and motivational effects—would help an organism behave in ways that would make the occurrence of desirable or evolutionarily advantageous states of affairs (e.g., acquiring food or a mate) more likely to occur and undesirable or evolutionary disadvantageous states of affairs (e.g., starvation, failing to find a mate, being injured/killed by a predator, or aggressive conspecific) less likely to occur.[9] As such, there is good evolutionary reason to think that the tendency to experience anticipatory affective states would have been selected for in humans and other species.

Why would these advantages not accrue to organisms that tended to experience (contextually appropriate) affective states upon imagining past events? Our answer here appeals to the causal asymmetry—the fact that, so far as we know, causation operates forward in time but not backward (for discussion, see Horwich, 1987, Ch. 8). Put more simply, our present actions can affect the future but not the past. Given the existence of the causal asymmetry, changes in an organism's behavior or motivation resulting from emotions activated by imagining or retrospecting past events would not have any effect on the past states of affairs in question. They would not help the organism to act so as to make desirable past states of affairs more likely to have occurred or undesirable past states of affairs less likely to have occurred. As such, there would be no adaptive advantage conferred on organisms disposed to experience strong emotion when imagining past states of affairs, as compared to organisms not so disposed, and the tendency to experience such emotions would not have been selected for.[10]

---

[9] For a striking modern demonstration of this, see Damasio and colleagues' work showing that the failure to activate appropriate affective states is a major reason for the profound decision-making deficits of individuals with ventromedial prefrontal cortex damage (e.g., Bechara et al., 1994, 1996).

[10] Indeed, to the extent that it would result in physiological and motivational states that were inappropriate to present and future demands, and in the use of time and energy on ineffectual efforts to alter past states of affairs, the tendency to experience such emotions would have been selected against.

Box 3   A Note on Mental Mechanisms and Evolution

Note that I am not suggesting that the PF asymmetry is the result of a dedicated suite of affective and valuation mechanisms that evolved solely for dealing with past/future choice. This explanation of the asymmetry would be very difficult to sustain given the lack of independent reason to believe in the existence of such mechanisms. More generally, such explanations—often couched in terms of domain-specific "modules" that evolved with the specific purpose of serving the function one is trying to explain (see, e.g., Cosmides and Tooby, 1997; Haidt and Joseph, 2007)—are one of the hallmarks of the explanatorily impotent evolutionary psychology (Buller and Hardcastle, 2007; Suhler and Churchland, 2011) that I am keen to avoid.

Instead, my starting point is the fact that mechanisms for valuation and choice among alternatives (e.g., food sources, mates, locations for nests/burrows/etc.) exist, in more or less sophisticated forms, in nearly all animals, and certainly in primates. In many, and perhaps most, species, these mechanisms' deployment is limited to temporally and spatially proximate alternatives. However, as the cognitive capacities of humans and their evolutionary forbears became more and more developed, it would have become possible to represent not only alternatives that were near in time and space, but also to represent temporally and spatially distant alternatives (e.g., that a grove of currently barren trees would be full of fruit in a few months' time). Once these more sophisticated representational capacities were in place, the preexisting mechanisms for valuation and choice among proximate alternatives could be applied to alternatives that were not immediately present (including not only temporally and spatially distant situations, but also counterfactuals).

With this explanation of the emergence of the ability to choose between non-immediate alternatives in place, one can then ask whether there is any evolutionary reason to expect the valuations of alternatives in the future or past to be differentially weighted. The answer is that considerations of the evolutionary functions of emotion, in combination with the existence of other asymmetries, provide strong reason to expect selection for the preferential valuation of future, as compared to past, benefits and costs. As such, our explanation does not make the implausible claim that there evolved some wholly new, dedicated mechanism for valuing past and future. Instead, it is more limited in scope, aiming to explain why extant valuation and choice mechanisms were deployed in certain ways once cognitive and representational capacities reached a point where temporally distal alternatives could be entertained.

Using two well-known and deep asymmetries, the causal and knowledge asymmetries, we have witnessed some ways in which they contribute to the affect asymmetry. Because of their physical origins, I suspect that the causal and knowledge asymmetries are in some sense prior to the affect asymmetry. However, it wouldn't surprise me if it turns out that some judgments of causation or knowledge sometimes hang on one's emotions too, thereby letting affect contribute to some causal and knowledge judgments. Whatever the exact genesis, what's important for us is this suite of asymmetries.

12.6 Explaining the PF Asymmetry

Why might we care about future headaches but not past headaches? Can the causal–knowledge–affect asymmetry complex explain this? It turns out that it can. Earlier I expressed the hope that the PF asymmetry can be given the kind of empirical attention that the PD asymmetry receives. Recently it has. In this work it is shown that the PF asymmetry exists and moreover that it is importantly related to the affect asymmetry just described.

The foundation of my account is a series of experiments by Caruso et al. (2008) that, for the first time in the psychological literature, explicitly aims to investigate the PF asymmetry. The general design of the experiments was as follows. Subjects read pairs of stories describing two events, one of which occurred a certain amount of time in the future and one of which occurred an equal amount of time in the past.[11] They were then given probes that involved placing a value on each of the two events.

In the first of the five experiments comprising the study, subjects read a story asking them to imagine that they would spend (or had spent) five hours on a computer data-entry task one month in the future (or one month in the past) and to indicate how much money they thought would be fair compensation for their completion of the task. In line with the hypothesis that people place a greater value on events in the future than on identical events in the equidistant past, subjects indicated, on average, that they should receive *twice* as much money for performing the task in the future as in the past ($125.04 versus $62.20), a highly significant difference (Fig. 12.2).[12]

Having demonstrated the PF asymmetry's existence, Caruso, Gilbert, and Wilson sought to tease out its causes. Most relevant for our purposes are the third and fourth of the five experiments they report. In the former, they hypothesized that subjects' differential valuation of past and future events might be due to differences in their affective responses when they imagine the events. To test this, Caruso and colleagues gave subjects a story in which they were asked to imagine that they had agreed to help their neighbor move out of his apartment. Paralleling the experiment described above, subjects received two variants of the story, one in which they helped the neighbor move one week in the past and one in which they would help him move one week in the

---

[11] Following standard psychological methodology, the order in which subjects received this pair of stories, as well as the others described below, was counterbalanced (i.e., some subjects read the future variant first, while others read the past variant first).

[12] Subjects were also asked, for both the future and past conditions, to rate (using a seven-point scale) how difficult they expected the work to be (or to have been) and how qualified they thought they would be (or would have been) to perform it. Subjects gave the past and future conditions the same difficulty and qualification ratings, ruling out differences in these variables as the cause of the differences in their responses concerning fair compensation for the work (Caruso et al., 2008).

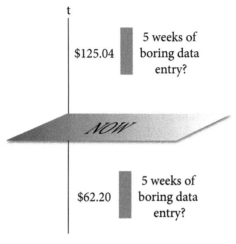

Figure 12.2 PF asymmetry

future. The story stipulated that the neighbor, to thank the subject for her help, would afterwards give her a coupon that would allow her to go to a website and select a bottle of wine from a list that included descriptions and the price of the wine. The researchers found that subjects asked to imagine helping the neighbor in the future chose a bottle of wine that was 38% more expensive than the bottle of wine they selected when given the past version of the story ($162.24 versus $117.96, where the wine prices ranged from $10 to $400). As before, the difference in subjects' compensation decisions was not due to differences in the anticipated difficulty, length, or satisfaction of the task, all of which were given similar ratings in the past and future conditions.

After subjects had selected the bottle of wine in each condition, they were asked to rate (again using a seven-point scale) how "tired, stressed, and dreadful" (Caruso et al., 2008, p. 798) they felt upon imagining the event. This revealed a past/future asymmetry in subjects' affective responses to the stories: subjects who imagined helping their neighbor move in the future felt significantly more tired, stressed, and dreadful than subjects who imagined helping the neighbor move in the past. These ratings were then averaged to create an "index of negative affect" (Caruso et al., 2008, p. 798). Using statistical techniques, the researchers found that this index of negative affect *fully mediated* the effect of temporal location on valuation.

Similar findings were obtained in the fourth of the five experiments, with an additional twist. Subjects were given the past and future versions of the data-entry story described above. However, some received versions in which they were asked to imagine themselves doing the work and indicate their own fair compensation, while others were asked to imagine a random member of the local community doing the work and indicate that person's fair compensation. As in the first experiment, subjects asked to imagine that they were performing the work indicated that they should receive significantly more money for the future condition than the past condition. Subjects asked to imagine that another individual was performing the work, however, indicated that the individual should receive the same compensation in the past and future conditions.

Why were the subjects' compensation judgments temporally asymmetrical in the "self" condition but temporally symmetrical in the "other" condition? The data suggest that differences in affective response are again responsible. In addition to the compensation question, subjects in the fourth experiment were asked to indicate how stressed they felt upon thinking about the task. The subjects asked to imagine themselves performing the task displayed an affective asymmetry, reporting greater stress in the future than the past condition. By contrast, the subjects who received stories in which another person was performing the task reported the same amount of stress in both conditions. As in the third experiment, statistical analysis revealed that this difference in affect fully mediated the interactive effect of past versus future, on the one hand, and self- versus other-relevance, on the other.

The affect asymmetry, itself plausibly connected to the causal and knowledge asymmetries, seems to explain the PF asymmetry. While Caruso et al. (2008) is merely one study—and its exclusive reliance on self-report measures of affect must be regarded as a significant methodological shortcoming—one general point in favor of their hypothesis is its consilience with a large body of evidence indicating that affect is critical not only in paradigmatically "emotional" responses such as experiencing strong fear at the sight of a predator, but also in valuation and decision-making generally (for discussion, see Damasio, 1994; Damasio et al., 1996; Panksepp, 1998; Thagard, 2006). To take just two well-known examples out of many, the reactivation of negative affective states have been found to be important in helping individuals avoid choosing options that have been associated with negative outcomes in the past (Bechara et al., 1996), while positive affective states have been shown to be important in reinforcing and motivating cooperative behavior (King-Casas et al., 2005). Moreover, it is known that emotions also play an integral role in the valuation of different options and other aspects of decision-making (see Damasio, 1994; Damasio et al., 1996; Panksepp, 1998; Thagard, 2006). Perhaps it even plays a role in moral decisions (see Haidt and Joseph, 2007 and Suhler and Churchland, 2011). As we'll see, these points connect up nicely with similar explanations of similar phenomena.[13]

---

[13] Against an earlier version of this argument Hare writes, "But the argument does not work. Whatever the capabilities of my swampy ancestors, I think that I am well capable of focusing my practical attention on future things over which I have control without being future-biased" (Hare, 2009, p. 512). No matter how confident Hare is in his abilities, it takes more than introspective judgments about one's own abilities to counter evolutionary (or other) explanations of bias. Typically biases are unconscious tendencies, so pointing out that one "doesn't notice" them isn't much of an objection. Furthermore, an evolutionary argument hopes to explain the general tendency and doesn't purport to claim that one always must operate with future bias.

12.7 Other Temporal Biases

Feelings of relief and dread are only one type of temporally asymmetric emotion. Virtually all of our emotions display a temporal bias one way or the other. I'm sad that a particular film is over, happy that a basketball game is soon, outraged about a past injustice, and so on. Not only emotions and desires are temporally biased, however. Not surprisingly, given the prevailing asymmetries in our world, we inherit a suite of temporally asymmetric behaviors and judgments. Psychologists have only recently started studying these. Some turn out to be quite surprising and to teach us important lessons. Let me provide some examples based on actual experiments:[14]

- *The future is more intentional.* An elderly woman named Gertrude kills her husband by giving him the wrong pills. Did she do it on purpose? Evidence is given both for and against this hypothesis: on the one hand, Gertrude's eyesight is poor and the medicine labels are small, but on the other, the marriage is rocky and the husband has a large life insurance policy. Imagine that you are on a jury and are asked whether the action was intended. One group of participants in the study was told the action was a past one (a month ago) and another that it is a future action (a month from now). Burns et al. (2012) found that subjects in the future condition setup felt much more strongly that Gertrude killed her husband intentionally (Fig. 12.3). Not surprisingly, they were also more upset at Gertrude's actions and felt that she deserved greater punishment.

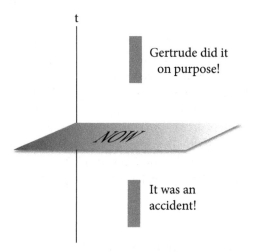

Figure 12.3 The Gertrude case

[14] Beware: you probably won't find this bias in yourself as you reflect on the examples. People tend to enforce consistency, so you may "erase" the bias because you're having both future and past conditions described to you in each case. The actual experiments, when relevant, avoid this by testing only one time condition per group.

- *... more potently willed.* Now turn your attention to Peter who did 60 pushups, 15 more than his previous personal best. How large a factor did willpower play in his feat, in comparison to physical strength and chance factors? One group was told that Peter performed the 60 pushups a year ago, another that he does them a year from now. Subjects asked the question where Peter performed pushups in the future felt that his willpower is proportionally more important than subjects told that it happened in the past. Further studies have refined this asymmetry, showing that it's not necessarily that we think we have *more* willpower in the future, but that what willpower we have is *more potent* in the future (Plaks, 2012; Helzer and Gilovich, 2012).

- *... more likely.* You roll a die and make a guess about the outcome. How confident are you? How much money would you bet on your choice? Interestingly, this too depends on whether one *has already* rolled the die, but not looked, or is *about to* roll the die. Many experiments reveal that subjects will bet more if the roll happens in the future instead of the past, e.g., Rothbart and Snyder (1970). Brun and Teigen (1990) show that we even believe guesses about the sex of a child are more likely when the birth is in the future. See also Helzer and Gilovich (2012). In general, the future is seen as more likely than the past (Hanko, 2007; Hanko and Gilovich, 2008).

- *... more rational and prototypical.* How rational were you in the past compared to how rational you will be in the future? And how emotional have you been and will you be? People rate their past selves as having stronger emotional capacities and weaker rational capacities, but their future selves the opposite (O'Brien, 2013). People also feel that the future will unfold in a more prototypical way than the past. That is, "prospection more than retrospection is grounded in scripts, schemas, stereotypes, and other prototypical mental representations of what people, places, and events are typically like" (Kane et al., 2012).

- *... more praiseworthy, more blameworthy.* As already mentioned, a soft drink company runs a trial of a vending machine that raises the price of a can of soda when the temperature is hot. How fair is this machine? Told that the trial would happen in a month, subjects were much more outraged than were subjects told that the trial happened a month ago. Other cases show that past transgressions are typically viewed as more morally acceptable than the same transgression in the future. Future transgressions are more blameworthy than past ones, all else being equal. And the same for good deeds, for here future charitable donations are viewed as more praiseworthy than past ones. See Caruso (2010) for more.

- *... and brighter, lighter, and closer.* How far away does Valentine's Day seem to you? People differ markedly if asked a week before Valentine's Day rather than a week after. In what is aptly called the temporal Doppler effect, the future feels closer to us now than the past does (Caruso et al., 2013). When asked to literally represent the future, which colors would you use? Bright ones or dark ones? Or if asked to represent the future with elements, would you use light ones or heavy ones? Helzer et al. (2011) show that we tend to represent the future as bright and light in comparison with the dark and heavy past.

The reader may worry about possible confounds in many of these. He or she would be right to worry—also about one's intuitive judgments when facing the thought experiments of philosophers, such as Parfit's hospital case with which we began the chapter. Given that the experiments (empirical or thought) take place in a world which is so profoundly temporally asymmetric, there are too many variables to control. For instance, we know that the soda machine trial in the past didn't cause widespread adoption of such machines, whereas we don't have that assurance about the future trial. The better experiments control, as in the Caruso study, some of the more obvious variables, e.g., whether subjects felt the past work was as difficult as the future work. But given the sweeping nature of time, it's probably impossible to get them all.

Those worries aside, I take it that the combined effect of these studies paints a very interesting picture, a picture of a large and deep suite of temporal biases pervading our lives. These are not simply emotions and desires, but judgments about intention, rationality, likelihood, and more. Furthermore, these biases really matter in potentially important situations. If the above is correct, all manner of legal decisions are subject to this bias. For example, a jury subject to this bias would award greater compensation for future suffering than for the same amount of past suffering; companies, governments, and individuals should just do controversial acts and let people find out about them afterwards, for that will engender less condemnation than announcing it and then doing it; and so on.

One other important point arises in the context of Prior's argument. Zimmerman seemed to want an account of our bias regarding headaches that not only *explained why we have it* but that *vindicated* the bias, i.e., that made the bias rational. I mentioned that this is hardly standard procedure in psychology. That said, I suppose it is preferable, all else being equal, if we can be charitable to ourselves. In light of our new awareness of temporal biases, however, I take it that making a blanket assertion or expectation that they should all emerge as rational or "make sense" is off the table. Some of the above biases are outlandish. Should a tensed metaphysics rationalize or make apt our wanting to punish future crimes more than past crimes, dice rolls being more likely in the future than past, and so on? It's neither clear why one would assume that nor how a tensed metaphysics could possibly handle all these cases.

Studying the rationality of these biases is a subject for another time. As a sneak preview, however, I assume the answer will be much like that for the PD asymmetry, namely, that whether it's rational or not is highly contextual. Many studies in that field show that Bayesian updaters immersed in a world with hazards and uncertainties scattered a certain way should apply PD discounting. But of course it's possible that the hazards and uncertainties are scattered in different ways, and when that is so, then the tendency, whose genesis makes sense, is inappropriate. As an analogy, consider the feeling of hunger. Its genesis makes perfect sense. The feeling is a great shortcut for us.

We don't need to calculate our caloric intake or usage or time since last eating. All of those computations are offloaded onto the feeling. Its genesis is not surprising. That doesn't mean, however, that we always feel hungry when we need food. A peek at my waistline proves this false.

12.8 Explaining Other Time Biases

Research in psychology is uncovering a wide variety of sometimes surprising time biases: we feel that the future is more likely, more predictable, more intentional, more potently willed, brighter and lighter, and so on. Can we trace these asymmetries back to the causal, knowledge, or affective asymmetries? If so, not only would we have a much richer explanation of the headache phenomenon than provided by the tensed theory, but we would also have an explanation that unifies much more phenomena.

Lack of space and skill prevent a full treatment here. Tracing every known past/future time bias back to the causal–knowledge–affect suite would be a long process and require new research. What I want to point out, however, is that the claim that many of these phenomena trace their origins to one or more of these asymmetries is not implausible. Indeed, many *already* seem explained by what we've shown. Experiments have and could tell us the answer in many cases.

For instance, why compensate people more for future harms suffered than past harms suffered, all else being equal? And why think future generosity is better than past generosity, all else being equal? Since we care more about the future, the explanation of these cases is straightforward. Since the future intensifies affect, the future matters more; hence future suffering deserves more compensation and future generosity demands more praise. To bring in our example of a headache, suppose reading this book causes you headaches, one headache per chapter (I hope not). Then I've caused you past headaches, and assuming you're willing to press on, will cause you a future one too. What compensation do you deserve for my crimes? Since future headaches are worse than past headaches, then I owe you more compensation than I do for all the past headaches I cause. And the reverse is true for any pleasure the book causes. Experiment shows that in these cases—compensation, generosity, and so on—an index of affect again largely mediates these biases (Caruso, 2010).

Less direct are explanations of some of the other biases. From experiment we know that, all else being equal, the future as compared to the past tends to seem

- more willed
- more emotion-driven
- more predictable and prototypical
- more likely
- more pre-meditated.

The key to explaining these, it seems to me, again begins with the knowledge and causal asymmetries. In Fischhoff and Beyth's (1975) study of hindsight bias they note that people implicitly interpret the past with a sense of "creeping determinism," the thought that past events are in some sense inevitable. Why do they do this? I suspect that much of the explanation is due to the knowledge asymmetry. When we look back at some event, thanks to our greater knowledge, we can fill in all the little details. Why did I buy that candy bar last night while I was walking home? As it happened it seemed an utterly spontaneous act. But knowing what I know now, I can see all the little factors pushing me in that direction: the hunger, the expectation of a too-healthy meal, the bakery being closed, the shifty character in front of the grocery with better alternatives, and so on. I can also see all that now hangs on the candy bar purchase: the sugar crash, the milk carton now empty, and more. Buying a candy bar seems locked in like a puzzle piece in a grand puzzle; removing it is conceptually difficult. Looking to the future, however, I don't enjoy anything like that level of detailed knowledge. I don't know how hungry I'll be, where the shifty characters will be located, or anything like that. Combine this knowledge asymmetry with the asymmetry of causal agency, the fact that there is nothing I can do about the past anyway, and we begin to understand the origin of the creeping determinism associated with the past.

This creeping determinism appears connected to the power of the will. Here is Helzer and Gilovich (2012):

> When they look back, people are hard pressed to see the role played by the will in shaping the outcomes they received. In contrast, when looking forward to an open, unrestricted future, people see the will as a powerful force guiding their behavior. Of course, a person's future attempts to exert will power are likely to run up against the same roadblocks as those experienced in the past. Those unexpected snags and distractions that interrupted the person's best-laid plans last week may very well show up in one form or another next week. Blind to this reality, people exaggerate the importance of the will as a decisive cause of tomorrow's outcomes, and, as a result, are overconfident about the future. (p. 1244)

All of that information about the past dilutes the perceived power of the will. Aiming at the future, by contrast, we don't have so much to go on, so the will stands out. That's why the will should work better next time, for now we can't see all those shop closings and so on that make it disappear. Since the will is a cause, and our causal agency is only in one direction, the asymmetry of causal agency can only amplify this effect.

These considerations motivate a sketch of why we might expect the above biases. The future seems more willed because the will stands out more against our background knowledge in the future than past. And since the future but not past is controllable, amplification will make us regard the future as more will-driven and therefore more rational. That's why those past New Year's diets didn't work but this next one will. The future is also more prototypical because I have less to go on when imagining it. I extrapolate from the present, read the present in the future, in a way I'm not tempted to in the past because I have so much information about it. For the same reason it's more predictable. It's more likely to go according to plans because that is all I have to go on. If I could see all those unexpected snags and defeaters of my plans, as I can toward the past, then I wouldn't regard it as so predictable.

Why is the future more likely? Why am I more confident about the results of future coin tosses than past ones? It might be because the future appears more predictable, but I suspect the greater contribution is a bit sillier. It's not so much that we believe the future is more likely, as I understand the data, but that we believe our guesses about unknown future outcomes will be better than guesses about unknown past outcomes. Many authors suggest that it's almost as if subjects believe they have a magic power to influence the die. That magic power is an exaggerated effect of the will. That plus the causal asymmetry give us our target: my will obviously is impotent with respect to the past roll of the die, but possibly not with respect to the future one.

What about Gertrude? Unlike in the previous cases, where we simply have my speculation, here we already have some empirical research along the same lines as Caruso et al.'s explanation of the PF asymmetry. There we saw that negative affect mediated the asymmetry in question. Burns et al. (2012) tested whether this is also so with the Gertrude case and found that when they included an index of negative affect in their model, "the effect of temporal perspective dropped to nonsignificance." They also asked the opposite, whether an index of intention ratings mediated the index of negative affect and found that it did. We don't, therefore, know whether one causes the other, only that there is a bidirectional correlation between the two. But we do have something very close to what we have in the case of the PF asymmetry case studied above. Moreover, I think that we can speculate on other reasons why Gertrude's action might be viewed as more pre-meditated when in the future. We're inveterate mind readers. In planning, it's crucial that we figure out what other people will do. People are unpredictable. Figuring out people's intentions is our way of fitting their future behavior into predictable patterns. Given that others act only in the future, like us, we therefore have more reason to understand their behavior as intentional towards the future than towards the past. Further, just as my will is viewed as more potent in the future, so is Gertrude's will. And against a background where usually the effect of the will is diluted in the past, we have another reason to view future actions as more intentional than past actions. For all these reasons, future Gertrude "did it."

I don't know whether explanations along these lines will hold up. They don't seem implausible. But let me be clear. I'm *not* saying that we can *always* explain every case directly in terms of the causal, knowledge or affect asymmetries. We're virtually certain about the next time Halley's comet will arrive, but that doesn't mean we think that future event is fixed; nor do we think the lives of Etruscan individuals are open, even though we know virtually nothing about them. Given the whopping number of differences between the past and future—recall not just this chapter's material, but also the previous chapter's on the flow—it would be stunning if organisms didn't adopt a heuristic that paints the entire future one way and the entire past the other way. As Caruso (2010) puts the point:

> I suggest the most compelling possibility is that reactions to future events are overlearned responses to the natural environment. Because a primary function of emotion is to prepare

organisms for action (Frijda, 1986), and because organisms can typically act on future events more successfully than past events, the emotional bias toward the future may be an overgeneralized response to future situations even when these situations are not actually under one's control.   (pp. 619–20)

Parfit (1984) and Van Boven and Ashworth (2007) make similar claims about overgeneralizing based on the frequency of future events being controllable and unknown and past events being known and uncontrollable. Hence we paint with a broad brush— sometimes literally, as when we color the future yellow and the past brown—and adopt a picture whereby the past is dead and the future open. That's why Etruscan lives are settled even if unknown and Halley's comet's next visit is open even if known.

12.9 Conclusion

Philosophers of time have attempted to use the existence of the PF asymmetry to argue for strong conclusions about the metaphysics of time. They contend that the PF asymmetry is a mark in favor of "tensed" metaphysical models of time, for they believe that tensed models explain the asymmetry and detensed ones have not. The chapter demonstrates that neither contention is correct. The tensed models "explain" the asymmetry by simply positing a primitive connection between the posited metaphysical "arrow" and our preferences. Not only is this something the opposition can just as easily posit, but it is hardly much of an explanation. There is no hint as to why our attitudes were shaped the way they were because of this arrow. In the absence of new posits, it tells us little or nothing about what to expect regarding different time biases. Like an *elan vital* posited to explain life, these arrows name a problem more than they provide a solution. By contrast, the "Humean" solution outlined here offers a much richer and satisfying explanation.

A sign of its richness is the fact that it has more or less morphed into a scientific research program in psychology and biology. The theory is still developing, but it is fruitful, unifying, and suggests novel predictions and theory. For instance, the above discussion might lead us to make several testable conjectures. We saw that the PF asymmetry holds in the first person situation but disappears in the third person case. Given the postulated role of emotions in this asymmetry, one might suspect that the PF asymmetry can be made to appear in the third person case if the third person is emotionally tied to the subject. I place a premium on my future labor, but I value your past and future labor alike (assuming you are a stranger). Do I treat my *children* like that too? This question, plus plenty of speculations in the previous section, are easily testable. In addition, I believe Caruso et al.'s hypothesis about affect and the temporal value asymmetry is ripe for more rigorous investigation using tools such as skin conductance response and functional magnetic resonance imaging (fMRI). Together they will help us hone the Humean explanation of our temporal asymmetries into a more exact scientific hypothesis. When evaluated along typical dimensions of theory choice in science, there is simply no contest between the two.

In sum, by marshaling research from a wide variety of fields, I hoped to develop a theory of our temporal biases. Of course we have much to learn. But it is possible to now see in outline how the causal and knowledge asymmetries leave a trail of biological and psychological asymmetries in their wake. Because our temporal biases have sweeping implications, ranging from the level of individual decisions about personal finances to international debates over social policy, it is of paramount importance that we understand the roots of these asymmetries. The next step is to assess which of these biases are rational and which aren't. I'll leave that for future work.

13

Moving Past the ABCs of Time

Readers familiar with philosophy of time may be puzzled. Not only have I set up the problem of time in an unfamiliar way, but I'm rarely using the same tools or language as analytic metaphysics. Where are the relational and monadic temporal properties, the A-predicates and B-predicates, the token reflexive and date analyses of tense, and the discussion of truth-makers for past statements? For heaven's sake, the reader may object, after all these pages he hasn't come clean on whether he is a presentist or eternalist, or even whether he believes the future exists! Worse still, I've helped myself to all manner of science, yet not once do I discharge my duty to purge the science of tensed concepts. What gives?

It is true that this book doesn't reflect the literature in analytic philosophy of time in a straightforward way. I move away from some of the analytic categories commonly in use not to be difficult but rather in the firm belief that this is the way forward in philosophy of time. That doesn't mean that my approach is irrelevant to mainstream philosophy of time, or vice versa. My approach is a slight modification of the program, not a wholesale substitute, and its motivations often lay in philosophy of time. In what follows, let me explain why I feel that philosophy of time, especially analytic metaphysics of time, can achieve less than is commonly supposed as it is currently practiced. I'll give a biased view of the history of the field and subject its main assumptions to various challenges. The upshot is that I view the field as having produced scores of metaphysical models of what time might be like but very little in showing how any of them provide good plausible explanations of temporal phenomena. Once that point is made, I'll then single out some ways in which traditional work bears on the current approach and produce a vision of a way forward.

13.1 Analytic Philosophy of Time: A Potted and Biased History

Although philosophers have studied time for as long as there have been philosophers (think Aristotle, Augustine, Leibniz, Kant, Husserl, and so on), the approach in recent analytic metaphysics can be traced more or less to McTaggart (1908). McTaggart argued that A-properties were essential to time, but that they were self-contradictory.[1] This provocative conclusion attracted the attention of philosophers. Few defended McTaggart's full argument, but a debate began between so-called A-theorists and B-theorists. The A-theory holds that temporal A-properties are fundamental and B-relations derivative, whereas the B-theory holds temporal B-properties to be fundamental and A-properties derivative.

The dispute primarily focused on the semantics of tense. This concentration is not surprising, for when philosophy of time became popular Anglo-American philosophy was entering its phase of ordinary language analysis. I'll leave the detailed sociology and history to those better able to tell it. Suffice to say, philosophy of time became about the question: can we substitute B-predicates for A-predicates in a statement while leaving its meaning invariant? If not, it was concluded that time *itself* is tensed, that there is something "out there" corresponding to the predicates "now," "past," and "future." A-theorists thus became philosophers who hold that that A-predicates are semantically irreducible to B-predicates. B-theorists obligingly played the stooge by offering ever more ingenious B-theoretic translations of statements with A-predicates. Given what we've learned about indexicals in philosophy of language, we now know the B-theorist translations were doomed to fail. Indexical language was not to be reduced to non-indexical language.

Over time the "A-theory" and "B-theory" became associated with a large grab-bag of theses. Nowhere is this better seen than in Gale's 1968 canonical text in philosophy of time. There the "A-theory" is defined as having four distinct doctrines, including (1) the semantic irreducibility of A-predicates to B-predicates, (2) temporal becoming (the idea that the future is unreal but becoming real, that future events acquire the A-property now-ness), (3) serious ontological differences between past and future (e.g., past is fixed, future open), and (4) the idea that change requires an A-series (see Fitzgerald, 1985). The theory thus lumps together two or three distinct metaphysical positions with one or two semantic positions, depending on how one understands the change dispute. The A-theory became a real mongrel. What's particularly odd about it is that (1)–(4) are logically independent of one another—at least, on a natural reading—and yet (1) carried almost all of the argumentative burden. Secure (1), it was thought, and (2)–(4) followed. Now we scorn arguments whose premises are entirely about language and whose conclusions are metaphysical. It's not clear how to get *anywhere* from (1).

Although other areas of philosophy have long recognized that there is a yawning gap between language and the world, the message spread (and is spreading[2]) slowly in philosophy of time. Since twentieth-century analytic philosophy as a whole often drew

---

[1] 1908 was a "rough year for Time" (Savitt, 1994). Not only did it witness McTaggart arguing that it is unreal, but that same year Minkowski announced that "Henceforth space by itself, and time by itself, are doomed to fade away into mere shadows" (Minkowski, 1908).

[2] Dyke (2003) chronicles three recent attempts in the area to draw metaphysical conclusions from arguments with only linguistic premises.

metaphysical conclusions from arguments with only linguistic premises, philosophy of time perhaps may be forgiven for this transgression. We can be especially sympathetic if we recognize that the framing of the main debate and its timing made it especially susceptible to this error. In any case, because B-theorists grew tired of losing the argument so quickly, given how easily (1) is secured, the project was refashioned in the 1980s: B-theorists were allowed to "win" if they produced B-kosher statements that had the same truth conditions as the A-predicate version, even if they didn't have exactly the same meanings. Now we are doing metaphysics, it was thought. If you could not produce B-theoretic truth conditions of statements with A-predicates in them, then there must be honest to goodness A-properties *in the world*. The A-versus B-debate lived on, but it became about whether A-facts or B-facts were *ontologically basic* rather than about whether one type of language was *semantically basic*.

This official methodological improvement didn't change much on the ground. The battle now occurred over various B-theoretic schemas for truth conditions. Token-reflexive, date analysis, and other alternatives were proposed. The question became whether these analyses were missing something or not. Yet all the pitfalls associated with these analyses have exact counterparts in the same project of giving truth conditions for spatial and personal indexicals too. Why the battle has something in particular to do with time was never clear. Even today, the mighty Routledge *Philosophy of Time* two-volume set devotes the bulk of the work (seventeen chapters!) to such analyses of tensed statements.

The question of what follows from (1) has a counterpart in this re-framed debate. Replace (1) with its metaphysical counterpart, (1*), that the A-properties or B-properties are ontologically basic. What do we learn from this? (2), (3), and (4) don't follow from (1*) any more than they did from (1)! The ultimate trouble with the A versus B debate is that neither position tells us much that is specific about time. Suppose A-properties exist. What would that tell us? It tells us a little: the truth-values of tensed statements change with time. In this fact A-theorists might say they've discovered time flow. But truth-values changing with time is a rather paltry thing: presumably many metaphysical models would get changing truth-values, and it tells us as much about truth as about time. Crucially, it doesn't tell us that the past or future are real or unreal, that time is branching, that time is discrete or dense or continuous, that we know more about the past than the future, or much else. None of this is at all surprising, given that the theory is ultimately characterized in terms of a linguistic category. It's like being told that a particle accelerator revealed the existence of something represented by a *noun*. This is too coarse-grained a category to tell us much about the world, and correspondingly, to make its investigation tractable. We know that quarks are ontologically more basic than protons because we have a powerful theory, vindicated by experiment, telling us how quarks compose protons, and not vice versa. But A-facts and B-facts?

As the questions were changing, so too was what counted as evidence for a theory of time. Once space and time were treated more or less on a par methodologically—consider their treatment in Newton, Helmholtz, Poincaré, for example. Evidence for a theory of time was like evidence for a theory of space, namely, theoretical applications of the ordinary scientific method (to the extent there is one). The dispute between Newton and Leibniz over whether time is substantival hung, ultimately, on what physical theory best explained the phenomena. With the "linguistic turn" and the debate's framing, however, contemporary philosophy altered the principal method of attack on the two: time was understood through language, space through science.

This division was evident every time I visited my university's library. Before draconian fiscal cutbacks, UC San Diego's Geisel Library (Geisel of "Dr. Seuss" fame) was divided between a Humanities & Social Science Library on the left and a Science & Engineering Library on the right. Fetching a philosophy of time book, I had to go left; fetching a philosophy of space (or spacetime) book, I had to go right. Study of the two basic modes of extension—or one if there is only spacetime—was divided by library.

Don't blame the librarians, however, for when one opens the books one sees that they warrant separation. Books in philosophy of space are filled with reflection on science, drawing on a tradition including Berkeley, Kant, Helmholtz, Newton, Leibniz, Maxwell, Poincaré, Einstein, and contemporary scientists. Books in philosophy of time, by contrast, are dominated by the semantics of time, drawing on a tradition dating to Augustine but mostly bypassing the above figures, and practiced almost exclusively by non-scientists. That the standards of evidence have changed during this makeover isn't surprising. Though metaphysicians profess to be following the same methodology as theoretical scientists, with philosophy of time moving so far away from science its evidential standards changed too. The result today is, well, downright odd. To find out whether Bigfoot exists, you look around in the woods and your best theory tells you the evidence is insufficient. To find out whether dark matter exists, you look at your most powerful theory of indefinitely many gravitational events, plus tons of auxiliary hypotheses, see that there is nothing better, and conclude that dark matter must exist, even if it's not directly observed. But to infer features of the nature of time we wonder whether the "date analysis" truth conditions leave something out?[3]

Today many philosophers of time scoff at those past muddles. The debate has moved on. Instead of skirmishing about tenses, they keep the ontological squarely before them by debating the question of what events *exist* for other events. Existence draws the very lines of debate in philosophy of time: "eternalists" believe past, present, and future events all "equally" exist, "possibilists" believe that past and present events exist, and "presentists" believe that only present events enjoy this lofty status (Fig. 13.1). Arguably the A/B debate is orthogonal to the present one (Parsons, 2002). The fixation on semantics is firmly behind us. Nothing is more ontological than existence!

---

[3] Imagine a two-volume set, counterpart to the Routledge piece mentioned above, devoted to philosophy of space or spacetime with seventeen chapters devoted to analysis of "here" or "here-now" There would be a riot.

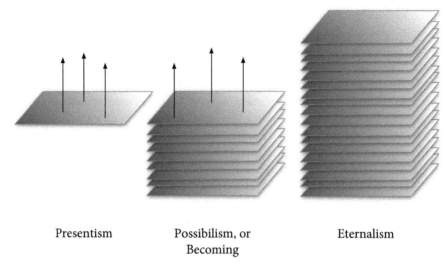

| Presentism | Possibilism, or Becoming | Eternalism |

Figure 13.1 Tensed theories

This trifold division is of course only the beginning. We can ask whether the distinguished present is thin or thick (instantaneous or not), whether it moves continuously or through discrete jumps, whether the so-called truth-maker principle holds, and if so, how it holds (itself tensed or not), and much more. The literature now offers increasingly byzantine models of time that ultimately differ only over the truths of what exists as of when. Not only do we have presentism, possibilism, and eternalism, but we have thick and thin presentism, priority presentism, existence presentism, moving spotlights, Meinongian presentism, fundamentally tensed presentism, scores of branching models, and absolute becoming (forward and reverse). With such a proliferation of models and a set of stock worries, e.g., the truth-maker objection, it is no wonder that a few years ago the prestigious *Australasian Journal of Philosophy* put out a call to its editors for more referees in philosophy of time. Submissions on presentism and related topics had exceeded reviewing capacity.

13.2 The Explanatory Challenge

I don't have a deep-seated conviction that one metaphysical model of time is right to the exclusion of the others, nor that they are notational variants (although they certainly *might* be—see Callender, 2000; Meyer, 2013, Ch. 9; Savitt, 2006; Dorato, 2006). Nor am I against the development of models of time. Yet I do think the current focus in analytic metaphysics of time has often lost sight of the ball. The models being developed don't do much explanatory work and the standards of evidence for or against a particular model are insensitive to the ultimate goal of philosophy of time. What would count as evidence of thick skipping presentism (a theory with a non-instantaneous present that doesn't evolve continuously)? Would the world seem like a stop-motion film? No, for we don't experience the non-existent moments. What would count as evidence of the non-existence of the future? Does it imply that we have a sense of free will? No, for physical laws still constrain us, and we can't phenomenologically detect the gap between the physically and metaphysically necessary; moreover, plenty of models of the sense of agency are compatible with a "block" universe.

The contrast with dark matter is illustrative. Dark matter is posited to explain a host of discrepancies between observed masses of various objects and their gravitational effects. It is, in a sense, metaphysical. As with time, no shortage of models of dark matter exist: WIMPs, WILPs, MACHOs, RAMBOs, axions, and more have all been posited. Many strike one as outlandish. Currently few are observationally distinguished from one another. Viewed from afar, the situation looks similar to the metaphysics of time. Up close, however, large differences emerge. First, some models are weeded out by observations, but more importantly, they all have *possible* observations in mind. Second, they are expected to meet a high evidential standard. One can posit so-called "warm" dark matter, but if particle physics has no independent need for such particles (those with mass between roughly 300 eV to 3000 eV) the theory is seen as suffering as a result. In the metaphysics of time, by contrast, no effort is directed at independent checks from other theories, using the models to unify the theories, or finding observational consequences. Just as linguistic issues such as the ineliminability of tense and ontological debates over the basicness of A-facts fail to elucidate much about time, so too do the "existence debates" fail to meet the usual standards employed when evaluating models of the world.

What happened, I suspect, is that metaphysics to some degree, and philosophy of time to a larger degree, found itself increasingly threatened by science.[4] Physics made do without a distinguished present, a flow or fundamental differences between the past and future. Not only did it muddle through, but it proved to be spectacularly successful. Psychology too turns out not to need skipping presents or other metaphysical extravagances to explain phenomena. With each new theory and new success the pressure on metaphysics of time increased. The assault was indirect. It was an attack of indifference. Neither physics nor psychology nor any other science required any of the apparatus of the metaphysics of time when tackling the temporal. Any proposed metaphysical addition that mattered, e.g., a preferred frame, seemed to conflict with science, for reasons to posit it were always lacking according to the usual standards of evidence in science.

[4] Here is a great quote that I learned from John Earman: "I am inclined to think that McTaggart's complete lack of acquaintance with contemporary natural science was in certain respects a great advantage to him as a philosopher. The recent advances in physical theory have been so important and spectacular that they have only too obviously 'gone to the heads' of some eminent physicists, and have encouraged them and the public to believe that their pronouncements on technical philosophical problems, for which they have no special training or aptitude, are deserving of serious attention" (Broad, 1933).

Eventually philosophy of time hit upon the perfect response. It bought itself *permanent immunity* from the threats of science by making philosophy of time about the one property *guaranteed* to generate no inconsistency with science: naked existence itself. What distinguishes the present moment from others? It is the only one that exists, say presentists, or the edge of what exists, says possibilists. What does the flow of time consistent in? The changing nature of what exists. Why might the past be fixed and the future open? The past and present exist, but the future does not, according to possibilists.[5]

The trouble with existence is that it explains everything and nothing. The existence of particular *kinds* of objects can explain. The existence of atoms would explain all of the data to which the physicist Jean Perrin famously points in the early twentieth century. What is really doing the work in this explanation, however, is not the atoms' existence, but their nature, i.e., their specific causal powers. Perrin's case for atoms was a fight over transmission speeds, specific heats, and other details of the models. To revert to a more old-fashioned terminology, without specific *essentia*, their *esse* explained nothing. And that's generally the way it is with explanation. Not all explanation is causal, but an awful lot of it is. Since naked existence doesn't cause anything, focusing on this feature doesn't allow for the bread-and-butter of explanation, causal explanation.

Philosophy of time buys itself immunity from science because existence is not a property, or if it is, it is not a helpful one for explanation of first-order events. In Kant and Hume's opinion, enshrined in the now-standard Frege–Russell view, existence is not a property or predicate of individuals (e.g., Hume, 2000). The reason is that it imparts no information about it's subject. An existing hundred dollars and a non-existing hundred dollars are the same amount of currency, the same color, the same weight, and more. According to Frege, the existential quantifier captures all that there is to capture about the nature of existence, as existence imparts neither denotation nor significance. Existence is a function from first-order concepts to truth-values. Objects aren't the sorts of things that exist (or not), strictly speaking, for "[i]t is a logical category mistake to ascribe existence to objects" (McGinn, 2000, p. 19). For something to exist is simply for some propositional function to have an instance. To say that the future exists, therefore, means that there are true sentences of the form "*a* is future," where *a* is the name of some future event. Clearly, to say that some event exists at some time does not change the nature of any object or event one jot. Existence is a predicate of predicates, so it can't enter into the rough and tumble first-order world of explanation.

---

[5] One might object that presentism and possibilism additionally posit a primitive direction of time: the present updates in one direction according to presentism and the growing block grows (as opposed to shrinks) according to possibilism. Fair enough. However, if a primitive temporal direction is doing all the explanatory work, then that is orthogonal to the eternalism debate. An eternalist can just as easily posit a primitive direction to time, and in fact, many have (Weingard, 1977; Christensen, 1993; Maudlin, 2002).

Perhaps the standard Fregean treatment of existence isn't correct.[6] Maybe existence is a property, the property that all existents share, just as feline is a property all feline objects have. That may be, but the ghost of Hume and Kant's objection that existence "makes no addition" to an object still threatens. Any account of existence as a property must contend with the awkward question of what it adds to an individual. Plainly, the right reply is to say that it adds existence to the individual and no more.

Santayana, by contrast, saw in existence the "strain and rumble of the universal flux," neo-Thomists see existence as a Perfection, and the Existentialists see it as absurd. Now these strains, rumbles, perfections, and absurdities may well alter the character of an experience and they might even play a role in a causal explanation. If we could test for, say, rumble, and find the simultaneity plane that rumbles, then we have a scientific thesis on our hands. These ideas, however, are non-starters:

> Existence must be such that any explicit doctrine that does not impute too much to it will impute much too little. An essence is so nearly nothing without its existence that Existence must be nearly everything; and yet there is so little left of the existent when we abstract from the essence that Existence evaporates to almost nothing.   (Williams, 1962, p. 752)

As soon as we spell out what existence's "ontic voltage" (Williams' phrase) consists in, we immediately say too much, for there may be existents without any such property and arguably non-existents with such properties.

Hume's famous point that existence is not observed is widely accepted (Miller, 2002). Given his flair for making such points, it's worthwhile hearing Williams again:

> That existence is not observable must be tested by everyone for himself... Now, I respect the person who stares at a doorknob, for example, which is not doing anything in particular, and thinks he can see that it is at any rate existing. I acknowledge for myself that I hardly know what to look for, and may well lack the requisite intuition. In spite, however, of a sympathetic eye for such ontological gems as universals, relations, and classes, I don't discover any Existence, and I think that Hume was right on this score. What one observes is not the existing of the knob but just the knob.   (1962, 751)

The point is correct but it shouldn't be restricted to observation. Bricker (2004), following Williams, asks how you know *you* exist, if existence is a property. No matter what you can observe, you still have the same parents, the same house, the same color shirt, the same mental state, and so on. It's not as if you are incomplete in some way, like Sherlock Holmes, who neither has nor doesn't have a freckle on his left shoulder. Notice that if existence changed the character of experience in any way, à la Santayana, we would not ask this question. Since it makes sense to ask, we see that existence, even if a property, does not affect the character of any experience. As Williams writes, "Existence evaporates to almost nothing." Not only is it unobservable, but it is causally impotent too.

---

[6] See Williams (1951), McGinn (2000), and Salmon (1987).

To drive this point home, consider the thought that the non-existence of the future *explains* either the attitude asymmetry discussed in Chapter 12 or the sense of freedom toward the future discussed in Section 11.2. The metaphysical model of the "growing block" includes a genuine existential temporal asymmetry: the past but not future exists and is growing. Perhaps it can be used to explain the attitude asymmetry or sense of freedom? Let's slightly modify an example from Dainton (2001) to display why this hope is a vain one.

For simplicity, restrict attention to three representative chunks of time in Broad's growing block model.

Broad states that t_j is later than t_i just in case the sum total of reality at t_j is greater than the sum total of reality at t_i. In the growth depicted in Fig. 13.2, we therefore ought to understand time as progressing from left to right. Call this the *block-arrow*. Does this existential asymmetry explain the attitude asymmetry, the sense of freedom, or other temporally asymmetric phenomena? It's hard to see how it could. Call the direction picked out by memories, causation, counter-factual dependence, knowledge, attitude, and so on the *world-arrow*. Focus, for instance, on memory. Then the world-arrow orders events roughly as follows: time t_j is later than t_i just in case there are memories of events from t_i at t_j and not vice versa. In Fig. 13.2, where a bigger arrow depicts later memories, we again read time as passing from left to right.

Absent any hypothesized connection between the world and block arrows, however, there is no reason for the two arrows to align. The scenario in Fig. 13.3 is therefore possible.

If we use the world arrow to convey temporal direction—from smallest to biggest arrow—then this world is a *shrinking block*, not growing block. In it, one's memories of one's childhood are memories of events that haven't existed yet and our anticipations are of what has already existed.

Let's now add a few non-trivial, rough but plausible assumptions. First, assume that the memory arrow is a crucial component of both the attitude asymmetry and the sense of freedom. The knowledge asymmetry is probably not the whole story in either phenomenon, but it certainly seems to be part of it. Assume also that memory supervenes, ultimately, upon the matter configuration at a time or times. Then we're

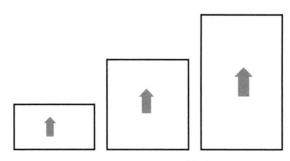

Figure 13.2 Growing block

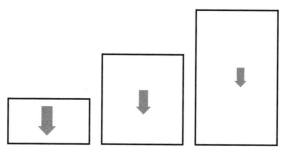

Figure 13.3 Shrinking block

in trouble. Since there is no reason to believe that the direction of accumulating memories is the direction of growing reality, there is no reason to believe that the block arrow has anything to do with explaining the attitude asymmetry or the sense of freedom. The theory is explanatorily inert.

One could posit a connection between the two arrows. Then misalignment couldn't happen. What would the connection be? A primitive one? A claim to the effect that I can only have particular memories at some time if the contents of my memories already existed according to the block arrow? With no independent reason to sub-scribe to this, this posit is akin to positing dormative virtues to explain the causal powers of narcotics. It's hard to see how anything but a primitive connection will do. Since existence doesn't imply anything about the nature of event, the contents of time slices are completely unmoored. The non-existence of the future by itself isn't explanatory. Pointing out that ghosts lack existence is useful in certain contexts. Yet if we really wish to explain the eerie creaking noise in the night one needs to roll up one's sleeves, climb into the attic, and find the open shutter or noisy critters.

By taking ownership of existence, philosophy of time protected itself from conflict with science and its explanations. Experiments pose no risk, for existence can't be seen. Theories don't either, for existence has no causal powers and can't disrupt the core posits of a theory. Existence, after all, makes no addition to an object. At the small price of adding surplus structure to a scientific theory—and since existence is nearly nothing, the surplus isn't large—the field makes itself orthogonal to scientific pursuits. As a bonus, philosophy of time purchases disciplinary autonomy for the field. Meta-physics is under threat from science and other quarters. The rise of metametaphysics as a field is a symptom of the worries many have when thinking about the warrant for many metaphysical claims. The conservative reaction is to find some specifically metaphysical domain over which philosophers have charge, one not encroached upon by science (see Callender, 2011a). Making philosophy of time the science of existence is the perfect tool. Almost by definition, existence will play no role in science, so philosophy of time will never be threatened by those grubby scientists with their labs, experiments, data, and elaborate theories. And since philosophers are all trained in logic, we already have all the tools we need to study time.

The irony about this maneuver—and this is the *big problem*, the point of this section—is that creating this safety zone around time leaves philosophers of time unable to do their original job. The metaphysical models are posited, after all, to help explain temporal phenomenology and related claims, e.g., that we can't change the past. Don't worry about that preferred frame, we are told, the frame isn't detectable or causally potent—we're doing metaphysics, not physics. When faced with the temporal phenomena, however, philosophers of time *want* this invisible frame to do some explanatory work. They want to use it to explain why we feel as we do, why the past is fixed and the future open, and so on. But this assumes that we have astonishing perceptual powers. Somehow we human beings are able to detect what the most powerful physical detectors and all the other sciences cannot! The irony is that the metaphysics is posited to explain the temporal phenomenology, yet in defending it from objections, philosophers make it unable to do its original job.[7] The supposed effects of "tensed existence" are the manifestations of the temporal phenomena themselves, which in turn are the basis for inferring the concept in the first place. This circular reasoning offers pseudo-explanation, and may deceive us into believing we have explained some aspect of temporal phenomena when in fact we have only labeled our ignorance.[8] Time's ontic voltage can't light the smallest bulb.

13.3 The ABCs of Physics

Earlier I said that the approach taken here is a modification of standard practice in philosophy of time, not something wholly new. We can appreciate this if we compare the A-series with manifest time and the B-series with physical time. Strictly speaking they are quite different, although there are crucial similarities. It is for the latter reason that I view the current approach as a more "naturalized" version of some recent work in philosophy of time, e.g., Mellor (1998), than something unconnected with past work.

Let's remind ourselves of McTaggart's time series. The C-series tells us that events are ordered. It tells us that events M, N, O, and P are "in the order M, N, O, P. And they are therefore not in the order M, O, N, P, or O, N, M, P, or in any other possible order" (McTaggart, 1908, 461–2). McTaggart doesn't delve into the mathematical details of order theory, but it's clear that he means that the C-series provides a betweenness relation among triples of events, e.g., that N is between events M and O.

---

[7] Miller (2013) makes this point nicely: "The key challenge for both growing-blockism and presentism is, as I see it, resolving objections in a way that does not undermine the very motivation for accepting either of these views. . . . The issue they face is not that their views cannot be made consistent with sTr, but rather, that the most naturalistically and scientifically respectable ways of doing so radically undermine the motivations for either view. For in making the privileged present empirically undetectable, it becomes very difficult to see how the presence of such a present could be the explanation for our temporal phenomenology, the very thing that motivates both views to posit a privileged present in the first place." Dieks (2016) also presses this point sharply.

[8] These sentences are a paraphrase of Keating's rejection of vitalism in Keating (2002).

The B-theory differs from the C-theory in the addition of a direction. B-relations are two-place asymmetric, irreflexive, and transitive relations. Examples include "event x is earlier than event y" and "event x is later than event y." From these one can define relations such as "event x is simultaneous with event y." B-relations, in McTaggart's example, allow us to distinguish that a series went from event M to N to O to P, and not from P to O to N to M. So if we pick a B-relation, say, earlier than, then it may pick out objectively the order M, N, O, P, where M is earlier than N, N earlier than O, and O earlier than P, and not P, O, N, M. Not much is made of this, but B-relations also encode metrical relations. Relations like "event M is two days earlier than N" are B-relations, and they provide information not only about the direction but also the duration between events M and N.

Finally, the A-series is the one that McTaggart felt embodied true change. In this series, events begin as future, then become present, and then past. Examples of A-properties are the one-place relations "is present or now," "is past" and "is future." McTaggart felt that one needed the A-series to turn the C-series into a B-series; that is, to acquire its directionality, the events in a C-series had to successively have and then lose A-properties. The A-series also encodes metrical ideas: for example, the one-place relation "The year 2000 is ten years past" is an A-property.

Manifest time and the A-series are not identical. The former has many properties that don't follow from the latter. For example, the openness of the future and fixity of the past doesn't follow from changing A-properties, so far as I can see, but they are part of the manifest image. Manifest time also includes dozens of other temporal features, such as (arguably) that time is absolute, that time is indivisible, and much more.

That said, both conceptions of time do share one very important feature: *the objectification of the egocentric temporal distinction of {past, present, future}*. It was McTaggart's great insight to highlight this aspect of our conception of time. His A, B, and C distinctions also turn out to be useful in developmental psychology's study of time acquisition. Although not identical, McTaggart's A-series does grab an important distinction made in manifest time.

Physical time and the B-series are not identical either. Two picky points demonstrate this. First, due to physical time's lack of fundamental directionality, C-time is probably closer to physical time than B-time is. B-time is intrinsically directed, whereas most models of physical time allow that time is fundamentally time reversal invariant; that is, the laws of nature don't care about the direction of time. Obviously, this point is hostage to whatever is the correct theory of fundamental physics. It is anyway picky, as mentioned, for it simply invites switching to C-relations. Second, relativity posits *spatiotemporal* relations, not temporal relations, so strictly speaking relativity contains no B-relations or C-relations. Clearly this is an important difference, although here too it is a little picky because one feels that there should be counterparts to B-relations in relativity theory.

As with manifest time and A-properties, there is a kernel of truth in the thought that "B-time" is more or less physical time. First, both the B-series and physical time plausibly lack the egocentric temporal distinctions of past, present, and future. Neither physics nor B-relations divide the world into past, present, and future. Second, with enough assumptions, updating, and imagination, the *earlier than*, *later than*, and *simultaneous with* relations can recover a surprising amount of structure. The quick way to see this is to notice that, with some updating, the earlier and later than relations will pick out the same net of relations as the *is causally connectible to* relation. Then one need only remind oneself of how powerful the causal theory of time was as developed by Robb, Reichenbach, Grünbaum, Winnie, and others. Recall that Robb and others derived the core structure of Minkowski spacetime from the relation of being causally connectible by a light pulse. Even better, Malament (1977b) shows that, up to conformal factors, one can recover all past- and future-distinguishing spacetimes from the causal precedence relation. So using something akin to B-relations, it is possible to recover an awful lot of spacetime structure. For reasons like this, there is indeed substantial overlap between physical time and a natural heir to B-theoretic time. That said, the causal program failed when extended to relativity in full. Relativistic time (and spacetime) is a lot richer than the net of all *earlier than* and *later than* relations. Even if we limit ourselves to spacetimes where the causal program succeeds, still I think that only one bent on self-defeat would purposefully obscure all of one's resources by packaging them together in a strange ill-suited vocabulary of B-relations. One would never think of doing the same in solving Eddington's "two tables" problem. We shouldn't in the two times problem either.

13.4 Eliminating Tense?

This book has not discharged its alleged obligation to explain the temporal phenomena purged of "tensed" concepts. I try to explain why creatures would model the world with a flowing present and past/future asymmetry. My account uses a mix of philosophy supplemented with biology, physics, cognitive science, and psychology. Not once do I attempt to purify these sciences of any extra-physical notions of time they may employ. Perhaps worse, I often appeal to the knowledge and causal arrows of time in my explanations, but again, I offer no sanitized versions of either. For all I have said, it may be that these sciences require time to be modeled as manifest time. Am I simply begging the question?

I hope not. And my hope isn't unreasonable. Regarding the causal and knowledge asymmetries, there do exist versions of these arrows cleansed of any extra-physical temporal metaphysics. Albert (2000), Loewer (2012), and Kutach (2011) offer some of the more sophisticated ways of understanding the causal and knowledge arrows; these are perfectly compatible with a "tenseless" metaphysics. Indeed, if Albert and Loewer's "imperialistic" picture of statistical mechanics is correct, then all of the special sciences are probabilistic corollaries of an "untensed" probabilistic posit over a special initial condition. Concepts that may seem to involve some aspects of manifest time, such as fitness in natural selection, would be cashed out in terms of this probabilistic posit over initial conditions (which itself could be understood "tenselessly"). If this imperialistic view is correct, then we have a kind of transcendental argument that the above two arrows plus all of the special sciences appealed to here are grounded in a "tenseless" foundation. If incorrect, then we would have to go through the arrows and sciences on a case by case basis. Even on this approach I'm optimistic, as most of the sciences employed here simply use ordinary clock time in their models, a temporal series easily shown to be compatible with physical time.

Be that as it may, I happily concede the point: I have not *shown* that such a cleansing could be done. Consequently, for all that I have argued in the book, it's possible that some new temporal metaphysics is necessary to underwrite the causal and knowledge arrows and perhaps also the various scientific theories used. Fine. Nothing I have done refutes any particular metaphysical model of time. Instead what I have done is give them less to do. Now, rather than going from (say) tensed attitudes about headaches to presentism, one must go via psychology to presentism; instead of going from alleged instances of feeling passage to presentism, one ought to tie presentism somehow to the self; and so on. If the explanations offered here or improved ones become compelling, that is, acquire theoretical and empirical successes, the metaphysician of time will not be free to ignore these explanations. Inevitably this will push back their explanation. To not threaten the science, they must now point out the "hidden" presentist assumptions behind each scientific and philosophical triumph. No one is doing that now, and hence I don't feel the need to burden myself with offering a "cleansed" version of every science I use.

13.5 Conclusion

There is a lesson here for philosophy of time. As the retreats pile up, presentism and other models of time will eventually be out of work. Stuck with existence as their all-purpose means of explanation, metaphysicians of time will never offer rich explanations of phenomena on par with what science provides. As our explanations of the present, flow, and the past/future division grow more sophisticated and testable, presentism and the others will be in a position much like vitalism was in a century ago. Vitalism, the theory that life was due to an *elan vital*, was never *refuted*. Instead it was made unemployed because redundant. As knowledge grew it simply had less to do. Even the demonstration of chemical processes that created organic compounds from inorganic matter didn't falsify the theory—it just made it more contrived. My hunch is that presentism et al. will share this fate. The good news, however, is that recognizing this frees us to tackle time with many more methods and tools.

14

Putting It All Together

Manifest time, I've argued, arises as a natural by-product of our attempt to successfully navigate the many obstacles thrown in our way. We're bombarded by signals from all sides. Did the signals originate in the same event or from different ones? The signal speeds differ, have different features, and impinge on different sensory modalities. Out of all this chaos we must somehow assemble experiences of roughly contemporaneous events. Getting it right has costs (binding thunder with lightning would take too long), as does getting it wrong (not binding speech with lip movement would hamper communication). Evolution shaped these experiential windows over millions of years. Due to hard physical constraints and many softer environmental factors, these evolved psychological presents emerge as widely shared amongst us, even if around the edges they vary widely. Widespread intersubjective agreement about this assembled experiential present tricks us into thinking that the present is objective, a global feature of the universe itself, when it is not.

To keep up with our busy environments, these psychological presents must regularly update themselves. More than that, organisms like us must develop a sophisticated sense of self. Primitive organisms need to know not to eat their arms when hungry and what body parts they control. Long-range planning is a major key to our species' success. We therefore develop not only our spatial boundaries but temporal ones too. For this we develop a sense of an enduring self. We are social creatures. To engage with ourselves and others, we are constantly telling stories about this self, representing it as having goals, as having a particular history, and more. Drawing on temporally asymmetric memories as the ingredients of these narratives, at each moment we regard ourselves as the leading edge of this extended object, this agent acting in the world. Because the story keeps going, this self thinks it is growing through time. This, I claim, is tantamount to conceiving of time as flowing.

Drawing a line between the past and future, we invest quite different properties to each region. This is a natural reaction to the vast physical, causal, and knowledge asymmetries flooding our universe. Evolution will not design creatures that waste energy on trying to cause the past to change when that is effectively impossible for creatures like us. Nor will it make us care about it as much, so we are rigged with a temporally asymmetric affect structure. Chapter 12 showed how deep this wiring goes. Combined with our sense of agency, it's not at all surprising that we paint the future as open and ripe with possibility and the past as dead and settled.

The above theory, in outline, is the explanation I offer of why manifest time emerges in creatures embedded in a world like ours. It is a perfectly sensible way for organisms to model the world, implicitly or explicitly, given the challenges we face. With the above theory in place, we are now in a position to answer the question D. C. Williams asks in his classic 1951 article, "The myth of passage":

> What does the theory [of the manifold] allege except what we find, and what do we find that is not accepted and asserted by the theory? Suppose a pure intelligence, bred outside of time, instructed in the nature of the manifold and the design of the human space-time worm, with its mnemic organization and the strands of world history which flank it, and suppose him incarnated among us: what could he have expected the temporal experience to be like except just about what he actually discovers it to be?

Assume Williams' imagined intelligence is implemented as our IGUS. Embed this organism in the world described. Stand back and ask how it will represent time. Do we expect something resembling manifest time to emerge in its model of the world? While duly acknowledging that I've only scratched the surface of understanding manifest time, I think that the answer is a resounding Yes. We would expect such a creature to divide the world up into past, present, and future, to update this division, to invest deep significance to the past, present, and future, and to believe that time flows. Given the constraints it faces, these emerge as exceedingly natural reactions.

Let me conclude with a few observations.

14.1 Common Structure

The explanations of all three features of manifest time share a common structure. To begin with, notice that the explanations of presentness, the flow, and the past/future asymmetry all rely on physical differences between time and space discussed in the first half of this book. The explanation of the present requires the one-dimensionality of time, the fact that we are always timelike-directed, the mobility asymmetry, signal transmission speeds, plus higher-level facts about the physical arrow of time. Depending upon the present, the flow and past/future asymmetry consequently rely on all of these physical facts as well. In addition, the explanation of why we temporally decenter appeals to a self, and the reason for that is due to our inability to decenter via movement as when we spatially decenter. In other words, many of the same physical features causing us to be "stuck in the now" also cause a situation where only an enduring self would allow temporal decentering. The past/future asymmetry turns out to depend heavily upon the causal and knowledge asymmetries; if you believe that these are connected in some way to the massive thermodynamic or radiation asymmetries of our world, as I do, then that feature also depends crucially upon some physical features. Call all of these physical features the *physical hooks*.

The physical hooks are nowhere near sufficient for explaining manifest time. The hooks don't give us the special features of environments that we need. So we require

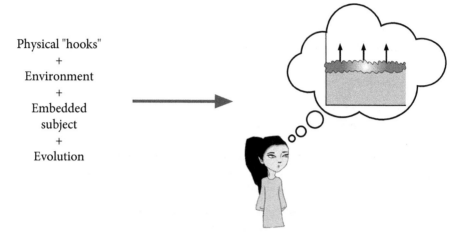

Physical "hooks"
+
Environment
+
Embedded
subject
+
Evolution

Figure 14.1 Explanatory schema

facts about the initial and boundary conditions of our situation. Together the physical hooks plus environment are still nowhere enough. They won't tell us that evolution would wire our emotions certain ways, define the macroscopic scale for creatures with our observational acuity, and give us narrated selves crawling up our worldlines. That demands embedding a subject in this situation and figuring out how it interacts with other subjects and the rest of the physical world. Evolution, for instance, shapes the windows of simultaneity, the boundaries of the self, and trains our emotions. Only when these ingredients are added do we have something approaching an explanation of manifest time.

For each feature of manifest time, the explanatory schema has roughly the form shown in Fig. 14.1.

This schema, I suspect, is what is needed if we're going to answer Einstein's worry in a Carnapian spirit. Seeing the many ingredients necessary, it's no wonder that an explanation has been so long in the waiting. Trying to explain the "flowing now illusion" with physics or metaphysics alone is like trying to bake a cake with only flour.

14.2 A Unified Flowing Now

The full story of the emergence of manifest time is surely more unified than that given here. In Chapter 1, I announced that for the sake of exposition and analysis I would tease apart manifest time into its separate core aspects, namely, a privileged present, flow, and direction. I noted, however, that psychologically and theoretically, manifest time possesses a unity not respected by the violence of this decomposition. Perhaps this is most obvious when we consider the thought that we probably wouldn't believe in a distinguished present if we didn't also update the present regularly. Certainly also the fixity of the past and openness of the future depends on the now slicing the world in two. Since our three features of manifest time all evolve and develop as part of an overall response to the internal and external world, we wouldn't expect these particular features to be all that separable from one another.

As our simultaneity windows evolve—as organisms "decide" what signals to bind together as now—so too is the past/future asymmetry evolving. Discounting the past and distant future has costs and benefits too, yet since the past and future are defined relative to the now, it's biologically naive to think of these conceptions as arising independently of one another. Same goes with memory, the self, and the sense of agency. Although we've taken manifest time apart, presumably the true story fills in many connections.

Having fragmented manifest time, I want to briefly highlight how some of the different parts are in fact intimately connected. They are bound not merely by logic but also via various mechanisms. These are small peeks at the unified character of manifest time. Let me mention two "high-level" connections, and then three surrounding a particular "low-level" perceptual effect.

- *The self and headaches.* In Chapter 12, I played up the temporal asymmetry of attitudes but I downplayed Prior's worry about making sense of the headache being *over*. That worry was, I said, simply the general one about expressing indexical language in non-indexical terms, a worry I dealt with in Chapter 9 regarding the present. That answer was fine, so far as it went, but we can now do better. A "self" crawls up your worldline. At each moment you feel that your self has made it to that moment in time. There is a huge difference between a future and past headache according to this self. Unlike the future one, the past headache is *part of the store of information* from which you built the story of your self. You're uncertain about that future headache and that uncertainty will trigger emotions in an asymmetric manner. But more than that, if my story of passage is correct, there really is a sense in which *you* haven't experienced the future headache yet but have the past one. The past one is *part* of what you now are, whereas the future one isn't. Interacting with the temporal asymmetries mentioned, this will lead you to feel that the past headache is not only fixed and settled but also *over*. That headache is already part of the story.

- *Narrative and openness.* Note also that the past headache is not only over, but it is "fixed" whereas the future one is "open." I appealed to the narrative self as a way of accounting for why we believe time itself to be dynamic. I think it can do more than that. Via its sense of agency and the sense of agency's connection to perceived freedom, I believe that it is an important part of why we feel the future is open. We can also see how it contributes to our sense of the fixity of the past. Lloyd (2008) makes the point that the autobiographer is constantly transforming the past into something that is fixed or necessary to fit the developing story, which is always told from the present tense. We are "impos[ing] a pattern of necessity on the fragments" (p. 264) that constitute our lives, cemented in place to fit into the narrative. Narration may not be the key to unlocking why we believe the past is settled and future open, but I suspect that it plays a supporting role.

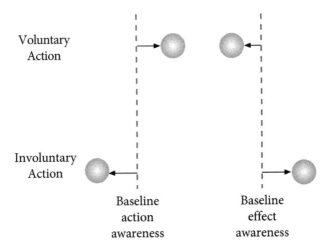

Figure 14.2 Intentional binding

The perceptual effect known as *intentional binding* connects many of the features discussed together. Put most generally, this effect vividly and robustly displays connections between our *organization of the world into a time series* and the degree to which *we take ourselves to cause external events*. In particular: intentionality shortens the perceived time interval between actions and their sensory outcomes (Fig. 14.2). Experiments by Haggard et al. (2002) first revealed this phenomenon, which has subsequently been replicated many times and across many sensory modalities. In the original experiment, participants pressed keys that created, after a fixed lag of 250 ms, audio tones. Using a so-called Libet clock, subjects were asked to mark the clock time of the key press or the audio tone, but not both in the same trial. The other variable of relevance is that some of the key presses were voluntary and others involuntary (induced by transcranial magnetic stimulation). Together these cases demonstrated a remarkable pattern: subjects perceived a sensory effect closer in time if they believed that they were the cause of that effect. Both the perceived time of the key pressing and the sensory effect seemed to shift toward one another. Perceived authorship of an effect thus bound it perceptually closer in time to the now. We use temporal contiguity to determine causation, as Hume thought, but we also use causation and agency to determine temporal contiguity. Overall, the effect is so robust that some have used intentional binding as a measure of authorship, i.e., how much one feels they caused an external effect.

- *Agency and the present.* In Chapters 9 and 10 we looked at the construction of an objective-seeming present, witnessing how we turn an assembly of discordant signals into a coherent now. Despite the confusing assault on our senses, we manage to organize a set of phenomena onto a canvas of space and time. In Section 11.2, I appealed to the sense of self and its sense of agency. It is the self that exists wholly in the moment. And it is the self that feels a sense of agency when outcomes conform with predictions. While the sensory effect under consideration in intentional binding is later and not present, the larger point is that one's organization of phenomena in time can hang in part on one's sense of agency. Here there is a connection between the self, who takes herself to be an agent authoring many events in the world, and the temporal organization of phenomena discussed in Chapter 9.

- *Intentional binding and the future.* Recent work may also connect intentional binding to our belief that the future is open. Barlas and Obhi (2013) modified the experimental protocol to give participants many, few, or no options on which buttons to press. When presented with many options ("greater freedom"), the binding effect was strongest, and when presented with no options ("no freedom"), the binding effect was least strong. Here agency, time, and freedom all seem linked (although how and why are open questions).

- *Intentional binding and past/future bias.* It's also possible to find connections between intentional binding and the thesis of the last chapter. Emotions, most now think, exist thanks to evolution and its need to prepare organisms for action. The sense of agency is viewed similarly. The thought is that it motivates goal-directed behavior. It's perhaps not so surprising to find that one can link the valence of emotions to the sense of agency via intentional binding. Do you cause that bad outcome? No! That good outcome? Yes! Not only do we take ownership of the good, but we literally bring it closer to ourselves in time, just as we literally make distant the bad (Yoshie, 2013).

As mentioned, these are simply a few peeks at some noteworthy connections amongst the different strands of manifest time. I have no doubt that readers can discover many more.

14.3 Animals

Responding to many of the same pressures as we have, many non-human animals probably employ a similar model of time. They too need to organize the temporal structure of their environment, and they too suffer from the same temporal asymmetries as we do. Indeed, I believe that we already have a small amount of evidence that this is right. Plenty of studies have been performed on the temporal discounting of non-human animals, such as various primates and even jays. Do cotton-top tamarins suffer from an immediacy bias like we do? And if so, what is its mathematical form? These questions can be answered with experiment (for some discussion and references, see Santos and Rosati, 2015). Whatever the particular answers here, notice that temporal discounting is properly conceived of as *tensed*; that is, the discounting is measured from the now or present, not from a particular clock or calendar time. If this is so– and I admit that these inferences are fraught with speculation—it shows that some non-human animals are at least carving up the world into the tensed tripartite division of {past, present, future} like we do.

Few animals will have such a rich sense of self, freedom, etc., as we do, so we can expect plenty of differences. Perhaps time doesn't flow in any interesting sense to the lobster. We can certainly expect differences in the psychological presents of many creatures. Some animals live in the sea, where the speed of sound is five times faster than in air, and some animals (giraffes) have 3 m nerves from their feet to their brains. Given the variance even among people, plus the different environments and pressures on animals, it would be startling if the temporal integration windows of non-human animals didn't vary a great deal. Perhaps, as I've mentioned, the giraffe has not only a long neck but a long now (More et al., 2013). When it comes to an implicit model of time in non-human creatures, I suspect we'll find what we've always found from the animal kingdom: a huge spectrum of surprising differences and commonalities.

14.4 An Illusion?

Is the flowing now an illusion? Those who deny that manifest time tracks real properties of physical or metaphysical time are often said to hold that presentness and passage are illusions. Some advocates of a mind-dependence theory of becoming embrace this label. Passage and the now are not "out there" in the world, but rather are a kind of misapprehension due to our cognitive and perceptual limitations. Is this the right way to understand the present theory?

Not really. Many features of experience have been described in this book. On a proper accounting—which the interested reader can readily work out—*none* of them turn out to be illusions. An illusion is defined, roughly, as a perceptual experience of an object where the object perceptually appears to have properties it doesn't actually have. In the famous Müller–Lyer illusion, for instance, one line looks longer than the other despite the two lines actually being the same length. The current theory allows that some features associated with the now and passage are part of our phenomenology and denies that others are. Often this division is controversial. But it turns out that the former are not illusions because they are not generally misapprehensions. Candidates for temporal phenomenology included, for example, sensed motion, sensed synchrony, and sensed duration. But these generally track genuine motion, genuine synchrony, and genuine duration. Of course matters can go awry, but no more so here than elsewhere in perception, and we don't classify all perception as illusory. The other aspects of the flowing now are not illusions because they are not features of perceptual experience but instead part of our implicit model of the world, e.g., time flow. Strictly speaking, nothing we counted as an experience of the now or of passage turned out to be illusory.

Little hangs on this point. In some very loose and coarse-grained sense, manifest time might be called an illusion without any harm done. However, for many of its aspects, it's a bit like calling our impression of a shape an illusion, and that seems wrong.

14.5 Conclusion

What is the value of philosophical speculation about time? My own view is similar to Michael Friedman's answer to this question more generally about philosophy. The job of philosophy, he thinks, is to provide meta scientific perspectives that open doors to new possibilities for science. Friedman gives historical examples, such as the philosophical discussion of geometry and space from Kant through Helmholtz and Poincaré making Einstein's discoveries possible. Clark Glymour, in a blog post,[1] provides more contemporary examples, such as Putnam's work leading to computational learning theory and Lewis' work on convention answering how meaning could arise amongst communicators. One can also point to work by Fodor and Chomsky providing the theoretical backdrop to a move from behaviorism to cognitive psychology.

I cannot accomplish anything as significant. Aspirationally, however, the present work shares this kind of methodological goal. It is metascientific in the sense that it situates the problem of time within a long-standing philosophical discussion, tackles problems not dealt with by any single science, and appeals to work across a wide variety of sciences. Explaining manifest time falls between the cracks of physics and psychology, broadly speaking. Even when one science is concerned, the questions presently asked are outside the norm. Without the proposed theoretical framework, physicists probably wouldn't explore "mixed" initial value problems that push data in a spatial direction. Psychologists wouldn't apply their theories of the self to the problem of time's passage. Only a philosopher would ask these questions. Yet for one with a clear idea of the problem of time in mind, they turn out to be natural questions for anyone wanting to understand time.

My goal has been to stimulate modest new theories and questions, some of which I hope lead to testable claims. The Hume–Parfit "causal" answer to Prior's "thank goodness" objection already has led to a thriving and interesting research program on time biases in psychology. My variant on this program suggests further testable hypotheses, e.g., that emotional attachment may mediate temporal bias shift from first to third person. My theory of the flow of time suggests various correlations that should show up in childhood development amongst independent diagnostic measures of self, memory, and kinds of temporal aptitude (e.g., decentering). Looking at the world sideways in mathematical physics may trigger new work on time's role in physics. Work on the subjective present suggests a framework for watching perceived simultaneity evolve in various game theoretic ways and also throughout the animal kingdom in different environments. I also lay down various challenges to interpretations of quantum gravity seeking to either eliminate or rescue time.

Should we be satisfied with these advances? I haven't answered the "big" questions. Do future events "already" exist? If the question makes sense, I'm a bit skeptical that its

---

[1] http://choiceandinference.com/2011/12/23/in-light-of-some-recent-discussion-over-at-new-apps-i-bring-you-clark-glymours-manifesto/.

answer will explain much. Does my failure to address this question prohibit progress on time? On this question I have an opinion much like that recently expressed by the computer scientist Scott Aaronson:

> whenever it's been possible to make definite progress on ancient philosophical problems, such progress has almost always involved a [kind of] "bait-and-switch." In other words: one replaces an unanswerable philosophical riddle Q by a "merely" scientific or mathematical question Q', which captures part of what people have wanted to know when they've asked Q. Then, with luck, one solves Q'... [T]his process of "breaking off" answerable parts of unanswerable riddles, then trying to answer those parts, is the closest thing to philosophical progress that there is.

Aaronson presumably has questions of logic and computability in mind, but I think it's often the case that philosophical progress is won this way. This picture fits Friedman's understanding of progress too, as often it's by asking some metascientific question Q that one comes across the solvable Q'.

To some extent that is what we have done here. Philosophy asks whether the future is real, whether time flowing is coherent, and much more. These are the big Q questions. Consideration of these leads to the smaller Q' questions. We begin asking why only the present seems real and end up answering smaller queries about the various physical, biological, and psychological facts that lead to a creature feeling stuck in a moment. We ask why we dread future but not past headaches and answer a related question about how emotions and more are tied to the causal and knowledge asymmetries. We ask how time differs from space and answer a related question of whether laws of nature can be productive deterministic in a non-temporal direction. Many Q' questions await replies. With luck, some of these new Q' will turn out to be fruitful to ask.

Sometimes answering the Q' questions leads one to reflect on the big Q questions and see them as less interesting. The big Q question that motivated biological vitalism, the theory that posited an elan vital, demanded an explanation for what animates dead matter. Biochemistry then made this question less interesting. Whether viruses are alive became more a terminological dispute than a mystery to be solved. The same may be the case with some of the big Q questions in philosophy of time. The eternalism debate may itself be eternal. While we wait for the verdict, I suggest we tackle some of the low hanging fruit dangling before us, as I've only scratched the surface of the philosophical applications of the burgeoning science of time.

Bibliography

Adelson, E. H. and Bergen, J. R. (1985). Spatiotemporal energy models for the perception of motion. *Journal of the Optical Society of America*, 2(2):284–99.

Aharonov, Y. and Albert, D. Z. (1981). Can we make sense out of the measurement process in relativistic quantum mechanics? *Physical Review D*, 24:359–70.

Alais, D. and Burr, D. (2004). The ventriloquist effect results from near-optimal bimodal integration. *Current Biology*, 14(3):257–62.

Albert, D. Z. (1992). *Quantum Mechanics and Experience*. Harvard University Press, Cambridge, MA.

Albert, D. Z. (2000). *Time and Chance*. Harvard University Press, Cambridge, MA.

Albertazzi, L. (1999). The time of presentness. A chapter in positivistic and descriptive psychology. *Axiomathes*, 10(1): 49–73.

Ames, K. A. and Straughan, B. (1997). *Non-Standard and Improperly Posed Problems*. Academic Press, New York.

Ananthaswamy, A. (2012). Time flows uphill for remote Papua New Guinea tribe. *New Scientist*, 2867:14.

Andersen, H. and Grush, R. (2009). A brief history of time-consciousness: Historical precursors to James and Husserl. *Journal of the History of Philosophy*, 47(2):277–307.

Anderson, E. (2007). Emergent semiclassical time in quantum gravity. i. mechanical models. *Classical and Quantum Gravity*, 24:2935–78.

Andersson, L., Barbot, T., Béguin, F., and Zeghib, A. (2012). Cosmological time versus CMC time in spacetimes of constant curvature. *Asian Journal of Mathematics*, 16(1):37–88.

Anonymous (2005). So much more to know. *Science*, 309(5731):78–102.

Arageorgis, A. (2012). Spacetime as a causal set: Universe as a growing block? Unpublished.

Arnold, D. H. and Yarrow, K. (2011). Temporal recalibration of vision. *Proceedings of the Royal Society B: Biological Sciences*, 278(1705):535–8.

Arntzenius, F. and Greaves, H. (2009). Time reversal in classical electromagnetism. *The British Journal for the Philosophy of Science*, 60(3):557–84.

Arntzenius, F. and Maudlin, T. (2002). Time travel and modern physics. In Callender, C., editor, *Time, Reality, and Experience*, pages 169–200. Cambridge University Press, Cambridge.

Arntzenius, F. and Maudlin, T. (2009). Time travel and modern physics. In Zalta, E. N., editor, *The Stanford Encyclopedia of Philosophy*. Spring 2009 edition. http://plato.stanford.edu/archives/spr2009/entries/time-travel-phys/.

Asgeirsson, L. (1937). Über eine Mittelwertseigenscaff von Losungen homogener linearer partieller Differentialgleichungen 2. Ordnungmit konstanten Koeffizienten. *Mathematische Annalen*, 113:321–46.

Aspect, A., Grangier, P., and Roger, G. (1981). Experimental tests of realistic local theories via Bell's theorem. *Physical Review Letters*, 47:460–3.

Atance, C. M. and O'Neill, D. K. (2005). The emergence of episodic future thinking in humans. *Learning and Motivation*, 36(2):126–44.

Audoin, C. and Guinot, B. (2001). *The Measurement of Time: Time, Frequency, and the Atomic Clock*. Cambridge University Press, Cambridge.

Aveni, A. (2002). *Empires of Time: Calendars, Clocks, and Cultures*. University Press of Colorado, Boulder, CO.

Bach, K. (1994). Conversational implicature. *Mind & Language*, 9:124–62.

Baker, D. (2007). Measurement outcomes and probability in everettian quantum mechanics. *Studies in History and Philosophy of Science Part B*, 38(1):153–69.

Balashov, Y. (2005). Times of our lives: Negotiating the presence of experience. *American Philosophical Quarterly*, 42:295–309.

Banks, T. (1985). TCP, quantum gravity, the cosmological constant and all that.... *Nuclear Physics B*, 249(2):332–60.

Bar-Anan, Y., Wilson, T. D., and Gilbert, D. T. (2009). The feeling of uncertainty intensifies affective reactions. *Emotion*, 9(1):123–7.

Barbour, J. B. (1993). Time and complex numbers in canonical quantum gravity. *Physical Review D*, 47:5422–9.

Barbour, J. B. (1999). *The End of Time: The Next Revolution in Physics*. Oxford University Press, Oxford.

Barlas, Z. and Obhi, S. S. (2013). Freedom, choice and the sense of agency. *Frontiers in Human Neuroscience*, 7(514).

Barrett, J. A. (1999). *The Quantum Mechanics of Minds and Worlds*. Oxford University Press, Oxford.

Barrett, T. (2015). Spacetime structure. *Studies in History and Philosophy of Modern Physics*, 51:37–53.

Bayne, T. (2008). The phenomenology of agency. *Philosophy Compass*, 3(1):182–202.

Bechara, A., Damasio, A. R., Damasio, H., and Anderson, S. W. (1994). Insensitivity to future consequences following damage to human prefrontal cortex. *Cognition*, 50:7–12.

Bechara, A., Tranel, D., Damasio, H., and Damasio, A. R. (1996). Failure to respond automatically to anticipated future outcomes following damage to prefrontal cortex. *Cerebral Cortex*, 6:215–25.

Beck, J. (1966). Effect of orientation and of shape similarity on perceptual grouping. *Perception & Psychophysics*, 1:300–2.

Bell, J. S. (1964). On the Einstein Podolsky Rosen paradox. *Physics*, 1(3):195–200.

Bell, J. S. (1987a). Are there quantum jumps? In Bell (1987b), chapter 22, pages 201–12.

Bell, J. S. (1987b). *Speakable and Unspeakable in Quantum Mechanics*. Cambridge University Press, Cambridge.

Bell, J. S. (1987c). Speakable and unspeakable in quantum mechanics. In Bell (1987b), pages 169–72.

Belot, G. (2011). *Geometric Possibility*. Oxford University Press, Oxford.

Belot, G. and Earman, J. (2001). Pre-Socratic quantum gravity. In Callender and Huggett (2001), chapter 10, pages 213–55.

Benussi, V. (1913). *Psychologie der Zeitauffassung*. Winter: Heidelberg.

Bergmann, P. G. (1961). Observables in general relativity. *Reviews of Modern Physics*, 33:510–14.

Bergson, H. (1922). *Durée et Simutanéité*. Clinamen Press Ltd., Manchester.

Berndl, K., Dürr, D., Goldstein, S., and Zanghì, N. (1996). Nonlocality, Lorentz invariance, and Bohmian quantum theory. *Physical Review A*, 53:2062–73.

Bishop, R. C. (2003). On separating predictability and determinism. *Erkenntnis (1975–)*, 58(2):169–88.

Block, R. A. (1990). Models of psychological time. In Block, R. A., editor, *Cognitive Models of Psychological Time*, pages 1–35. Lawrence Erlbaum Associates, Mahwah, NJ.

Boenke, L., Deliano, M., and Ohl, F. (2009). Stimulus duration influences perceived simultaneity in audiovisual temporal-order judgment. *Experimental Brain Research*, 198(2–3): 233–44.

Bohm, D. and Hiley, B. J. (1993). *The Undivided Universe: An Ontological Interpretation of Quantum Theory*. Routledge, London.

Bojowald, M., Morales-Técotl, H. A., and Sahlmann, H. (2005). Loop quantum gravity phenomenology and the issue of Lorentz invariance. *Physical Review D*, 71:084012.

Bondi, H., van de Burg, M. G. J., and Metzner, A. W. K. (1962). Gravitational waves in general relativity. VII. Waves from axi-symmetric isolated systems. *Proceedings of the Royal Society of London A*, 269:21–52.

Boolos, G. S. and Jeffrey, R. C. (1989). *Computability and Logic*. Cambridge University Press, Cambridge, 3rd edition.

Boroditsky, L. (2000). Metaphoric structuring: understanding time through spatial metaphors. *Cognition*, 75(1):1–28.

Boroditsky, L., Schmidt, L., and Phillips, W. (2003). Sex, syntax and semantics. In Gentner, D. and Goldin-Meadow, S., editors, *Language in Mind: Advances in the Study of Language and Thought*, pages 61–78. The MIT Press, Cambridge, MA.

Borstnik, N. M. and Nielsen, H. B. (2002). Why nature has made a choice of one time and three space coordinates? *Journal of Physics A: Mathematical and General*, 35(49):10563–73.

Bourne, C. (2004). Becoming inflated. *The British Journal for the Philosophy of Science*, 55(1): 107–19.

Bricker, P. (2004). Discussion – McGinn on non-existent objects and reducing modality. *Philosophical Studies*, 113(3):439–51.

Brightwell, G., Dowker, F., Garcia, R., Henson, J., and Sorkin, R. (2003). "Observables" in causal set cosmology. *Physical Review D*, 67:084031.

Brink, D. O. (2011). Prospects for temporal neutrality. In Callender (2011a), pages 353–81.

Broad, C. D. (1923). *Scientific Thought*. Routledge & Kegan Paul, London.

Broad, C. D. (1933). *Examination of McTaggart's Philosophy. Volume 1*. Cambridge University Press, Cambridge.

Broad, C. D. (1938). *Examination of McTaggart's Philosophy. Volume 2*. Cambridge University Press, Cambridge.

Broad, C. D. (1959). A reply to my critics. In Schilpp, P., editor, *The Philosophy of C. D. Broad*. Open Court, New York.

Broekaert, J. B. (2007). A Lorentz-Poincaré type interpretation of the weak equivalence principle. *International Journal of Theoretical Physics*, 46(6):1722–37.

Brown, H. R. (2001). The origins of length contraction: I. the FitzGerald–Lorentz deformation hypothesis. *American Journal of Physics*, 69(10):1044–54.

Brun, W. and Teigen, K. H. (1990). Prediction and postdiction preferences in guessing. *Journal of Behavioral Decision Making*, 3(1):17–28.

Buller, D. J. and Hardcastle, V. G. (2007). Evolutionary psychology, meet developmental neurobiology: Against promiscuous modularity. *Brain and Mind*, 1(3):307–25.

Burns, Z. C., Caruso, E. M., and Bartels, D. M. (2012). Predicting premeditation: Future behavior is seen as more intentional than past behavior. *Journal of Experimental Psychology: General*, 141(2):227–32.

Butterfield, J. (1984). Seeing the present. *Mind*, 93:161–76.

Butterfield, J. (2013). On time in quantum physics. In Bardon, A. and Dyke, H., editors, *A Companion to the Philosophy of Time*, pages 220–41. Wiley-Blackwell, Chichester.

Callahan, J. (2000). *The Geometry of Spacetime*. Springer, New York.

Callender, C. (2000). Shedding light on time. *Philosophy of Science*, 67:S587–S599.

Callender, C. (2007). The emergence and interpretation of probability in Bohmian mechanics. *Studies in History and Philosophy of Science Part B: Studies in History and Philosophy of Modern Physics*, 38(2):351–70.

Callender, C. (2008). The common now. *Philosophical Issues*, 18(1):339–61.

Callender, C. (2010). Is time an illusion? *Scientific American*, June:58–65.

Callender, C. (2011a). Metaphysics and philosophy of science. In French, S. and Saatsi, J., editors, *Continuum Companion to the Philosophy of Science*, pages 33–54. Continuum, London.

Callender, C., editor (2011b). *The Oxford Handbook of Philosophy of Time*. Oxford University Press, Oxford.

Callender, C. (2011c). The past histories of molecules. In Beisbart, C. and Hartmann, S., editors, *Probabilities in Physics*, chapter 4, pages 83–114. Oxford University Press, Oxford.

Callender, C. and Cohen, J. (2010). Special sciences, conspiracy and the better best system account of lawhood. *Erkenntnis*, 73(3):427–47.

Callender, C. and Huggett, N., editors (2001). *Physics Meets Philosophy at the Planck Scale: Contemporary Theories in Quantum Gravity*. Cambridge University Press, Cambridge.

Callender, C. and Weingard, R. (1996). Time, Bohm's theory, and quantum cosmology. *Philosophy of Science*, 63(3):470–4.

Callender, C. and Weingard, R. (1997). Trouble in paradise? Problems for Bohm's theory. *The Monist*, 80(1):24–43.

Campbell, J. (1994). *Past, Space and Self*. MIT Press, Cambridge, MA.

Campbell, J. (1997). The structure of time in autobiographical memory. *European Journal of Philosophy*, 5(2):105–18.

Campbell, S. (1990). Circadian rhythms and human temporal experience. In Block, R. A., editor, *Cognitive Models of Psychological Time*, pages 101–18. Lawrence Erlbaum Associates, Mahwah, NJ.

Čapek, M. (1991). *The New Aspects of Time: Its Continuity and Novelties*. Boston Studies in the Philosophy of Science. Kluwer Academic Publishers, Dordrecht.

Caponigro, I. and Cohen, J. (2011). On collection and covert variables. *Analysis*, 71(3):478–88.

Carnap, R. (1963). Carnap's intellectual biography. In Schilpp, P., editor, *The Philosophy of Rudolph Carnap*, pages 3–84. Open Court, La Salle, IL.

Carroll, J. W. and Markosian, N. (2010). *An Introduction to Metaphysics*. Cambridge University Press, Cambridge.

Carroll, S. (2003). *Spacetime and Geometry*. Addison-Wesley, Boston, MA.

Cartan, É. (1923). Sur les variétés a connexion affine et la théorie de la relativité généralisée. *Annales Scientifiques de l'Ecole Normale Supérieure*, 40:325–412.

Cartan, É. (1924). Sur les variétés a connexion affine et la théorie de la relativité généralisée. *Annales Scientifiques de l'Ecole Normale Supérieure*, 41:1–25.

Caruso, E. M. (2010). When the future feels worse than the past: A temporal inconsistency in moral judgment. *Journal of Experimental Psychology: General*, 139(4):610–24.

Caruso, E. M., Gilbert, D. T., and Wilson, T. D. (2008). A wrinkle in time: Asymmetric valuation of past and future events. *Psychological Science*, 19(8):796–801.

Caruso, E. M., Van Boven, L., Chin, M., and Ward, A. (2013). The temporal doppler effect: When the future feels closer than the past. *Psychological Science*, 24(4):530–6.

Casasanto, D. (2008). Who's afraid of the big bad Whorf? Crosslinguistic differences in temporal language and thought. *Language Learning*, 58:63–79.

Castañeda, H.-N. (1966). 'He': A study in the logic of self-consciousness. *Ratio*, 7:130–57.

Castañeda, H.-N. (1967). Indicators and quasi-indicators. *American Philosophical Quarterly*, 4(2):85–100.

Chalmers, D. J. (1996). *The Conscious Mind: In Search of a Fundamental Theory*. Oxford University Press, Oxford.

Choquet-Bruhat, Y., Chruściel, P. T., and Martín-García, J. M. (2011). An existence theorem for the Cauchy problem on a characteristic cone for the Einstein equations. *Contemporary Mathematics*, 554:73–81.

Choquet-Bruhat, Y. and Geroch, R. (1969). Global aspects of the Cauchy problem in general relativity. *Communications in Mathematical Physics*, 14(4):329–35.

Chou, C. W., Hume, D. B., Rosenband, T., and Wineland, D. J. (2010). Optical clocks and relativity. *Science*, 329(5999):1630–3.

Christensen, F. M. (1993). *Space-Like Time: Consequences Of, Alternatives To, and Arguments Regarding the Theory that Time is Like Space*. University of Toronto Press, Toronto.

Christian, J. (2001). Why the quantum must yield to gravity. In Callender and Huggett (2001), chapter 14, pages 305–38.

Clifton, R. and Hogarth, M. (1995). The definability of objective becoming in Minkowski spacetime. *Synthese*, 103(3):355–87.

Cockburn, D. (1998). Tense and emotion. In Le Poidevin, R., editor, *Questions of Time and Tense*, pages 77–91. Clarendon Press, Oxford.

Cohen, J. (2009). *The Red and the Real*. Oxford University Press, New York.

Cohen, J. (2012). Computation and the ambiguity of perception. In Hatfield, G. and Allred, S., editors, *Visual Experience: Sensation, Cognition, and Constancy*, pages 160–76. Oxford University Press, New York.

Cohen, J. and Callender, C. (2009). A better best system account of lawhood. *Philosophical Studies*, 145(1):1–34.

Colonius, H. and Diederich, A. (2010). The optimal time window of visual-auditory integration: A reaction time analysis. *Frontiers in Integrative Neuroscience*, 4(11).

Cosmides, L. and Tooby, J. (1997). The modular nature of human intelligence. In Scheibel, A. and Schopf, J. W., editors, *The Origin and Evolution of Intelligence*, pages 71–101. Jones and Bartlett, Boston, MA.

Courant, R. and Hilbert, D. (1962). *Methods of Mathematical Physics*, volume II. Interscience, New York.

Craig, W. L. (1996). The new B-theory's *tu quoque* argument. *Synthese*, 107:249–69.

Craig, W. L. (1999). Tensed time and our differential experience of the past and future. *Southern Journal of Philosophy*, 37:515–37.

Craig, W. L. (2000). *The Tensed Theory of Time*. Kluwer Academic Publishers, Dordrecht.

Craig, W. L. and Smith, Q., editors (2008). *Einstein, Relativity, and Absolute Simultaneity.* Routledge, New York.

Craig, W. L. and Weinstein, S. (2009). On determinism and well-posedness in multiple time dimensions. *Proceedings of the Royal Society A: Mathematical, Physical and Engineering Science*, 465(2110):3023–46.

Crisp, T. M. (2008). Presentism, eternalism, and relativity physics. In Craig and Smith (2008), pages 262–78.

Cromer, R. F. (1971). The development of the ability to decenter in time. *British Journal of Psychology*, 62(3):353–65.

Cunningham, D., Billock, V., and Tsou, B. (2001). Sensorimotor adaptation to violations of temporal contiguity. *Psychological Science*, 12:532–5.

Dainton, B. (2001). *Time and Space.* McGill Press, Montreal.

Damasio, A. R. (1994). *Descartes' Error: Emotion, Reason, and the Human Brain.* G. P. Putnam and Sons, New York.

Damasio, A. R., Everitt, B. J., and Bishop, D. (1996). The somatic marker hypothesis and the possible functions of the prefrontal cortex [and discussion]. *Philosophical Transactions of the Royal Society of London. Series B: Biological Sciences*, 351(1346):1413–20.

Darmois, G. (1927). *Les équations de la gravitation einsteinienne*, volume Mémorial des Sciences Mathématiques. Gauthier-Villars.

Dautcourt, G. (1963). Matter and gravitational shock waves in general relativity. *Archive for Rational Mechanics and Analysis*, 13(1):55–8.

Deng, N. (2013). On explaining why time seems to pass. *The Southern Journal of Philosophy*, 51(3):367–82.

Dennett, D. C. (1989). The origins of selves. *Cogito*, 3:163–73.

Dennett, D. C. (1992). The self as a center of narrative gravity. In *Self and Consciousness: Multiple Perspectives*, pages 103–115. Lawrence Erlbaum Associates, Mahwah, NJ.

DeWitt, B. S. (1967). Quantum theory of gravity. I. The canonical theory. *Physical Review*, 160:1113–48.

Dieks, D. (2016). Physical time and experienced time. In Dolev, Y. and Roubach, M., editors, *Cosmological and Psychological Time*, volume 285 of Boston Studies in the Philosophy and History of Science, pages 3–20. Springer International Publishing, Heidelberg.

Dixon, N. and Spitz, L. (1980). The detection of auditory visual desynchrony. *Perception*, 9:719–21.

Dorato, M. (2006). The irrelevance of the presentism eternalism debate for the ontology of Minkowski spacetime. In Dieks, D., editor, *The Ontology of Spacetime*, pages 93–109, Elsevier, Amsterdam.

Dorato, M. (2011). The Alexandroff present and Minkowski spacetime: Why it cannot do what it has been asked to do. In Dieks, D., Gonzalo, W., Uebel, T., Hartmann, S., and Weber, M., editors, *Explanation, Prediction, and Confirmation*, pages 379–94, Springer, Dordrecht.

Dorato, M. (2015). Presentism and the experience of time. *Topoi*, 34(1):265–75.

Dorling, J. (1970). Energy-momentum conservation for collision, creation, and annihilation processes: Geometrical derivation from a very simple classical action principle. *American Journal of Physics*, 38(8):1023–8.

Dowker, F. (2004). Quantum gravity phenomenology, Lorentz invariance and discreteness. *Modern Physics Letters*, A19:1829–40.

Dowker, F. (2014). The birth of spacetime atoms as the passage of time. Unpublished.

Dray, T., Ellis, G., Hellaby, C., and Manogue, C. (1997). Gravity and signature change. *General Relativity and Gravitation*, 29(5):591–7.

Dretske, F. I. (1962). Moving backward in time. *Philosophical Review*, 71(1):94–8.

Dummett, M. (1960). A defense of McTaggart's proof of the unreality of time. *The Philosophical Review*, 69(4):497–504.

Durgin, F. H. and Sternberg, S. (2002). The time of consciousness and vice versa. *Consciousness and Cognition*, 11(2):284–90.

Durkheim, E. (1912). *The Elementary Forms of Religious Life*. Oxford University Press, Oxford.

Dürr, D., Goldstein, S., Münch-Berndl, K., and Zanghì, N. (1999). Hypersurface Bohm-Dirac models. *Physical Review A*, 60:2729–36.

Dürr, D., Goldstein, S., and Zanghì, N. (1992). Quantum equilibrium and the origin of absolute uncertainty. *Journal of Statistical Physics*, 67(5–6):843–907.

Dyke, H. (2003). Temporal language and temporal reality. *The Philosophical Quarterly*, 53(212):380–91.

Eagleman, D. and Holcome, A. (2002). Causality and the perception of time. *Trends in Cognitive Science*, 6:323–6.

Eakin, P. J. (2008). *Living Autobiographically: How we Create Identity in Narrative*. Cornell University Press, Ithaca, NY.

Earman, J. (1972). Notes on the causal theory of time. *Synthese*, 24(1/2):74–86.

Earman, J. (1984). Laws of nature: The empiricist challenge. In Bogdan, R., editor, *D.M. Armstrong*, volume 4 of *Profiles*, pages 191–223. Springer, Netherlands.

Earman, J. (1986). *A Primer on Determinism*, volume 32 of The Western Ontario Series in Philosophy of Science. D. Reidel Pub. Co., Dordrecht.

Earman, J. (1995). *Bangs, Crunches, Whimpers, and Shrieks: Singularities and Acausalities in Relativistic Spacetimes*. Oxford University Press, Oxford.

Earman, J. (2006). Aspects of determinism in modern physics. In Butterfield, J., Earman, J., Gabbay, D. M., Thagard, P., and Woods, J., editors, *Handbook of the Philosophy of Science: Philosophy of Physics*, pages 1369–434. North Holland, Amsterdam.

Earman, J. (2008). Reassessing the prospects for a growing block model of the universe. *International Studies in the Philosophy of Science*, 22:135–64.

Earman, J. (2011). Sharpening the electromagnetic arrow(s) of time. In Callender (2011a), pages 33–54.

Earman, J. and Norton, J. D. (1998). Exorcist XIV: The wrath of Maxwell's demon. Part I. From Maxwell to Szilard. *Studies in History and Philosophy of Science Part B: Studies in History and Philosophy of Modern Physics*, 29(4):435–71.

Earman, J. and Norton, J. D. (1999). Exorcist XIV: The wrath of Maxwell's demon. Part II. From Szilard to Landauer and beyond. *Studies in History and Philosophy of Science Part B: Studies in History and Philosophy of Modern Physics*, 30(1):1–40.

Eddington, A. S. (1928). *The Nature of the Physical World*. Cambridge University Press, Cambridge.

Eijk, R. L. J., Kohlrausch, A., Juola, J. F., and Par, S. (2008). Audiovisual synchrony and temporal order judgments: Effects of experimental method and stimulus type. *Perception & Psychophysics*, 70(6):955–68.

Ellis, G. F. R. (2006). Physics in the real universe: Time and spacetime. *General Relativity and Gravitation*, 38(12):1797–824.

Emerson, R. W. (1836). *Nature*. James Munroe and Company, Boston, MA.

Ernst, M. (2005). A Bayesian view on multimodal cue integration. In Knoblich, G., Thornton, I., Grosjean, M., and Shiffrar, M., editors, *Human Body Perception From the Inside Out*, pages 105–31. Oxford University Press, New York.

Evans, G. (1982). *The Varieties of Reference*. Oxford University Press, Oxford.

Evans, V. (2003). *The Structure of Time: Language, Meaning and Temporal Cognition*. John Benjamins Publishing, Amsterdam.

Falk, A. (2003). Time plus the whoosh and whiz. In Jokić and Smith (2003), chapter 7, pages 211–50.

Fendrich, R. and Corballis, P. M. (2001). The temporal cross-capture of audition and vision. *Perception & Psychophysics*, 63(4):719–25.

Fine, K. (2005). Tense and reality. In *Modality and Tense: Philosophical Papers*. Oxford University Press, Oxford.

Fiocco, M. O. (2007). Passage, becoming and the nature of temporal reality. *Philosophia*, 35:1–21.

Fischhoff, B. and Beyth, R. (1975). I knew it would happen: Remembered probabilities of once-future things. *Organizational Behavior and Human Performance*, 13(1):1–16.

Fitzgerald, P. (1985). Four kinds of temporal becoming. *Philosophical Topics*, 13(3):145–77.

Fivush, R. (2001). Owning experience: The development of subjective perspective in autobiographical memory. In Moore, C. and Lemmon, K., editors, *The Self in Time: Developmental Perspectives*, pages 35–52. Psychology Press, Hove.

Fivush, R. and Haden, C., editors (2003). *Autobiographical Memory and the Construction of a Narrative Self*. Lawrence Erlbaum Associates, Mahwah, NJ.

Fleming, G. N. (1996). Just how radical is hyperplane dependence? In *Perspectives on Quantum Reality: Non-Relativistic, Relativistic, and Field-Theoretic*, pages 11–28. Kluwer Academic Publishers, Dordrecht.

Foster, J. G. and Müller, B. (2010). Physics with two time dimensions. eprint. arXiv:1001.2485 [hep-th].

Fraisse, P. (1984). Perception and estimation of time. *Annual Review of Psychology*, 35:1–36.

Freeman, E. D., Isper, A., Palmbaha, A., Paunoiu, D., Brown, P., Lambert, C., and Driver, J. (2013). Sight and sound out of synch: Fragmentation and renormalization of audiovisual integration and subjective timing. *Cortex*, 49:2875–87.

Frege, G. (1956). The thought: A logical inquiry. *Mind*, 65(259): 289–311.

Friedman, J. L. (1991). Spacetime topology and quantum gravity. In Ashtekar, A. and Stachel, J., editors, *Conceptual Problems of Quantum Gravity*, pages 539–72. Birkhäuser, Boston, MA.

Friedman, J. L. (2004). The Cauchy problem on spacetimes that are not globally hyperbolic. In Chruściel, P. T. and Friedrich, H., editors, *The Einstein Equations and the Large Scale Behavior of Gravitational Fields*, pages 331–46. Birkhäuser, Basel.

Friedman, M. (1983). *Foundations of Space-time Theories: Relativistic Physics and Philosophy of Science*. Princeton University Press, Princeton, NJ.

Friedrichs, K. (1927). Eine Invariante Formulierung des Newtonschen Gravitationsgesetzes und der Grenzüberganges vom Einsteinschen zum Newtonschen Gesetz. *Mathematische Annalen*, 98:566–75.

Frijda, N. H. (1988). The laws of emotion. *American Psychologist*, 43(5):349–58.

Frijda, N. H., Kuipers, P., and ter Schure, E. (1989). Relations among emotion, appraisal, and emotional action readiness. *Journal of Personality and Social Psychology*, 57:212–28.

Frisch, M. (2010). Does a low-entropy constraint prevent us from influencing the past? In Hüttemann, A. and Ernst, G., editors, *Time, Chance, and Reduction: Philosophical Aspects of Statistical Mechanics*, pages 13–33. Cambridge University Press, Cambridge.

Fuhrman, O., McCormick, K., Chen, E., Jiang, H., Shu, D., Mao, S., and Boroditsky, L. (2011). How linguistic and cultural forces shape conceptions of time: English and Mandarin time in 3D. *Cognitive Science*, 35(7):1305–28.

Fujisaki, W., Koene, A., Arnold, D., Johnston, A., and Nishida, S. (2006). Visual search for a target changing in synchrony with an auditory signal. *Proceedings of the Royal Society B: Biological Sciences*, 273(1588):865–74.

Fujisaki, W., Shimojo, S., Kashino, M., and Nishida, S. (2004). Recalibration of audiovisual simultaneity. *Nature Neuroscience*, 7:773–8.

Gale, R. M. (1968). *The Language of Time*. Routledge & Kegan Paul, London.

Galison, P. (2004). *Einstein's Clocks, Poincaré's Maps: Empires of Time*. W. W. Norton, New York.

Garabedian, P. R. (1998 [1964]). *Partial Differential Equations*. AMS Chelsea Publishing, Providence, RI.

Garipov, R. M. and Kardakov, V. B. (1973). The Cauchy problem for the wave equation with a nonspatial initial manifold. *Doklady Akademii Nauk SSSR*, 213(5):1047–50.

Geach, P. T. (1979). *Truth, Love, and Immortality*. Hutchinson, London.

Gell, A. (1992). *The Anthropology of Time: Cultural Constructions of Temporal Maps and Images*. Berg, Oxford.

Gell-Mann, M. (1994). *The Quark and the Jaguar*. Freeman, San Francisco, CA.

Gentner, D. (2001). Spatial metaphors in temporal reasoning. In Gattis, M., editor, *Spatial Schemas and Abstract Thought*, chapter 8, pages 203–22. The MIT Press, Cambridge, MA.

Geroch, R. (1970). Domain of dependence. *Journal of Mathematical Physics*, 11:437–49.

Geroch, R. (1977). Prediction in general relativity. In Earman, J., Glymour, C., and Stachel, J., editors, *Foundations of Space-Time Theories*, volume 8, pages 81–93. University of Minnesota Press, Minneapolis, MN.

Geroch, R. (1981). *General Relativity from A to B*. University of Chicago Press, Chicago, IL.

Geroch, R. (1996). Partial differential equations of physics. In Hall, G. S. and Pulham, J. R., editors, *General Relativity: Proceedings of the Forty Sixth Scottish Universities Summer School in Physics, Aberdeen, July 1995*, pages 19–60. CRC Press, New York.

Geroch, R. (2011). Faster than light? In Plaue, M., Rendall, A. D., and Scherfner, M., editors, *Advances in Lorentzian Geometry: Proceedings of the Lorentzian Geometry Conference in Berlin*, volume 49 of AMS/IP Studies in Advanced Mathematics, pages 59–70. American Mathematical Society, Providence, RI.

Ghirardi, G. C., Rimini, A., and Weber, T. (1986). Unified dynamics for microscopic and macroscopic systems. *Physical Review D*, 34:470–91.

Gleick, J. (2011). *Genius: The Life and Science of Richard Feynman*. Open Road Media, New York.

Gleitman, L. and Papafragou, A. (2005). Language and thought. In Holyoak, K. and Morrison, B., editors, *The Cambridge Handbook of Thinking and Reasoning*, pages 633–61. Cambridge University Press, Cambridge.

Gödel, K. (1949). A remark on the relationship between relativity theory and idealistic philosophy. In Schilpp, P., editor, *Albert Einstein: Philosopher-Scientist*, volume 4 of Library of Living Philosophers, pages 555–62. Open Court, La Salle, IL.

Gogberashvili, M. (2000). Brane-universe in six dimensions with two times. *Physics Letters B*, 484(1–2):124–8.

Gold, T. (1967). *The Nature of Time*. Cornell University Press, Ithaca, NY.

Goldie, P. (2000). *The Emotions: A Philosophical Exploration*. Oxford University Press, Oxford.

Goldman, A. I. (1967). A causal theory of knowing. *The Journal of Philosophy*, 64(12):357–72.

Gourgoulhon, É. (2012). *3 + 1 Formalism in General Relativity: Bases of Numerical Relativity*. Springer, Berlin Heidelberg.

Greene, B. (2004). *The Fabric of the Cosmos: Space, Time, and the Nature of Reality*. A. A. Knopf, New York.

Greensite, J. (1993). Dynamical origin of the Lorentzian signature of spacetime. *Physics Letters B*, 300(1–2):34–7.

Griffiths, P. E. (1997). *What Emotions Really Are: The Problem of Psychological Categories*. University of Chicago Press, Chicago, IL.

Grünbaum, A. (1968). *Modern Science and Zeno's Paradoxes*. Allen & Unwin, London.

Grünbaum, A. (1973). *Philosophical Problems of Space and Time*, volume XII of Boston Studies in the Philosophy of Science. Reidel, Dordrecht.

Grush, R. (2007). Time and experience. In Müller, T., editor, *The Philosophy of Time*. Klosterman, Frankfurt.

Hadamard, J. (1902). Sur les problèmes aux dérivées partielles et leur signification physique. *Princeton University Bulletin*, pages 49–52.

Hafele, J. C. and Keating, R. E. (1972). Around-the-world atomic clocks: Observed relativistic time gains. *Science*, 177(4044):168–70.

Haggard, R., Clark, S., and Kalogeras, J. (2002). Voluntary action and conscious awareness. *Nature Neuroscience*, 5:382–5.

Haidt, J. and Joseph, C. (2007). The moral mind: How five sets of innate intuitions guide the development of many culture-specific virtues, and perhaps even modules. In Carruthers, P., Laurence, S., and Stich, S., editors, *The Innate Mind*, volume 3: Foundations and the Future, pages 367–92. Oxford University Press, New York.

Hanko, K. (2007). On once and future things: A temporal asymmetry in judgments of likelihood. PhD thesis, Cornell University.

Hanko, K. and Gilovich, T. (2008). On once and future things: A temporal asymmetry in judgments of likelihood. Paper presented at the Society for Consumer Psychology.

Hansteen, R. W. (1968). Visual latency as a function of stimulus onset, offset, and background luminance. PhD thesis, Tulane University.

Hansteen, R. W. (1971). Visual latency as a function of stimulus onset, offset, and background luminance. *Journal of the Optical Society of America*, 61(9):1190–5.

Hardcastle, V. G. (2008). *Constructing the Self*. John Benjamins Publishing, Amsterdam.

Hare, C. (2007). Self-bias, time-bias, and the metaphysics of self and time. *The Journal of Philosophy*, 104(7):350–73.

Hare, C. (2009). *On Myself, and Other, Less Important Subjects*. Princeton University Press, Princeton, NJ.

Harris, L. R., Harrar, V., Jaekl, P., and Kopinska, A. (2010). Mechanisms of simultaneity constancy. In Nijhawan, R. and Khurana, B., editors, *Space and Time in Perception and Action*, chapter 15, pages 232–53. Cambridge University Press, Cambridge.

Hartle, J. B. (2005). The physics of now. *American Journal of Physics*, 73(2):101–9.

Hawking, S. W. (1969). The existence of cosmic time functions. *Proceedings of the Royal Society of London A*, 308(1494):433–5.

Hawking, S. W., King, A. R., and McCarthy, P. J. (1976). A new topology for curved space-time which incorporates the causal, differential, and conformal structures. *Journal of Mathematical Physics*, 17(2):174–81.

Hawking, S. W. and Penrose, R. (1996). *The Nature of Space and Time*. Princeton University Press, Princeton, NJ.

Hawley, K. (2015). Temporal parts. *The Stanford Encyclopedia of Philosophy*, Winter 2015 Edition.

Hayden, B. (2015). Time discounting and time preferences: A critical review. *Psychonomic Bulletin & Review*.

Helzer, E. G. and Gilovich, T. (2012). Whatever is willed will be: A temporal asymmetry in attributions to will. *Personality and Social Psychology Bulletin*, 38(10):1235–46.

Helzer, E. G., Hanko, K. C., and Gilovich, T. (2011). Common associations to the past and future and their role in the past/future attributional asymmetry. In preparation.

Hestevold, H. S. (1990). Passage and the presence of experience. *Philosophy and Phenomenological Research*, 50:537–52.

Hilgevoord, J. and Atkinson, D. (2011). Time in quantum mechanics. In Callender (2011a), pages 647–62.

Hirsh, I. J. and Sherrick, J. E. (1961). Perceived order in different sense modalities. *Journal of Experimental Psychology*, 62:423–32.

Hoagland, H. (1933). The physiological control of judgments of duration: Evidence for a chemical clock. *The Journal of General Psychology*, 9(2):267–87.

Hoefer, C. (2011). Time and chance propensities. In Callender (2011a), pages 68–90.

Hoerl, C. (2009). Time and tense in perceptual experience. *Philosophers' Imprint*, 9(12):1–18.

Hoerl, C. (2014). Do we (seem to) perceive passage? *Philosophical Explorations*, 17:188–202.

Holcombe, A. (2009). Seeing slow and seeing fast: two limits on perception. *Trends in Cognitive Science*, 13(5):216–21.

Holcombe, A. (2013). The temporal organization of perception. In Wagemans, J., editor, *The Oxford Handbook of Perceptual Organization*, chapter 37. Oxford University Press, Oxford.

Home, D. (1997). *Conceptual Foundations of Quantum Mechanics–an Overview from Modern Perspectives*. Plenum, New York.

Horgan, T. (2007). Mental causation and the agent-exclusion problem. *Erkenntnis*, 67:183–200.

Horwich, P. (1987). *Asymmetries in Time: Problems in the Philosophy of Science*. MIT Press, Cambridge, MA.

Huggett, N. (2006). The regularity account of relational spacetime. *Mind*, 115(457):41–73.

Huggett, W. J. (1960). Losing one's way in time. *Philosophical Quarterly*, 10(40):264–7.

Hughes, R. I. G. (1992). *The Structure and Interpretation of Quantum Mechanics*. Harvard University Press, Cambridge, MA.

Hume, D. (2000). *A Treatise of Human Nature*. Oxford University Press, Oxford.

Ismael, J. (2007). *The Situated Self*. Oxford University Press, Oxford.

Ismael, J. (2011). Decision and the open future. In Bardon, A., editor, *The Future of the Philosophy of Time*, chapter 8, pages 149–68. Routledge, London.

Itin, Y. and Hehl, F. W. (2004). Is the Lorentz signature of the metric of spacetime electromagnetic in origin? *Annals of Physics*, 312(1):60–83.

Jackson, F. (1986). What Mary didn't know. *Journal of Philosophy*, 83:291–5.

Jammer, M. (2006). *Concepts of Simultaneity: From Antiquity to Einstein and Beyond*. The Johns Hopkins University Press, Baltimore, MD.

Jeans, J. (1936). "Man and the universe," Sir Stewart Alley Lecture. In *Scientific Progress*, pages 13–38. The McMillan Company, New York.

Jokić, A. and Smith, Q., editors (2003). *Time, Tense, and Reference*. MIT Press, Cambridge, MA.

Kaistrenko, V. M. (1975). On the Cauchy problem for a second order hyperbolic equation with data on a timelike surface. *Siberian Mathematical Journal*, 16(2):306–8.

Kane, J., Van Boven, L., and McGraw, A. P. (2012). Prototypical prospection: future events are more prototypically represented and simulated than past events. *European Journal of Social Psychology*, 42(3):354–62.

Kaplan, D. (1979). On the logic of demonstratives. *Journal of Philosophical Logic*, 8(1):81–98.

Kassam, K. S., Gilbert, D. T., Boston, A., and Wilson, T. D. (2008). Future anhedonia and time discounting. *Journal of Experimental Social Psychology*, 44(6):1533–7.

Keetels, M. and Vroomen, J. (2012). Perception of synchony between the senses. In Murray, M. M. and Wallace, M. T., editors, *The Neural Bases of Multisensory Processes*, chapter 9. CRC Press, Boca Raton, FL.

Keating, J. C. (2002). The meaning of innate. *Journal of the Canadian Chiropractic Association*, 46(1):4–10.

Keren, G. and Roelofsma, P. (1995). Immediacy and certainty in intertemporal choice. *Organizational Behavior and Human Decision Processes*, 63:287–97.

Kern, S. (2003). *The Culture of Time and Space, 1880–1918*. Harvard University Press, Cambridge, MA.

Kiefer, C. (2012). *Quantum Gravity*, volume 155 of International Series of Monographs on Physics. Oxford University Press, Oxford.

Kiernan-Lewis, D. (1991). Not over yet: Prior's "thank goodness" argument. *Philosophy*, 66: 241–3.

King, D. (2004). Two-dimensional time: Macbeath's "Time's Square" and special relativity. *Synthese*, 139(3):421–8.

King-Casas, B., Tomlin, D., Anen, C., Camerer, C. F., Quartz, S. R., and Montague, P. R. (2005). Getting to know you: Reputation and trust in a two-person economic exchange. *Science*, 308(5718):78–83.

Kitcher, P. (2003). *Science, Truth, and Democracy*. Oxford University Press, New York.

Klein, S. B. (2001). A self to remember: A cognitive neuropsychological perspective on how self creates memory and memory creates self. In Sedikides, C. and Brewer, M., editors, *Individual Self, Relational Self, and Collective Self*, pages 25–46. Psychology Press, Philadelphia, PA.

Klein, S. B. and Ganagi, C. E. (2010). The Multiplicity of self: Neuropsychological evidence and its implications for the self as a construct in psychological research. *Annals of the New York Academy of Sciences*, 1191: 1–15.

Klein, S. B., German, T. P., Cosmides, L., and Gabriel, R. (2004). A theory of autobiographical memory: Necessary components and disorders resulting from their loss. *Social Cognition*, 22: 460–90.

Klein, S. B. and Nichols, S. (2012). Memory and the sense of personal identity. *Mind*, 121(483):677–702.

Klibanov, M. and Rakesh (1992). Numerical solution of a time-like Cauchy problem for the wave equation. *Mathematical Methods in the Applied Sciences*, 15(8):559–70.

Koenderink, J., Richards, W., and van Doorn, A. J. (2012). Space-time disarray and visual awareness. *i-Perception*, 3(3):159–65. doi: 10.1068/i04905as.

Köhler, W. (1947). *Gestalt Psycholgy: An Introduction to New Concepts in Modern Psychology*. Liveright, New York.

Komar, A. (1965). Foundations of special relativity and the shape of the Big Dipper. *American Journal of Physics*, 33(12):1024–7.

Kopinska, A. and Harris, L. R. (2004). Simultaneity constancy. *Perception*, 33(9):1049–60.

Kutach, D. (2011). The asymmetry of influence. In Callender (2011a), pages 247–75.

Lamarque, P. (2004). On not expecting too much from narrative. *Mind and Language*, 19: 393–408.

Lautman, A. (1946). Le problème du temps. *Actualités Scientifiques et Industrielles*, 1012: 23–46.

Le Poidevin, R. (2003). Why tenses need real times. In Jokić and Smith (2003), chapter 9, pages 305–24.

Le Poidevin, R. (2007). *The Images of Time*. Oxford University Press, Oxford.

LeDoux, J. E. (1995). Emotion: Clues from the brain. *Annual Review of Psychology*, 46(1):209–35.

Leray, J. (1953). *Hyperbolic Differential Equations*. Institute for Advanced Study, Princeton, NJ.

Levine, R. (1998). *A Geography of Time: On Tempo, Culture, and the Pace of Life*. Basic Books, New York.

Lewis, D. (1973). *Counterfactuals*. Blackwell, Oxford.

Lewis, D. (1979). Counterfactual dependence and time's arrow. *Noûs*, 13(4):455–76.

Lewis, D. (1994). Humean supervenience debugged. *Mind*, 103:473–90.

Lewis, P. (2016). *Quantum Ontology: A Guide to the Metaphysics of Quantum Mechanics*. Oxford University Press, Oxford.

Lloyd, G. (2008). Shaping a life: Narrative, time, and necessity. In Mackenzie, C. and Atkins, K., editors, *Practical Identity and Narrative Agency*, chapter 12, pages 255–68. Routledge, New York.

Loewer, B. (2001). Determinism and chance. *Studies in History and Philosophy of Science Part B*, 32(4):609–20.

Loewer, B. (2004). Humean supervenience. In Carroll, J., editor, *Readings on Laws of Nature*, pages 176–206. University of Pittsburgh Press, Pittsburgh, PA.

Loewer, B. (2007a). Counterfactuals and the second law. In Price, H. and Corry, R., editors, *Causation, Physics, and the Constitution of Reality: Russell's Republic Revisited*, chapter 11, pages 293–326. Clarendon Press, Oxford.

Loewer, B. (2007b). Laws and natural properties. *Philosophical Topics*, 35:313–28.

Loewer, B. (2012). Two accounts of laws and time. *Philosophical Studies*, 160(1):115–37.

Lorentz, H. A. (1900). Considerations on gravitation. *Proceedings of the Royal Netherlands Academy of Arts and Sciences*, 2:559–74.

Lorentz, H. A. (1927). *Problems of Modern Physics*. Ginn, Boston.

Lowe, E. J. (2009). *Serious Endurantism and the Strong Unity of Human Persons*. In Honnefelder, L. Schick, B., and Runggaldier, E., editors, *Unity and Time in Metaphysics*, pages 67–82. Walter de Gruyter Inc., Berlin.

Lucas, J. R. (1973). *A Treatise on Time and Space*. Methuen, London.

Lucas, J. R. (1986). The open future. In Flood, R. and Lockwood, M., editors, *The Nature of Time*. Basil Blackwell, Oxford.

Lucas, J. R. (1998). Transcendental tense II. *Aristotelian Society Supplementary Volume*, 72: 29–43.

Lucas, J. R. (1999). A century of time. In Butterfield, J., editor, *The Arguments of Time*, pages 1–20. Oxford University Press, Oxford.

Lucas, J. R. (2008). The special theory and absolute simultaneity. In Craig and Smith (2008), pages 279–90.

Ludlow, P. (1999). *Semantics, Tense, and Time: An Essay in the Metaphysics of Natural Language*. MIT Press, Cambridge, MA.

Luk, J. (2011). On the local existence for the characteristic initial value problem. eprint. arXiv:1107.0898 [gr-qc].

MacBeath, M. (1993). Time's square. In Le Poidevin, R. and MacBeath, M., editors, *The Philosophy of Time*, Oxford Readings in Philosophy, chapter 11, pages 183–202. Oxford University Press, Oxford.

McCabe, G. (2005). The Standard Model of particle physics in other universes. eprint. http://philsci-archive.pitt.edu/2218.

McCall, S. (1969). Time and the physical modalities. *The Monist*, 53(3):426–46.

McCormack, T. and Hoerl, C. (1999). Memory and temporal perspective: The role of temporal frameworks in memory development. *Developmental Review*, 19:154–82.

McCormack, T. and Hoerl, C. (2008). Temporal decentering and the development of temporal concepts. *Language Learning*, 58:89–113.

McCoy, C. (2016). Philosophical implications of inflationary cosmology. Master's thesis, University of California, San Diego, CA.

McGinn, C. (2000). *Logical Properties*. Clarendon Press, Oxford.

Maclaurin, J. and Dyke, H. (2002). "Thank goodness that's over": The evolutionary story. *Ratio*, 15(3):276–92.

McTaggart, J. M. E. (1908). The unreality of time. *Mind*, 17(68):457–74.

Malament, D. (1977a). Causal theories of time and the conventionality of simultaneity. *Noûs*, 11(3):293–300.

Malament, D. (1977b). The class of continuous timelike curves determines the topology of spacetime. *Journal of Mathematical Physics*, 18(7):1399–404.

Malament, D. (1977c). Observationally indistinguishable space-times. In Earman, J., Glymour, C., and Statchel, J., editors, *Foundations of Space-Time Theories*, volume VIII of Minnesota Studies in Philosophy of Science, pages 61–80. University of Minnesota Press, Minneapolis, MN.

Malament, D. (1986). Gravity and spatial geometry. In Barcan Marcus, R., Dorn, G. J. W., and Weingartner, P., editors, *Logic, Methodology and Philosophy of Science VII*, Studies in Logic and the Foundations of Mathematics, pages 405–11. Elsevier Science, Amsterdam.

Malament, D. (2007). Classical relativity theory. In Butterfield, J. and Earman, J., editors, *Philosophy of Physics*, volume A. Elsevier, Amsterdam.

Malament, D. (2012). *Topics in the Foundations of General Relativity and Newtonian Gravitation Theory*. University of Chicago Press, Chicago, IL.

Malotki, E. (1983). *Hopi Time: A Linguistic Analysis of the Temporal Concepts in the Hopi Language*. Mouton Publishers, New York.

Manchak, J. (2008). Is prediction possible in general relativity? *Foundations of Physics*, 38(4): 317–21.

Maroney, O. J. E. (2005). The (absence of a) relationship between thermodynamic and logical reversibility. *Studies in History and Philosophy of Science Part B: Studies in History and Philosophy of Modern Physics*, 36(2):355–74.

Maroney, O. J. E. (2010). Does a computer have an arrow of time? *Foundations of Physics*, 40(2):205–38.

Massimi, M. (2007). Saving unobservable phenomena. *The British Journal for the Philosophy of Science*, 58(2):235–62.

Maudlin, T. (1994). *Quantum Non-Locality and Relativity: Metaphysical Intimations of Modern Physics*. Blackwell, Cambridge, MA.

Maudlin, T. (1996). Space-time in the quantum world. In Cushing, J. T., Fine, A., and Goldstein, S., editors, *Bohmian Mechanics and Quantum Theory: An Appraisal*, pages 285–307. Springer, Boston, MA.

Maudlin, T. (2002). Remarks on the passing of time. *Proceedings of the Aristotelian Society*, 102(3):237–52.

Maudlin, T. (2007). *The Metaphysics Within Physics*. Oxford University Press, Oxford.

Maudlin, T. (2010). I–Time, topology and physical geometry. *Aristotelian Society Supplementary Volume*, 84(1):63–78.

Mauk, M. and Buonomano, D. (2004). The neural basis of temporal processing. *Annual Review of Neuroscience*, 27:307–40.

Mayo, B. (1961). Objects, events, and complementarity. *The Philosophical Review*, 70(3):340–61.

Mayo, B. (1976). Space and time re-assimilated. *Mind*, 85(340):576–80.

Meiland, J. W. (1966). Temporal parts and spatio-temporal analogies. *American Philosophical Quarterly*, 3(1):64–70.

Mellor, D. H. (1981). *Real Time*. Cambridge University Press, New York.

Mellor, D. H. (1998). *Real Time II*. Routledge, London.

Mellor, D. H. (2001). The time of our lives. *Royal Institute of Philosophy Supplement*, 48:45–59.

Merleau-Ponty, M. (1945). *Phénoménology de la perception*. Gallimard, Paris.

Meyer, U. (2013). *The Nature of Time*. Clarendon Press, Oxford.

Michon, J. (1978). The making of the present: A tutorial review. In Requin, J., editor, *Attention and Performance*, volume 7, chapter 6, pages 89–111. Lawrence Erlbaum Associates, Mahwah, NJ.

Miller, B. (2002). Existence. In Zalta, E. N., editor, *The Stanford Encyclopedia of Philosophy*. Summer 2002 edition. http://plato.stanford.edu/archives/sum2002/entries/existence/.

Miller, K. (2013). Presentism, eternalism, and relativity physics. In Dyke, H. and Bardon, A., editors, *A Companion to the Philosophy of Time*, pages 345–64. John Wiley and Sons, Ltd., Chichester.

Milne, E. A. (1948). *Kinematic Relativity*. Oxford Unversity Press, Oxford.

Minkowski, H. (1908). Raum und Zeit. *Physikalische Zeitschrift*, 10:75–88.

Mlodinow, L. and Brun, T. (2014). Relation between the psychological and thermodynamic arrows of time. *Physical Review E*, 89:052102.

Misner, C. W., Thorne, K. S., and Wheeler, J. A. (1973). *Gravitation*. Macmillan, New York.

Mollon, J. D. and Perkins, A. J. (1996). Errors of judgement at Greenwich in 1796. *Nature*, 380:101–2.

Montare, A. (1988). Further learning effects of knowledge of results upon time estimation. *Perceptual and Motor Skills*, 66:579–88.

Monton, B. (2006). Presentism and quantum gravity. In Dieks, D., editor, *The Ontology of Spacetime*, volume 1 of Philosophy and Foundations of Physics, chapter 14, pages 263–80. Elsevier, Amsterdam.

Moore, C., Lemmon, K., and Skene, K., editors (2001). *The Self in Time: Developmental Perspectives*. Lawrence Erlbaum Associates, Mahwah, NJ.

More, H. L., O'Connor, S. M., Brøndum, E., Wang, T., Bertelsen, M. F., Grøndahl, C., Kastberg, K., Hørlyck, A., Funder, J., and Donelan, J. M. (2013). Sensorimotor responsiveness and resolution in the giraffe. *The Journal of Experimental Biology*, 216(6):1003–11.

Morein-Zamir, S., Soto-Faraco, S., and Kingstone, A. (2003). Auditory capture of vision: Examining temporal ventriloquism. *Cognitive Brain Research*, 17(1):154–63.

Mott, N. F. (1931). On the theory of excitation by collision with heavy particles. *Mathematical Proceedings of the Cambridge Philosophical Society*, 27:553–60.

Moutoussis, K. and Zeki, S. (1997). A direct demonstration of perceptual asynchrony in vision. *Proceedings of the Royal Society of London B*, 264:393–9.

Mozersky, M. J. (2006). A tenseless account of the presence of experience. *Philosophical Studies*, 129:441–76.

Mozersky, M. J. (2011). Presentism. In Callender (2011a), pages 122–44.

Mundy, B. (1986). The physical content of Minkowski geometry. *The British Journal for the Philosophy of Science*, 37:25–54.

Narlikar, J. V. (2002). *An Introduction to Cosmology*. Cambridge University Press, Cambridge.

Nesse, R. M. (1990). Evolutionary explanations of emotions. *Human Nature*, 1(3):261–89.

Nichols, S. (2008). Imagination and the I. *Mind & Language*, 23(5):518–35.

Nikolić, H. (2005). Relativistic quantum mechanics and the Bohmian interpretation. *Foundations of Physics Letters*, 18(6):549–61.

Norton, J. D. (2000). What can we learn about physical laws from the fact that we have memories only of the past? *International Studies in the Philosophy of Science*, 14:11–23.

Norton, J. D. (2005). Eaters of the lotus: Landauer's principle and the return of Maxwell's demon. *Studies in History and Philosophy of Science Part B: Studies in History and Philosophy of Modern Physics*, 36(2):375–411.

Norton, J. D. (2010). Time really passes. *Humana.Mente: Journal of Philosophical Studies*, 13:23–4.

Nùñez, R. and Cooperrider, K. (2013). The tangle of space and time in human cognition. *Trends in Cognitive Science*, 17(5):220–9.

Núñez, R. E., Cooperrider, K., Doan, D., and Wassmann, J. (2012). Contours of time: Topographic construals of past, present, and future in the Yupno valley of Papua New Guinea. *Cognition*, 124(1):25–35.

Núñez, R. E., Motz, B. A., and Teuscher, U. (2006). Time after time: The psychological reality of the ego- and time-reference-point distinction in metaphorical construals of time. *Metaphor and Symbol*, 21(3):133–46.

O'Brien, E. (2013). Emotional pasts, rational futures: Time perspective influences perceived and experienced affect. Presented at the 14th Annual Meeting of the Society for Personality and Social Psychology.

Oeckl, R. (2006). General boundary quantum field theory: Timelike hypersurfaces in the Klein-Gordon theory. *Physical Review D*, 73:065017.

Oeckl, R. (2007). Probabilites in the general boundary formulation. *Journal of Physics: Conference Series*, 67(1):012049.

Øhrstrøm, P. and Hasle, P. (2011). Future contingents. In Zalta, E. N., editor, *The Stanford Encyclopedia of Philosophy*. Summer 2011 edition. http://plato.stanford.edu/archives/sum2011/entries/future-contingents/.

Okasha, S. (2011). Optimal choice in the face of risk: Decision theory meets evolution. *Philosophy of Science*, 78(1):83–104.

Ornstein, R. E. (1969). *On the Experience of Time*. Penguin, Harmondsworth.

Overbye, D. (2013). A quantum of solace: Timeless questions about the universe. *The New York Times*.

Page, D. (1983). How big is the universe today? *General Relativity and Gravitation*, 15(2): 181–5.

Panksepp, J. (1998). *Affective Neuroscience: The Foundations of Human and Animal Emotions*. Oxford University Press, New York.

Parfit, D. (1984). *Reasons and Persons*. Clarendon Press, Oxford.

Parsons, J. (2002). A-theory for B-theorists. *The Philosophical Quarterly*, 52(206):1–20.

Paul, L. A. (2010). Temporal experience. *The Journal of Philosophy*, 107(7):333–59.

Paul, L. A. (2014). Experience and the arrow. In Wilson, A., editor, *Chance and Temporal Asymmetry*, pages 175–93. Oxford University Press, Oxford.

Penrose, R. (1963). Asymptotic properties of fields and space-times. *Physical Review Letters*, 10:66–8.

Penrose, R. (1968). Structure of space-time. In DeWitt-Morette, C. and Wheeler, J. A., editors, *Batelle Rencontres: 1967 Lectures in Mathematics and Physics*, pages 121–235. Benjamin, New York.

Penrose, R. (1979). Singularities and time-asymmetry. In Hawking, S. W. and Israel, W., editors, *General Relativity: An Einstein Centenary Survey*, pages 581–638. Cambridge University Press, Cambridge.

Penrose, R. (1989). *The Emporer's New Mind: Concerning Computers, Minds, and the Laws of Physics*. Oxford University Press, Oxford.

Perry, J. (1979). The problem of the essential indexical. *Noûs*, 13(1):3–21.

Perry, L. (2001). Time, consciousness, and the knowledge argument. In Oaklander, L. N., editor, *The Importance of Time*. Kluwer Academic Publishers, Dordrecht.

Phillips, I. B. (2013). The temporal structure of experience. In Lloyd, D. and Arstila, V., editors, *Subjective Time: the Philosophy, Psychology, and Neuroscience of Temporality*. MIT Press, Cambridge, MA. In press.

Plaks, J. (2012). Your will seems stronger in the future. *Psychology Today*.

Poincaré, H. (1913). The measure of time. In *The Foundations of Science (The Value of Science)*, pages 222–34. Science Press, New York.

Pooley, O. (2013). Relativity, the open future, and the passage of time. *Proceedings of the Aristotelian Society*, 113:321–63.

Pöppell, E. (1988). *Mindworks: Time and Conscious Experience*. Harcourt Brace Jovanovich, San Diego, CA.

Popper, K. R. (1982). *Quantum Theory and the Schism in Physics: From the Postscript to the Logic of Scientific Discovery*. Hutchinson, London.

Povinelli, D. (2001). The self: Elevated in consciousness and extended in time. In Moore, C. and Lemmon, K., editors, *The Self in Time: Developmental Perspectives*, pages 75–95. Lawrence Erlbaum Associates, Mahwah, NJ.

Povinelli, D. and Simon, B. (1998). Young children's reactions to briefly versus extremely delayed images of the self: Emergence of the autobiographical stance. *Developmental Psychology*, 43:188–94.

Povinelli, D. J., Landau, K. R., and Perilloux, H. K. (1996). Self-recognition in young children using delayed versus live feedback: Evidence of a develomental asynchrony. *Child Development*, 67:1540–54.

Povinelli, D. J., Landry, A. M., Theall, L. A., Clark, B. R., and Castille, C. M. (1999). Development of young children's understanding that the recent past is causally bound to the present. *Developmental Psychology*, 35(6):1426–439.

Price, H. (1996). *Time's Arrow & Archimedes' Point: New Directions for the Physics of Time*. Oxford University Press, Oxford.

Prigogine, I. (1980). *From Being to Becoming: Time and Complexity in the Physical Sciences*. W. H. Freeman and Co., Ltd., New York.

Prior, A. N. (1959). Thank goodness that's over. *Philosophy*, 34:12–17.

Prior, A. N. (1967). *Past, Present, and Future*. Clarendon Press, Oxford.

Prior, A. N. (1996). Some free thinking about time. In Copeland, B. J., editor, *Logic and Reality*. Clarendon Press, Oxford.

Prosser, S. (2006). Temporal metaphysics in Z-land. *Synthese*, 149(1):77–96.

Putnam, H. (1967). Time and physical geometry. *The Journal of Philosophy*, 64(8):240–7.

Rammsayer, T. H. (1999). Neuropharmacological evidence for different timing mechanisms in humans. *The Quarterly Journal of Experimental Psychology Section B*, 52(3):273–86.

Rao, R. P. N., Eagleman, D. M., and Sejnowski, T. J. (2001). Optimal smoothing in visual motion perception. *Neural Computation*, 13:1243–53.

Rasmusen, E. B. (2008). Some common confusions about hyperbolic discounting. eprint. doi:10.2139/ssrn.1091392.

Read, D. (2008). Intertemporal choice. In Koehler, D. J. and Harvey, N., editors, *Blackwell Handbook of Judgment and Decision Making*, pages 424–43. John Wiley & Sons, Hoboken, NJ.

Recanati, F. (1993). *Direct Reference: From Language to Thought*. Wiley-Blackwell, Hoboken, NJ.

Redhead, M. (1989). *Incompleteness, Nonlocality, and Realism: A Prolegomenon to the Philosophy of Quantum Mechanics*. Oxford University Press, Oxford.

Reichardt, W. (1969). *Processing of Optical Data by Organisms and Machines*. Academic Press, New York.

Reichenbach, H. (1956). *The Direction of Time*. UCLA Press, Berkeley, CA.

Reichenbach, H. (1957). *The Philosophy of Space and Time*. Dover, New York.

Reichenbach, H. (1969 [1924]). *Axiomatization of the Theory of Relativity*. University of California Press, Berkeley, CA.

Rendall, A. D. (1990). Reduction of the characteristic initial value problem to the Cauchy problem and its applications to the Einstein equations. *Proceedings of the Royal Society of London. A. Mathematical and Physical Sciences*, 427(1872):221–39.

Rendall, A. D. (1997). Existence and non-existence results for global constant mean curvature foliations. *Nonlinear Analysis: Theory, Methods & Applications*, 30(6):3589–98.

Rey, G. (1997). *Contemporary Philosophy of Mind: A Contentiously Classical Approach*. Blackwell, Cambridge, MA.

Ribeiro, P. L. (2010). Diamond-shaped regions as microcosmoi. arXiv:1010.5032 [gr-qc].

Ricoeur, P. (1984–6). *Time and Narrative*. Cambridge University Press, Cambridge.

Rideout, D. P. and Sorkin, R. D. (1999). Classical sequential growth dynamics for causal sets. *Physical Review D*, 61:024002.

Rideout, D. P. and Sorkin, R. D. (2000). A classical sequential growth dynamics for causal sets. *Physical Review D*, 61:024002.

Rietdijk, C. W. (1966). A rigorous proof of determinism derived from the special theory of relativity. *Philosophy of Science*, 33(4):341–4.

Rindler, W. (1981). Public and private space curvature in Robertson-Walker universes. *General Relativity and Gravitation*, 13(5):457–61.

Robb, A. A. (1914). *A Theory of Space and Time*. Cambridge University Press, Cambridge.

Robson, A. J. and Samuelson, L. (2007). The evolution of intertemporal preferences. *American Economic Review*, 97(2):496–500.

Romney, G. (1977). Temporal points of view. *Proceedings of the Aristotelian Society*, 78:237–52.

Roseboom, W. and Arnold, D. H. (2011). Twice upon a time: Multiple concurrent temporal recalibrations of audiovisual speech. *Psychological Science*, 22(7):872–7.

Roseboom, W., Nishida, S., and Arnold, D. H. (2009). The sliding window of audio-visual simultaneity. *Journal of Vision*, 9(12).

Rothbart, M. and Snyder, M. (1970). Confidence in the prediction and postdiction of an uncertain outcome. *Canadian Journal of Behavioural Science*, 2(1):38–43.

Rovelli, C. (2009). "Forget time". eprint. arXiv:0903.3832 [gr-qc].

Rugh, S. and Zinkernagel, H. (2011). Weyl's principle, cosmic time and quantum fundamentalism. In Dieks, D., Gonzalez, W. J., Hartmann, S., Uebel, T., and Weber, M., editors, *Explanation, Prediction, and Confirmation*, volume 2 of *The Philosophy of Science in a European Perspective*, pages 411–24. Springer Netherlands, Dordrecht.

Saari, D. (1977). A global existence theorem for the four-body problem of newtonian mechanics. *Journal of Differential Equations*, 26:80–111.

Sachs, R. K. (1962). Gravitational waves in general relativity. VIII. Waves in asymptotically flat space-time. *Proceedings of the Royal Society of London A*, 270(1340):103–26.

Salmon, N. (1987). Existence. In Tomberlin, J., editor, *Philosophical Perspectives, 1: Metaphysics*, pages 49–108. Ridgeview Press, Atascadero, CA.

Santayana, G. (1942). *Realms of Being*. Cooper Square Publishers, New York.

Santos, L. and Rosati, A. (2015). The evolutionary roots of human decision making. *The Annual Review of Psychology*, 66(13):1–27.

Savitt, S. F. (1994). The replacement of time. *Australasian Journal of Philosophy*, 72(4): 463–74.

Savitt, S. F. (2000). There's no time like the present (in Minkowski spacetime). *Philosophy of Science*, 67:S563–S574.

Savitt, S. F. (2006). Presentism and eternalism in perspective. In Dieks, D., editor, *The Ontology of Spacetime*, volume 1 of *Philosophy and Foundations of Physics*, chapter 6, pages 111–27. Elsevier, Amsterdam.

Savitt, S. F. (2009). The transient *nows*. In *Quantum Reality, Relativistic Causality, and Closing the Epistemic Circle*, volume 73 of The Western Ontario Series in Philosophy of Science, pages 349–62. Springer, Dordrecht.

Savitt, S. F. (2012). Of time and the two images. *Humana Mente-Journal of Philosophical Studies*, 21.

Savitt, S. F. (2014). I [heart] [diamond]s. In *Philosophy of Science Association 24th Biennial Meeting*.

Schechtman, M. (2007). Stories, lives, and basic survival: A refinement and defence of the narrative view. In Hutto, D., editor, *Narrative and understanding persons*. Cambridge University Press, Cambridge.

Schechtman, M. (2011). The narrative self. In Gallagher, S., editor, *The Oxford Handbook of the Self*, pages 394–416. Oxford University Press, Oxford.

Scheier, C. R., Nijhawan, R., and Shimojo, S. (1999). Sound alters visual temporal resolution. *Investigative Ophthalmology & Visual Science*, 40(4):4169.

Schlesinger, G. (1975). The similarities between space and time. *Mind*, 84(334):161–76.

Schlesinger, G. N. (1991). E pur si muove. *The Philosophical Quarterly*, 41:427–41.

Schmidt, J. H. (1997). Classical universes are perfectly predictable! *Studies in History and Philosophy of Science Part B: Studies in History and Philosophy of Modern Physics*, 28(4): 433–60.

Schmidt, J. H. (1998). Predicting the motion of particles in Newtonian mechanics and special relativity. *Studies in History and Philosophy of Science Part B: Studies in History and Philosophy of Modern Physics*, 29(1):81–122.

Schuster, M. M. (1986). Is the flow of time subjective? *The Review of Metaphysics*, 39: 695–714.

Schutz, J. (1997). *Independent Axioms for Minkowski Space-Time*. Longman, Harlow.

Sellars, W. (1962a). Philosophy and the scientific image of man. In Colodny, R., editor, *Frontiers of Science and Philosophy*, pages 35–78. University of Pittsburgh Press, Pittsburgh, PA.

Sellars, W. (1962b). Time and the world order. In Feigl, H. and Maxwell, G., editors, *Minnesota Studies in the Philosophy of Science*, volume III, pages 527–618. University of Minnesota Press, Minneapolis, MN.

Shanks, N. (1991). Probabilistic physics and the metaphysics of time. *The South African Journal of Philosophy*, 10(2):37–42.

Shimony, A. (1993). The transient *now*. In *Search for a Naturalistic World View*, volume II, chapter 18, pages 271–87. Cambridge University Press, Cambridge.

Shimony, A. (1998). Implications of transience for spacetime structure. In Huggett, S. A., Mason, L. J., Tod, K. P., Tsou, S. T., and Woodhouse, N. M. J., editors, *The Geometric Universe: Science, Geometry and the Work of Roger Penrose*, pages 161–72. Oxford University Press, Oxford.

Shoemaker, S. (1963). *Self-Knowledge and Self-Identity*. Cornell University Press, Ithaca, NY.

Shoemaker, S. (1996). *The First-Person Perspective and Other Essays*. Cambridge University Press, Cambridge.

Shorter, J. M. (1981). Space and time. *Mind*, XC(357):61–78.

Sider, T. (2001). *Four-Dimensionalism: An Ontology of Persistence*. Oxford University Press, New York.

Sider, T. (2003). Against vague existence. *Philosophical Studies*, 114(1–2):135–46.

Sidgwick, H. (1907). *The Methods of Ethics*. Macmillan, New York, 7th edition.

Skow, B. (2007). What makes time different from space? *Noûs*, 41(2):227–52.

Skow, B. (2012). Why does time pass? *Noûs*, 46(2):223–42.

Smart, J. J. C. (1966). The river of time. In Flew, A., editor, *Essays in Conceptual Analysis 3*, pages 213–27. St. Martin's Press, London.

Smith, G. J. and Weingard, R. (1987). A relativistic formulation of the Einstein–Podolsky–Rosen paradox. *Foundations of Physics*, 17(2):149–71.

Smolin, L. (2006a). The case for background independence. In Rickles, D., French, S., and Saatsi, J., editors, *The Structural Foundations of Quantum Gravity*, pages 196–239. Oxford University Press, Oxford.

Smolin, L. (2006b). *The Trouble With Physics: The Rise of String Theory, the Fall of a Science, and What Comes Next*. Houghton Mifflin Harcourt, Boston, MA.

Smolin, L. (2013). *Time Reborn: From the Crisis in Physics to the Future of the Universe*. Houghton Mifflin Harcourt, Boston, MA.

Sobel, J. H. (1998). *Puzzles for the Will*. University of Toronto Press, Toronto.

Sorkin, R. D. (2006). Geometry from order: Causal sets. *Einstein Online*, 2:1007.

Sorkin, R. D. (2007). Relativity theory does not imply that the future already exists: A counterexample. In Petkov, V., editor, *Relativity and the Dimensionality of the World*, pages 153–61. Springer, Dordrecht.

Souzou, P. D. (1998). On hyperbolic discounting and uncertain hazard rates. *Proceedings of the Royal Society of London B*, 265:2015–20.

Spence, C. and Squire, S. (2003). Multisensory integration: Maintaining the perception of synchrony. *Current Biology*, 13(13):R519–21.

Stapp, H. P. (1977). Quantum mechanics, local causality, and process philosophy. *Process Studies*, 7(3):173–82.

Stein, H. (1968). On Einstein–Minkowski space-time. *The Journal of Philosophy*, 65(1):5–23.

Stein, H. (1991). On relativity theory and openness of the future. *Philosophy of Science*, 58(2): 147–67.

Stelmach, L. B. and Herdman, C. M. (1991). Directed attention and perception of temporal order. *Journal of Experimental Psychology: Human Perception and Performance*, 17(2):539–50.

Stephens, D. and Anderson, D. (2001). The adaptive value of preference for immediacy: When shortsighted rules have farsighted consequences. *Behavioral Ecology*, 12(3):330–9.

Sternberg, S. and Knoll, R. L. (1973). The perception of temporal order: Fundamental issues and a general model. In Kornblum, S., editor, *Attention and Performance*, volume IV, pages 629–85. Academic Press, New York.

Stetson, C., Cui, X., Montague, P. R., and Eagleman, D. M. (2006). Motor-sensory recalibration leads to an illusory reversal of action and sensation. *Neuron*, 51(5):651–9.

Stone, J. V., Hunkin, N. M., Porrill, J., Wood, R., Keeler, V., Beanland, M., Port, M., and Porter, N. R. (2001). When is now? Perception of simultaneity. *Proceedings of the Royal Society B: Biological Sciences*, 268:31–8.

Strawson, G. (2004). Against narrativity. *Ratio*, 17(4):428–52.

Strotz, R. H. (1956). Myopia and inconsistency in dynamic utility maximization. *Review of Economic Studies*, 23(3):165–80.

Sugita, Y. and Suzuki, Y. (2003). Audiovisual perception: Implicit estimation of sound-arrival time. *Nature*, 421:911.

Suhler, C. L. and Callender, C. (2012). Thank goodness that argument is over: Explaining the temporal value asymmetry. *Philosopher's Imprint*, 12(15).

Suhler, C. L. and Churchland, P. (2011). Can innate, modular "foundations" explain morality? Challenges for Haidt's moral foundations theory. *Journal of Cognitive Neuroscience*, 23(9):2103–16.

Swinburne, R. (2008). Cosmic simultaneity. In Craig and Smith (2008), pages 244–61.

Taylor, R. (1955). Spatial and temporal analogies and the concept of identity. *The Journal of Philosophy*, 52(22):599–612.

Taylor, R. (1959). Moving about in time. *The Philosophical Quarterly*, 9(37):289–301.

Taylor, R. (1983). Metaphysics. In *Metaphysics*, volume 3, Prentice Hall, Englewood Cliffs, NJ.

Tegmark, M. (1997). On the dimensionality of spacetime. *Classical and Quantum Gravity*, 14:L69–L75.

Thagard, P. (2006). *Hot Thought: Mechanisms and Applications of Emotional Cognition*. MIT Press, Cambridge, MA.

Thomas, E. A. C. and Weaver, W. B. (1975). Cognitive processing and time perception. *Perception & Psychophysics*, 17:363–7.

Thomson, J. J. (1965). Time, space, and objects. *Mind*, 74(293):1–27.

Thorne, K. S. (1994). *Black Holes and Time Warps: Einstein's Outrageous Legacy*. W. W. Norton, New York.

Tooley, M. (1997). *Time, Tense, and Causation*. Oxford University Press, Oxford.

Tooley, M. (2008). Two arguments for absolute simultaneity. In Craig and Smith (2008), pages 229–43.

Torre, S. (2011). The open future. *Philosophy Compass*, 6(5):360–73.

Trautman, A. (1965). Foundations and current problems of general relativity. In Deser, D. and Ford, K. W., editors, *Lectures on General Relativity*. Prentice Hall, Upper Saddle River, NJ.

Treisman, A. (1982). Perceptual grouping and attention in visual search for features and objects. *Journal of Experimental Psychology: Human Perception and Performance*, 8(2):194–214.

Treisman, M. (1963). Temporal discrimination and the indifference interval: Implications for a model of the "internal clock." *Psychological Monographs*, 77(13):1–31.

Tse, P. U., Intriligator, J., Rivest, J., and Cavanagh, P. (2004). Attention and the subjective expansion of time. *Perception & Psychophysics*, 66(7):1171–89.

Tulving, E. (1972). Episodic and semantic memory. In Tulving, E. and Donaldson, W., editors, *Organization of Memory*, pages 381–402. Academic Press, New York.

Unruh, W. G. and Wald, R. M. (1989). Time and the interpretation of canonical quantum gravity. *Physical Review D*, 40:2598–614.

Valentini, A. (2008). Hidden variables and the large-scale structure of spacetime. In Craig and Smith (2008), pages 125–55.

van Boven, L. and Ashworth, L. (2007). Looking forward, looking back: Anticipation is more evocative than retrospection. *Journal of Experimental Psychology: General*, 136(2): 289–300.

van Dam, H. and Ng, Y. (2001). Why 3+1 metric rather than 4+0 or 2+2? *Physics Letters B*, 520 (1–2):159–62.

Van der Burg, E. (2008). Pip and pop: Non-spatial auditory signals improve spatial visual search. *Journal of Experimental Psychology: Human Perception and Performance*, 34(5):1653–65.

Van der Burg, E. (2009). Poke and pop: Tactile-visual synchrony increases visual saliency. *Neuroscience Letters*, 450(1):60–4.

van Fraassen, B. (1985). *An Introduction to the Philosophy of Time and Space*. Columbia University Press, New York, 2nd edition.

van Santen, J. P. H. and Sperling, G. (1985). Elaborated Reichardt detectors. *Journal of the Optical Society of America*, 2(2):300–20.

Velleman, J. D. (2006). So it goes. The Amherst Lecture in Philosophy.

Vroomen, J. and de Gelder, B. (2004). Temporal ventriloquism: Sound modulates the flash-lag effect. *Journal of Experimental Psychology: Human Perception and Performance*, 30:513–18.

Wald, R. M. (1984). *General Relativity*. University of Chicago Press, Chicago, IL.

Wald, R. M. (1994). *Quantum Field Theory in Curved Spacetime and Black Hole Thermodynamics*. University of Chicago Press, Chicago.

Wald, R. M. (2009). The formulation of quantum field theory in curved spacetime. eprint. arXiv:0907.0416 [gr-qc].

Watzl, S. (2012). Silencing the experience of change. *Philosophical Studies*, 1–24.

Wearden, J. H. (2005). The wrong tree: Time perception and time experience in the elderly. In Duncan, J., Phillips, L., and McLeod, P., editors, *Measuring the Mind: Speed, Age, and Control*, pages 137–58. Oxford University Press, Oxford.

Webb, C. W. (1977). Could space be time-like? *The Journal of Philosophy*, 74(8):462–74.

Weber, B. and Chapman, G. (2005). The combined effects of risk and time on choice: Does uncertainty eliminate the immediacy effect? Does delay eliminate the certainty effect? *Organizational Behavior and Human Decision Processes*, 96:104–18.

Weingard, R. (1972). Relativity and the reality of past and future events. *The British Journal for the Philosophy of Science*, 23(2):119–21.

Weingard, R. (1977). Space-time and the direction of time. *Noûs*, 11(2):119–31.

Weist, R. (2002). Temporal and spatial concepts in child language: Conventional and configurational. *Journal of Psycholinguistic Research*, 31(3):195–210.

Welch, R. B., DuttonHurt, L. D., and Warren, D. H. (1986). Contributions of audition and vision to temporal rate perception. *Perception & Psychophysics*, 39(4):294–300.

Whitehead, A. (1929). *Process and Reality*. The Macmillan Co., New York.

Whitrow, G. (1980[1961]). *The Natural Philosophy of Time*. Oxford University Press, Oxford, 2nd edition.

Whorf, B. L. (1956). *Language, Thought, and Reality: Selected Writings*. MIT Press, Cambridge, MA.

Williams, D. C. (1951). The myth of passage. *Journal of Philosophy*, 48:457–72.

Williams, D. C. (1962). Dispensing with existence. *Journal of Philosophy*, 59:748–63.

Wilson, T. D., Centerbar, D. B., Kermer, D. A., and Gilbert, D. T. (2005). The pleasures of uncertainty: Prolonging positive moods in ways people do not anticipate. *Journal of Personality and Social Psychology*, 88(1):5–21.

Wilson, T. D. and Gilbert, D. T. (2005). Affective forecasting: Knowing what to want. *Current Directions in Psychological Science*, 14(3):131–4.

Wilson, T. D., Wheatley, T., Meyers, J. M., Gilbert, D. T., and Axsom, D. (2000). Focalism: A source of durability bias in affective forecasting. *Journal of Personality and Social Psychology*, 78(5):821–36.

Wüthrich, C. (2010). No presentism in quantum gravity. In Petkov, V., editor, *Space, Time, and Spacetime: Physical and Philosophical Implications of Minkowski's Unification of Space and Time*, pages 257–78. Springer, Berlin.

Wüthrich, C. (2013). The fate of presentism in modern physics. In Ciuni, R., Miller, K., and Torrengo, G., editors, *New Papers on the Present: Focus on Presentism*, pages 91–131. Philosophia Verlag, Munich.

Wüthrich, C. and Callender, C. (2015). What becomes of a causal set. *British Journal for the Philosophy of Science*, https://doi.org/10.1093/bjps/axv040.

Ynduráin, F. J. (1991). Disappearance of matter due to causality and probability violations in theories with extra timelike dimensions. *Physics Letters B*, 256(1):15–16.

Yoshie, M. and Haggard, P. (2013). Negative emotional outcomes attenuate sense of agency over voluntary actions. *Current Biology*, 23(20):2028–32.

Zahar, É. (2001). *Poincaré's Philosophy: From Conventionalism to Phenomenology*. Open Court, La Salle, IL.

Zampini, M., Guest, S., Shore, D., and Spence, C. (2005). Audio-visual simultaneity judgments. *Perception & Psychophysics*, 67(3):531–44.

Zelazo, P. and Sommerville, J. (2001). Levels of consciousness of the self in time. In Moore, C. and Lemmon, K., editors, *The Self in Time: Developmental Perspectives*, pages 229–52. Lawrence Erlbaum Associates, Mahwah, NJ.

Zimmerman, D. W. (2007). The privileged present: Defending an 'A-theory' of time. In Sider, T., Hawthorne, J., and Zimmerman, D., editors, *Contemporary Debates in Metaphysics*, pages 211–25. Wiley-Blackwell, Malden, MA.

Zimmerman, D. W. (2011). Presentism and the space-time manifold. In Callender (2011a), pages 163–244.

Index

Notes, figures, and boxes are indicated by an italic *n*, *f*, and *b* following the page number.

animated time 252
argument from experience 228
asymmetry
    affect 275
    boundary condition 133
    built-in 133
    causal-counterfactual 259
    knowledge 243, 244*n*, 247, 259, 271, 274, 276, 286, 298
    mechanisms for valuation and choice among alternatives 278*b*
    memory 244, 246, 247
    natural kind 22, 122, 133, 137, 148, 179
    past/future (PF), *see* past/future (PF) asymmetry
    proximal/distant (PD), *see* proximal/distant (PD) asymmetry
    solutions for 272
    strength, of 148*n*, 179
    temporal value asymmetry, explanation of 264
asymmetry of strength 148*n*, 179
asynchronous becoming 99

'base space', time as 38*f*
becoming
    absolute 246, 294
    asynchronous 99
    causal set theory (CST), and 99
    causet becoming 104
    collapses, via 94
    concept of 29, 56
    experience of 93
    frame of 92, 93, 94
    illustration of 134
    mind-dependence theory of 310
    present becoming past 246
    quantum 81
    restoration of 99
    space 138
    temporal 55, 81, 230, 291
    tensed theory of 269, 294
    time 138
bias, *see* temporal biases
binding time 144
Bohmian configuration space 91
boundary condition asymmetry 133

care
    caring 99, 100, 132, 155, 163, 163*n*, 248, 272, 273, 279, 285
    drop-off in 265
    need for 143*n*
    not caring 39, 46, 95, 99, 142, 163, 301, 304
    question of 20, 125
    time and 266*f*
Cartan metrics 33*n*
Cartan spacetime, *see* Newton–Cartan spacetime
Cartan theory, *see* Trautman–Cartan theory
Cauchy problems
    example of 162*f*
    illustration of 170
    sideways 164*f*
    solutions for 67, 68
    timelike 178
    well-posed 158, 165*n*, 166, 172, 176, 177, 178
Cauchy time 67, 73, 74, 75, 76, 79, 80, 84
causal-counterfactual asymmetry 259
causality, physical theories of time and space, and 128
causal order of flow of time 256*f*
causal set theory (CST)
    asynchronous becoming in 99
    basic kinematics of 100
    becoming and 99
    growth, treatment of 103
    manifest time, and 22, 99, 101, 102
    Minkowski spacetime, and 102, 104
    quantum gravity, and 100
    relativity theory, and 138
change, sensing of 238
classical duration, path-independence of 39
classical physics
    Cartan physics 32, 40
    ideal clocks 37, 49
    manifest time, and 37
    manifold set 32
    mass, inertial and gravitational 136
    massive bodies 53
    metrics 32, 40, 44, 46
    Newtonian physics 32
    proper time 47
    quantum mechanics, and 31
    relativity theory, and 31

classical physics (*cont.*)
  semiclassical time 112
  spacetime 32, 34, 52
  tangent spaces 33
  theories 32
  time and 31
  Trautman–Cartan theory 40
  vectors 33
classical spacetime 32, 34, 36, 38, 45, 58, 61,
     100, 114, 125, 213
clocks, ideal 37, 49
closed timelike curves 145
common now, *see* now
continuous metrics 78
coordination problem in quantum
     mechanics 90, 93*f*
CST, *see* causal set theory (CST)
cylinder spacetime 137*n*

'diamond presents' 64, 80
direction of motion 58, 194*n*
direction of space
  direction of time, and 22
  spacelike direction 139, 177
  spatial direction 134, 139, 311
direction of time
  binding time 148
  difference between time and space 132, 137
  direction of space, and 22
  non-temporal direction 312
  physical time 20, 21, 132
  physics and 122, 179
  preference for 142, 156
  primitive 296*n*
  quantum theory 81
  relativity theory 39, 40
  temporal direction 120, 132, 139, 142, 143,
     154, 155, 166, 167, 171, 174, 296*n*, 298
  timelike direction 151, 166, 171, 176, 177, 180
  which direction? 166, 166*f*
domain of dependence in spacetime 68
'domes' 63
'donut presents' 62*f*
duration
  felt duration, experience of 242
  path-independence of classical duration 39

Eddington's two tables 24
ego-moving perspective of time 254*f*
Einstein–Poincaré 'Radar' Synchronization 51*b*
EPR–Bell experiment 88
Euclidean metrics 44
Euclidean spacetime 128
existence, philosophy of time, and 296,
     299, 303
experience
  becoming, of 93

felt duration, of 242
manifest time argued from 228
presence of 182
present, of 72

felt duration, experience of 242
flowing now 21, 31, 35, 181, 238, 306, 310
flow of time
  acting self 259
  animated time 252
  apparantness of 1, 3, 270
  argument from experience 228
  background to 226
  causal order 256*f*
  change and 11
  conclusions about 262
  ego-moving perspective 254*f*
  existence 296
  IGUS, *see* IGUS
  manifest time 10, 14, 23, 206
  memory and 243
  now, of 22
  objective now, and 11
  passage, sense of 261
  phenomenology and 11
  physical time 31, 227, 238
  quantum theory, and 84
  relativity and 22
  self and 247, 252, 255, 259
  self in time 253*f*
  senses of 10
  temporal decentering 255, 257*f*
  theories of 226, 311
  timeline 256*f*
  time-moving perspective 254*f*
  undifferentiated past and future 255*f*
foliation
  choices of 143
  Minkowski spacetime 89, 101
  time and space 36*f*
fragmentation of time 135
Friedman–Lemaître–Robertson–Walker
     (FLRW) time 73
future
  now, and the 6
  open future, RPP argument against 52
  past/future (PF) asymmetry, *see* past/future
     (PF) asymmetry
  temporal biases about 282

Galilean spacetime 42, 81, 89, 124, 151
general relativity, *see* relativity theory
Gödel spacetime 60, 130, 143
growth
  causal set theory (CST), and 103
  post growth 109*f*

ideal clocks 37, 49
IGUS (information gathering and utilizing
      system)
   class of 232
   description of 232
   felt duration, experience of 242
   manifest time 27*f*
   memories and flow in relation 243
   outfitting of 237, 261
   proximity to eachother 235
   Reichardt-like motion detector 239*f*
   relativistic spacetime, in 26
   sense of passage 261
   sensing of motion and change 238
   specious present, experiencing of 240
   stuck in time 234
   world of 261

knowledge asymmetry 243, 244*n*, 247, 259, 271,
      274, 276, 286, 298

light, pages of 174
lightcone coordinates 123*n*
lightcone presents 61*f*
lightcone structure, relativity theory 45
Lorentzian metrics 44*n*, 124, 137
Lorentzian spacetime 172

Malament's treatment of spacetime 79*n*
manifest present, features of 50
manifest time
   argument from experience 228
   causal set theory (CST), and 22, 99,
      101, 102
   concept of 2, 4, 7, 304, 305, 306, 309
   Eddington's two tables 24
   explanation of 311
   flow of 10
   fragmentation of time 135, 307
   fuzziness of 231
   memory asymmetry, and 247
   mobility asymmetry 305
   now, and the 7
   objective global now, and 206
   past/future asymmetry 12
   phenomenology and 230
   physical time, and 7, 20, 23, 26, 49, 119, 157,
      180, 210, 300, 301, 310
   portrayal of present by 224
   presentness 184, 189
   Putnam's argument against 212
   quantum mechanics, and 21, 81
   quantum theory, and 81
   relativity theory, and 31, 67, 80,
      175, 176
   restoration of 22
   self and 248, 259, 260

special relativity, and 59
tensed and tenseless theories of time,
      and 182, 184
time as 27*f*
'two times' problem, and 23
universality of 14
maximal antichains 102
memories and flow in relation 243
memory asymmetry 244, 246, 247
metaphysics of time 181
   conservative variation 149*f*
   radical variation 150*f*
   variations 149
metrics
   Cartan 33*n*
   classical 32, 40, 44, 46
   continuous 78
   Euclidean 44
   FLRW 73
   Lorentzian 44*n*, 124, 137
   metrical structure of spacetime 122
   Minkowski spacetime 125
   non-degenerate 78
   relativistic 32, 43, 44
   Riemannian 44, 124
   spatial 61, 124, 125
   temporal 61, 124, 125, 140
   Wheeler–DeWitt equation 114
Minkowski spacetime
   causal set theory (CST), and 102, 104
   collapses in 87
   derivation of core structure of 302
   flat 48
   foliation of 89, 101
   general relativity, and 59
   Malament's treatment of 79*n*
   metric 125
   'no-go' theorem in 62
   non-invariant present in 72
   prediction in 160
   productive time, and 157
   relativistic quantum theory, and 81
   relativity theory, and 52
   Robb orthogonality in 63
   Robb's axiomatization of 40*n*
   special relativity, and 42
   tachyons 128
   tearing apart of 67, 80
mobility asymmetry
   constraint, as 234, 257
   feature of time 129, 130, 179
   fragmentation of time 135
   illustration of 131*f*
   importance 222
   manifest time 305
   manifold of events 147
   speculation about 156

motion
  Reichardt-like detector 239*f*
  sensing of 238

natural kind asymmetry 22, 122, 133, 137,
      148, 179
Newton–Cartan spacetime 32, 34*f*, 35, 41,
      124, 125, 127
non-degenerate metrics 78
now 210
  common 40, 55, 206, 216*f*
  common global now 65
  flowing now 21, 31, 35, 181, 238, 306, 310
  future and 6
  illusion of 3, 310
  manifest time, and 7, 206
  manufacturing of 213
  objective global now 206, 216
  past and 6
  perceived synchrony, PH case 210
  physical time, and 31, 181
  present and 6
  problem of 3
  time and space, and 9*f*, 308
  unified flowing now 306
  updating now 11

one-dimensionality of time 144
open future, RPP argument against 52

pages of light 174
pages of time 177
passage of time, sense of 261
past/future (PF) asymmetry
  choice and 278
  example of 280*f*
  explanation of 266, 272, 274, 279, 287, 288
  proximal/distant asymmetry, and 265, 270
path-independence of classical duration 39
perceived synchrony, PH case 210
phenomenology
  flow of time, and 11
  manifest time, and 230
philosophy of time
  analytic 290
  author's analytical approach 290, 300
  existence, focus on 296, 299, 303
  existence of 209
  focus of 223
  goal of 4, 294
  manifest time 49
  Minkowski spacetime 52
  presence of experience, problem of 182
  Prior's 'thank goodness that's over'
        argument 266
  quantum gravity 99
  questions in 312

relativity theory 42
science and 295
Temporal Knowledge Argument 185
tensed and tenseless theories 181
physical time
  aspects of 179
  change and 11, 229
  concept of 2, 19
  direction of time 132
  Eddington's two tables 24
  flow of time, and 227, 238
  fragmentation of time 135, 144
  informative strength 158
  manifest time, and 7, 20, 23, 26, 49, 119, 157,
        180, 210, 300, 301, 310
  now, and the 31, 181
  relativity theory, and 27, 31
  self and 249
  tensed and tenseless theories of time,
        and 303
  'two times' problem, and 23
physics
  A-, B- and C- series 300
  aspects of time in 122, 179
  classical, *see* classical physics
  philosophy of time, and 295
  sideways, *see* sideways physics
  time and space theories, challenges to 128
Poincaré, Jules Henri, *see* Einstein–Poincaré
      'Radar' Synchronization
pop out 204*f*
Popper, Karl 21, 84, 96
present
  becoming past 246
  'diamond presents' 64, 80
  'donut presents' 62*f*
  experience of, *see* experience of the present
  lightcone presents 61*f*
  manifest present, features of 50
  manifest time, presentness of 184, 189
  non-invariant 72
  now, and the 6
  pop out 204*f*
  portrayal by manifest time 224
  private presents 63*f*
  puny presents 62*f*
  simultaneity judgments 198*f*
  specious present 50*n*, 240
  stream–bounce illusion 200*f*
  subjective simultaneity, measurement
        of 197
  temporal integration window 193*f*
  temporal order judgments 199*f*
  temporal recalibration 194
  temporal ventriloquism 193
Prior's 'thank goodness that's over'
      argument 266

private presents 63*f*
proximal/distant (PD) asymmetry
  examples of 265
  explanation of 266, 284
  feature of time 270
  illustration of 266*f*
  past/future (PF) asymmetry, and 265, 279
puny presents 62*f*
Putnam's argument against manifest
    time 212

quantum becoming 81
quantum gravity
  canonical quantum gravity 98, 110, 116
  causal set theory (CST), and 100
  change and 11
  disappearing time in 110
  function of 21, 98
  initial value problems, and 174
  loop quantum gravity 98, 99, 112
  maximal antichains 102
  quantum gravitational time 98
  semiclassical time 112
  theory of 67, 85
  time 30, 118, 311
  timeless quantum gravity 112, 120
quantum mechanics
  coordination problem 90, 93*f*
  determinism 144
  function of 85
  interpretations of 87, 89, 117
  macroscopic scales 214
  manifest time, and 21, 81
  non-locality of 85, 86
  non-relativistic 81
  operation of 82
  probability 117, 128
  relativity theory, and 86, 89, 98
  spacetime 81
  tensed and tenseless theories of time,
    and 92
  time 99, 112, 115, 116, 117
  únitarity 128
  wavefunction collapse, and 94, 95
quantum theory
  Bohmian configuration space 91
  classical physics, and 31
  discovery of 81
  EPR–Bell experiment 88
  Everettian interpretation 117
  gravity 81, 98, 111
  Hamiltonian interpretation 111
  interpretations of 82, 117
  manifest time, and 81
  relativistic 81, 83
  relativity theory, and 19*n*, 86, 96, 98
  spacetime 81, 83

symmetric temporality of 95
tensed and tenseless theories of time, and 92
time 19, 81, 83, 98

Reichardt-like motion detector 239*f*
relativistic spacetime 26, 49, 65, 67, 70, 100, 105,
    106, 125, 160
relativity theory
  Cauchy time 67
  classical physics, and 31
  'diamond presents' 64, 80
  'domes' 63
  'donut presents' 62*f*
  Einstein–Poincaré 'Radar'
    Synchronization 51*b*
  general relativity
    'asynchronous becoming', and 104
    Cauchy problems 178
    discovery of 42
    formulation of 93, 105, 111
    gravity 41, 42
    Hamiltonian formulation of 111
    Lorentzian interpretation of 59
    manifest time, and 59
    meaning of 42
  lightcone presents 61*f*
  lightcone structure 45
  Lorentzian Time 57
  manifest time, and 31, 67, 80, 175, 176
  metrics 32, 43, 44
  Minkowski Spacetime 52
  outside Minkowski Spacetime 59
  private presents 63*f*
  puny presents 62*f*
  quantum theory, and 19*n*, 86, 96, 98
  relativity 42
  relativity of simultaneity 53
  restrictions on 76
  RPP argument 52
  spacetime torn apart 67
  special relativity
    causal theory, and 138
    discovery of 42, 51
    focus on 43, 52
    Lorentzian time 57
    manifest time, and 59
    maximal antichains, and 102
    spacetime 42, 52, 77
    'temporal becoming', and 55
  Stein's 'temporal becoming' theorem 55
  time 49
Riemannian metrics 44, 124
Rietdijk–Putnam–Penrose (RPP) argument
    against an open future 52
Robb's axiomatization of Minkowski
    spacetime 40*n*

science, philosophy of time, and 295
self, time and 247, 252, 253, 259
semiclassical time 112, 115*f*
sideways physics
    argument about 167
    aspects of time in 179
    background to 157
    Cauchy problems 158
    conclusions from 179
    illustration of 170, 171*f*
    implicit definition of time 171
    non-temporal direction 173
    pages of light 174
    pages of time 177
    possible worlds 164
    proposed theory methodology 166
    sideways Cauchy problem 164*f*
    telling of world's story sideways 177*f*
    transcendental determinism 158, 160*f*
simultaneity
    relativity of 53
    simultaneity judgments 198*f*
    subjective simultaneity, measurement of 197
space and time, *see* time and space
spacetime
    Cartan 125
    Cauchy time 67
    classical 32, 34, 36, 38, 45, 58, 61, 100, 114,
        125, 213
    cylinder 137*n*
    domain of dependence 68
    Euclidean 128
    Friedman–Lemaître–Robertson–Walker
        (FLRW) time 73
    Galilean 42, 81, 89, 124, 151
    Gödel 60, 130, 143
    Lorentzian 172
    metrical structure of 122
    Minkowski, *see* Minkowski Spacetime
    Newton–Cartan 32, 34*f*, 35, 41, 124, 125, 127
    relativistic 26, 49, 65, 67, 70, 100, 105, 106,
        125, 160
    restrictions on theories 76
    slices of 134*f*
    spacetime behaviour chart 72*b*
    tearing apart of 67, 69*f*
    thrown into 26
    Trautman spacetime 41, 77*n*
    'unique' time functions 72
spatial direction, *see* direction of space
spatial metrics 61, 124, 125
special relativity, *see* relativity theory
specious present 50*n*, 240
stability, physical theories of time and space,
        and 128
Stein's 'temporal becoming' theorem 55
stream–bounce illusion 200*f*
subjective simultaneity, measurement of 197

tachyons, physical theories of time and space,
        and 128
temporal biases
    description of 282
    explanation of 285
temporal decentering 255, 257*f*
temporal direction, *see* direction of time
temporal flow, *see* flow of time
temporal integration window 193*f*
Temporal Knowledge Argument 185
temporal metrics 61, 124, 125, 140
temporal order judgments 199*f*
temporal recalibration 194
temporal value asymmetry, explanation of
    affect asymmetry 275
    background to 264
    conclusions about 288
    Humean solution 272
    knowledge asymmetry 274
    mechanisms for valuation and choice among
        alternatives 278*b*
    other temporal biases
        description of 282
        explanation of 285
    past/future (PF) asymmetry 279
    Prior's 'thank goodness that's over'
        argument 266
    proximal/distant (PD) asymmetry 270
temporal ventriloquism 193
tensed and tenseless theories
    becoming, of 269, 294
    elimination of tense 302
    manifest time 182, 184
    philosophy of time 181
    physical time 303
    quantum mechanics 92
    quantum theory 92
    tensed theories, types of 294*f*
'thank goodness that's over' argument
        (Prior) 266
time
    analytic philosophy of 290
    author's analytical approach 290
    'base space', as 38*f*
    binding time 144
    Broad's growing block model 298
    care and 266*f*
    Cauchy time 67, 73, 74, 75, 76, 79, 80, 84
    common now 206
    dimensionality 126
    direction, *see* direction of time
    explanatory challenge 294
    flow, *see* flow of time
    FLRW time 73
    fragmentation of 135
    as 'great informer' 142
    manifest time, *see* manifest time
    metaphysics of 149, 181

now, and the, *see* now
one-dimensionality 144
pages of 177
philosophy of, *see* philosophy of time
physical time 19
physics, in 122
possibilities for 20*f*
present, experience of, *see* experience of the
    present
problem of 1
quantum mechanics, *see* quantum mechanics
quantum theory, *see* quantum theory
relativity theory, *see* relativity theory
semiclassical time 112, 115*f*
sideways physics 157
space and, *see* time and space
specialness of 21
system laws, and 138, 140
temporal value asymmetry 264
'two times' problem 23, 29, 30, 105, 127, 129,
    154, 302
unified theory of 304
'unique' time functions 72
time and space
  alikeness 22, 119, 292
  alternative choices in 278
  difference 10, 22, 23, 27, 77, 83, 119, 121,
    122, 123, 124, 132, 133, 137, 138, 142,
    151, 153, 156, 163, 179, 208, 221, 222,
    234, 305

discovery 60
foliation 36*f*
joining 121
lightcone coordinates 123*n*
Lorentzian time, and 57
'macroscopic' scales 261
mobility asymmetry 130, 131*f*
natural kind asymmetry 133
now, and the 308
physical theories, challenges to 128
recovery 34
relativity theory, and 29
separation 35, 37, 67, 127
time as 'base space' 38*f*
unique 35
wiggling in 221
time-moving perspective 254*f*
time stamps 213
Trautman–Cartan theory 40
Trautman spacetime 41, 77*n*
'two times' problem 23, 29, 30, 105, 127, 129,
    154, 302

undifferentiated past and future 255*f*
'unique' time functions 72
unitarity, physical theories of time and space,
    and 128

Wheeler–DeWitt equation 111, 112, 113, 114
wiggling in time and in space 221